T.M. Jenkins
and
W.R. Nelson

(Xtra "loan" copy)

Computer Techniques in Radiation Transport and Dosimetry

ETTORE MAJORANA INTERNATIONAL SCIENCE SERIES

Series Editor:
Antonino Zichichi
European Physical Society
Geneva, Switzerland

(PHYSICAL SCIENCES)

Volume 1   INTERACTING BOSONS IN NUCLEAR PHYSICS
               Edited by F. Iachello

Volume 2   HADRONIC MATTER AT EXTREME ENERGY DENSITY
               Edited by Nicola Cabibbo and Luigi Sertorio

Volume 3   COMPUTER TECHNIQUES IN RADIATION TRANSPORT
               AND DOSIMETRY
               Edited by Walter R. Nelson and Theodore M. Jenkins

Volume 4   EXOTIC ATOMS '79: Fundamental Interactions and Structure
               of Matter
               Edited by Kenneth Crowe, Jean Duclos, Giovanni Fiorentini, and
               Gabriele Torelli

Computer Techniques in Radiation Transport and Dosimetry

Edited by

Walter R. Nelson
CERN
Geneva, Switzerland
and
SLAC
Stanford, California

and

Theodore M. Jenkins
SLAC
Stanford, California

Plenum Press · New York and London

Library of Congress Cataloging in Publication Data

International School of Radiation Damage and Protection, 2d, Erice, Italy, 1978.
Computer techniques in radiation transport and dosimetry.

(Ettore Majorana international sciences series: Physical science; v. 3)
Includes index.
1. Shielding (Radiation)—Data processing—Congresses. 2. Shielding (Radiation)—Computer programs—Congresses. 3. Radiation dosimetry—Data processing—Congresses. 4. Radiation dosimetry—Computer programs—Congresses. I. Nelson, Walter Ralph, 1937- II. Jenkins, Theodore M. III. Title. IV. Series.
TK9210.I57 1978   621.48'028   79-20872
ISBN 0-306-40307-2

Proceedings of the Second International School of Radiation Damage and
Protection, held in Erice, Sicily, October 25—November 3, 1978

© 1980 Plenum Press, New York
A Division of Plenum Publishing Corporation
227 West 17th Street, New York, N.Y. 10011

All rights reserved

No part of this book may be reproduced, stored in a retrieval system, or transmitted, in any form or by any means, electronic, mechanical, photocopying, microfilming, recording, or otherwise, without written permission from the publisher

Printed in the United States of America

PREFACE

In October 1978, a group of 41 scientists from 14 countries met in Erice, Sicily to attend the Second Course of the International School of Radiation Damage and Protection "Ettore Majorana", the proceedings of which are contained in this book.

The countries represented at the School were: Brazil, Canada, Federal Republic of Germany, Finland, German Democratic Republic, Hungary, India, Italy, Japan, Spain, Sweden, Switzerland, United States of America, and Yugoslavia.

The School was officially sponsored by the Italian Health Physics Association, the Italian Ministry of Public Education, the Italian Ministry of Scientific and Technological Research, and the Sicilian Regional Government. In addition, administrative and technical support was received from the Stanford Linear Accelerator Center and from CERN.

The past 15 or so years have witnessed a significant development of computer methods in the science of radiation protection. The radiation transport codes associated with hadronic and electromagnetic cascades, reactor shielding, unfolding techniques, and gamma ray spectrum analysis have reached the state-of-the-art level, and the Erice Course aimed at presenting as comprehensive an overview of these programs as was possible within the allotted time span.

In addition, however, we are convinced by experience that the same computer programs are extremely valuable outside the immediate field of radiation protection. For example, in the period between the end of the Course and the publication of this book, the EGS and KASPRO programs have been combined and used at CERN in order to predict positron-electron yields from hadronic-electromagnetic cascade development in extended targets at 400 and 20 GeV, and the EGS and HETC programs have been combined at ORNL for similar problems. The CERN calculations are not for purposes of shielding and dosimetry, but are meant to be used in the design of the LEP and antiproton storage ring projects. Nevertheless, while the computer programs that were used for this particular problem have been developed, in most part, within radiation protection groups at CERN and SLAC, they have been used extensively for other things in addition to shielding and dosimetry. The same can be said of the many other computer programs developed elsewhere, i.e., at ORNL and Fermilab.

The lectures that comprise this book illustrate, therefore, not only the usefulness of various computer programs in radiation protection and damage, but show how they can be useful in related scientific fields, such as medical physics, detector design in high-energy physics, accelerator component design, cosmic ray physics, etc. The book therefore should have appeal to a wide scientific audience. Those interested in electromagnetic and/or hadronic cascade calculations will find it extremely valuable as a beginning reference on the subject.

As users of most of the programs presented in this book, we have seen a need to explain various facets from the point of view of the _user_ as opposed to the _writer_. Users have often become dissatisfied with their limited knowledge of the codes that they use; even more than that, the complexity of problems generated by users has dictated that they know the strengths and weaknesses of the codes. We hope that the method of presentation given at Erice, and consequently by this text, will enable users to better understand and appreciate these computer programs.

PREFACE

A number of people have aided us in our efforts to develop the Erice Course into the success it was, and in the publication of this book. We thank Professor Antonino Zichichi, Director of the "Ettore Majorana" Centre for Scientific Culture, for providing us the opportunity to have the Course. Dr. A. Gabriele and Miss P. Savalli were very much appreciated by all participants at Erice, and we are grateful to them. Our close friend and colleague, Dr. A. Rindi, Director of the School of Radiation Damage and Protection, formulated the initial idea for the Course. His help in conducting this Course and his advice as to its concept have been invaluable.

Finally, it is very difficult to coordinate the writing and typing efforts of twelve lecturers in seven countries such that a uniform and consistent book is created as a result. The editorial and typing staffs at CERN and SLAC deserve a special thanks. In particular, Mrs. K. Wakely and Mrs. Vascotto at CERN worked very hard on this book, and our co-worker, Mrs. Ruth Parker at SLAC, has been actively involved in all aspects from the inception of the Course.

17 April 1979

Walter R. Nelson
CERN and SLAC

T. M. Jenkins
SLAC

CONTENTS

INTRODUCTORY LECTURES

Lecture 1: Computer Methods in Radiation
          Protection as Viewed by a User   9
    Graham R. Stevenson

Lecture 2: The Physics of Radiation Transport   17
    Keran O'Brien

LOW ENERGY NEUTRON AND GAMMA-RAY PROGRAMS AND THEIR APPLICATIONS

Lecture 3: The Methods and Applications of Dis-
          crete Ordinates in Low Energy Neutron-
          Photon Transport (ANISN, DOT) Part I:
          Methods   59
    W. W. Engle, Jr.

Lecture 4: The Methods and Applications of Monte
          Carlo in Low Energy ($\leq$ 20 MeV) Neutron-
          Photon Transport (MORSE) Part I:
          Methods.   77
    T. A. Gabriel

Lecture 5: The Methods and Applications of Dis-
          crete Ordinates in Low Energy Neutron-
          Photon Transport (ANISN, DOT) Part II:
          Applications   97
    W. W. Engle, Jr.

Lecture 6: The Methods and Applications of Monte
          Carlo in Low Energy ($\leq$ 20 MeV) Neutron-
          Photon Transport (MORSE) Part II:
          Applications   99
    T. A. Gabriel

Lecture 7: The European Shielding Information
           Service - ESIS   121
   C. Ponti

Lecture 8: Cross Section Processing Codes and
           Data Bases (AMPX).   123
   W. W. Engle, Jr.

Lecture 9: Radiation Shielding Information
           Center and Biomedical Computing
           Technology Information Center   125
   T. A. Gabriel

Lecture 10: Approximate Methods in Reactor
            Shielding Calculations   127
   C. Ponti

ELECTROMAGNETIC CASCADE SHOWER PROGRAMS AND THEIR APPLICATIONS

Lecture 11: The Physics of Electromagnetic
            Cascade   141
   Keran O'Brien

Lecture 12: Solution of the Electromagnetic
            Cascade Shower Problem by Analog
            Monte Carlo Methods - EGS   173
   Walter R. Nelson

Lecture 13: Some Examples for the Application
            of the Monte Carlo Code EGS   197
   Herbert Dinter

Lecture 14: Calculation of the Average
            Properties of Electromagnetic Cascades
            at High Energies (AEGIS)   211
   A. Van Ginneken

Lecture 15: Electron Dosimetry Using Monte
            Carlo Techniques   223
   Keran O'Brien

Lecture 16: Application of EGS to Detector
            Design in High Energy Physics   239
   Walter R. Nelson

Lecture 17: Application of EGS and ETRAN to Prob-
            lems in Medical Physics and Dosimetry . . .   253
   Walter R. Nelson

HADRONIC CASCADE PROGRAMS AND THEIR APPLICATIONS

Lecture 18: Introduction to Hadronic Cascades   269
  Tony W. Armstrong

Lecture 19: Particle Production Models,
            Sampling High-Energy Multiparticle
            Events from Inclusive Single-
            Particle Distributions   279
  J. Ranft

Lecture 20: The Intranuclear-Cascade-
            Evaporation Model   311
  Tony W. Armstrong

Lecture 21: Calculation of the Average
            Properties of Hadronic Cascades
            at High Energies (CASIM)   323
  A. Van Ginneken

Lecture 22: The FLUKA and KASPRO Hadronic
            Cascade Codes   339
  J. Ranft

Lecture 23: The HETC Hadronic Cascade Code   373
  Tony W. Armstrong

UNFOLDING METHODS AND SPECTRUM ANALYSIS

Lecture 24: Unfolding Techniques for
            Activation Detector Analysis   389
  J. T. Routti and J. V. Sandberg

Lecture 25: Bremsstrahlung Spectrum Analysis
            by Activation Method (LYRA, DIBRE,
            REFUM)   409
  Takashi Nakamura

Lecture 26: Application of Activation-
            Spectrum Analysis Method to Shield-
            ing (TAURUS, LYRA, DIBRE, SAND-II) . . .   443
  Takashi Nakamura

Lecture 27: Activation Detectors and Their
            Gamma Spectrum Analysis   479
  M. J. Koskelo and J. T. Routti

INVITED PRESENTATIONS FROM STUDENTS AND SUMMARY LECTURE

Monte Carlo Calculation of Exposure Rates
      in Dwelling Rooms   501
  Laszlo Koblinger

Integral Equation for Radiation Transport –
      ASFIT   503
  V. Sundara Raman

Thermal Effects Induced by High Energy
      Protons in Target and Absorber
      Materials   507
  P. Sievers

Summary Lecture   511
  Graham R. Stevenson

Participants   513

Index   515

OPENING OF THE COURSE

Alessandro Rindi - Director of the School

Istituto Nazionale di Fisica Nucleare
00044 Frascati (Roma)
Italy

I am very happy to welcome you to the Ettore Majorana Centre by opening the Second Course of the School of Radiation Damage and Protection.

When we started this School with the First Course in 1975, I was a little afraid that it might be the last in this field. This is not to imply that the course was not technically excellent nor well received, but simply to state that the E. Majorana Centre aims to organize courses at the frontier of science and at the highest scientific level, and I questioned at that time whether others would identify exactly how Health Physics fit in with these precepts.

I have always tried during my career as a health physicist to show that our discipline is not just one of routine work; not just a job for "surveyors" who limit themselves to practical applications of knowledge and methods developed in other branches of science. On the contrary, I firmly believe that Health Physics is a field of research in its own right. We take advantage of the research of other fields, of course, but Health Physics also provides scientific results that can be original and can be used by other scientific disciplines at the same time. I believe that Health Physics has long since reached the status of a regular branch of science, competitive with many other scientific professions.

It has not been easy to gain acceptance of this point of view by members of the older branches of physics, such as the nuclear or high energy experimentalists and theoreticians---yet it is coming. The fact that we are opening the Second Course of this School, that most likely there will be a Third Course next year with more to follow, and that there are so many participants in this room today, clearly demonstrates that we have been successful in our efforts.

The development of computer programs to study the interaction of radiation with matter in order to predict dose, dose distribution, shielding, energy deposition, etc., is basic to Health Physics. However, many of these same programs are also extensively used for other applications beyond their primary aim, thereby typifying the impact and usefulness of the profession.

Ralph Nelson from SLAC is one of the members of this profession who certainly qualifies for what I would like to define as a radiation scientist. I am very pleased that he accepted directorship of this course; it may be just the first of a series in the field.

On behalf of myself and the Centre I want to thank him for the enthusiasm he has shown and the enormous efforts he has devoted to the organization of this course, and to wish the best of success.

INTRODUCTORY REMARKS

Walter R. Nelson - Director of the Course*
Health and Safety Division
CERN
Geneva, Switzerland

I would like to take a few minutes to tell you how this course originated. To begin with, I attended the 1975 Course at Erice along with a few of you that are here today. During that course, which dealt primarily with radiation protection technology at high energies, it was quite evident that computer techniques play a significant role in much of the work that we do in Health Physics. Subsequent to that time, I was approached by Dr. Rindi, who asked if I thought that a course on computer methods in Health Physics would be worth presenting at the Ettore Majorana Centre. The idea appealed to me, but I first wanted to make a survey among our colleagues throughout the world in order to determine exactly what type of course should be organized. We sent out over 200 questionaires to heads of departments and laboratories, to close colleagues, and to many others that we knew from their scientific publications. More than 20% responded, with all of those showing some degree of interest, and this made us quite happy and led to the decision to hold the course.

It was obvious from the comments and suggestions, however, that we would never be able to satisfy all interests with a single course.

---

* On leave of absence from SLAC (Stanford University), 1978/79.

Because the Health Physics profession is so diversified — its members coming from many branches of science — this came as no shock to us. What we finally decided, as a solution, was to organize a course on one of the predominant interests that were suggested by the survey, and to leave the other subjects to possible future courses.

In the 1975 Course, I was asked to give the keynote lecture to open the program. A number of us have had careers in Health Physics that have required us to interact with others outside our immediate profession. At SLAC, for example, we work in a very challenging way with high-energy physicists, not only in legislating what they can and cannot do, but also aiding in the design of their experiments and detectors. The title of my 1975 lecture was "The Role of the High-Energy Health Physicist", and I believe that I demonstrated in that talk the simple fact that many of the things that we do are of definite interest and value to high-energy physicists, and to many outside our profession as well. I hope that you also will become aware of this in the present course. We have people attending this course who are interested in hadron calorimetry at hundreds of GeV, who want to better understand radiation fields from medical accelerators used in the treatment of cancer, who would like to have better tools for designing reactor or hospital shielding, or who wish to expand their knowledge on computer techniques in analysing activation detectors. The list of interests is rather impressive, as I have been able to view this from my position as Director of the Course.

We have a good attendance for this course and the list of participants is truly international. I am pleased by this and I am sure that Professor Zichichi, the Director of the Centre, will be pleased too.

In keeping with the tradition of the Centre, I hope to see close contact between the participants. We have some excellent people giving lectures. These are individuals who have made significant contributions to a select group of computer codes that are at the "state of the art" level. We have an equally select group of students, many of whom could well have been lecturers at this course. I believe that we will all learn a lot from each other during the next nine days.

This course concerns itself with certain techniques that are used in radiation transport and in the unfolding of data, especially data obtained using activation detectors. In particular, it is a course on how to use existing codes and to use them in an intelligent way. Let me illustrate this by giving an example from experimental physics. A good experimentalist using modular electronics should learn how to use his equipment properly (i.e., as designed). To do this, he would most likely first read the instruction manual. He learns to recognize when things are not working correctly by making certain checks as he goes along. He then studies the results to see if they are reasonable and sound. And before he ever attempts to use his equipment, he performs "back-of-the-envelope" calculations to determine whether or not the choice of equipment is correct for the measurement at hand.

Using computer codes of the type to be described in this Course is no different from using experimental equipment. That is, you must know your codes: how they do what they do and their limitations. I believe it was John von Neumann who once remarked that the method of Monte Carlo was "experimental mathematics". We are here then, in this course, to learn how to do experimental mathematics in a scientifically intelligent way.

INTRODUCTORY LECTURES

LECTURE 1: COMPUTER METHODS IN RADIATION PROTECTION

AS VIEWED BY A USER

Graham R. Stevenson

Health and Safety Division
CERN
Geneva, Switzerland

ABSTRACT

Five elements in computer methods as applied to radiation protection are identified and discussed:

1. The user, from the person who never touches a keyboard to the person who has never made a fluence measurement.

2. The problem, analytical, iterative or Monte Carlo.

3. The objective, for multiple engineering runs, for single in-depth studies or for program development.

4. The interface, or how to get the data in and the answer out.

5. The solution, or is this really the right answer?

INTRODUCTION

The purpose of the first lecture in any conference, school, or seminar is to introduce the subject, set the tone and standard of the meeting, to give some penetrating insights into the philosophy of the various aspects of the subjects to be treated, and to do this without either stealing the subject matter of the succeeding lecturers or taking too long over it. The subject matter to be treated in this School is vast, ranging from the physics of the transport of electromagnetic and hadronic cascades in matter to the mathematical techniques used to make a complete theoretical study of the cascade possible, from the witchcraft or sorcery of spectrum unfolding to a study of the god-given fundamental cross-section data. The tone of the Course is automatically set by the impressive credentials of the lecturers here, who represent a significant fraction of the entire world's knowledge in the subjects under discussion. The task which is left for me today is to provide the "penetrating insights of philosophy". To do this I can only draw on my experience in the use of transport programs that it has been my lot to suffer during approximately the last seven years in the design and construction of the 400 GeV proton accelerator at CERN. The aim of my talk is to address myself as a student to the lecturers who follow me and to suggest to them what should be the object of all their hard work. They, the program writers, should be aware who the users of their programs really are; the mathematical method of solution should be the one best adapted to the problem -- one should not use a sledgehammer to crack a nut; it should be clear for which type of problem the program is optimized; the use of the program should not depend on sophisticated computer facilities which may only be available at the largest laboratories; finally, there must be a significant body of evidence to justify that a certain answer from a program is in fact the correct answer. I feel that in studying these five points in closer detail you will see computer methods used in radiation protection as viewed by at least one user.

THE USER

Probably the most important person involved in a computer program is the user. Without a user, program-writing can become mere self-gratification. The user of the program could well be its author; this state of affairs illustrates an important aspect in any computer program, namely the user must have a problem which he recognizes as unsolvable without the capacity of a computer. In many instances potential users have dismissed problems as unsolvable, because they were not aware that either a program existed to resolve their problem or that it would be worth their effort to write the required program themselves. This puts an obligation on authors to publicize their programs and to teach potential users how to run their programs. The author should not assume that the user of his program is as expert as he, the author, in computer techniques: the user may never have touched a keyboard before and everyone except the author who runs a program for the first time must be considered as inexperienced. Thus standard options should be built into all programs, but these options must be understandable and flexible enough to allow the experienced user to penetrate deeper into the problem. In this we come to the problem of communication between the person who has the problem and the author of the program. The author could well be a pure-mathematician, expert in multi-dimensional minimization, but who is not in the least interested in whether certain effluent water is a danger to the population. In addition, neither the person with the problem nor the author of the program may be sufficiently expert in the use of a GeLi gamma spectrometer to understand the problems of efficiency calibration, pile-up and base-line shift and dead-time effects. Only when the same person is an expert in problem solving, measurement, and programming does the communication problem disappear. It is by bringing together in Courses like the present one these three classes of people that this communication problem can be resolved.

THE PROBLEM

It is useful at this stage in this introductory talk to illustrate the capacity of the available computer techniques in solving the types of problems encountered in radiation protection, in the widest sense. At the outset of one's studies in physics or applied mathematics all the problems one met could be solved by purely analytical techniques and all the integrals in these problems were analytically integrable. The days of wine and roses. As an example of this simplest problem one could consider the creation of a radio-isotope in a mono-energetic neutron field, where the flux density is constant in time. The creation and subsequent decay of this isotope, even including the complexities of dilution of the isotope in a cooling water circuit are easily treated by the simplest of computer techniques -- the slide-rule. However, to repeat the calculation for more than one isotope, to introduce time variations in the neutron fluence rate or fluid transport or to vary production and decay parameters in a systematic way would so extend the slide-rule calculation in time so that it becomes more efficient to use a programmable desk calculator or computer. The example chosen above is one of direct calculation from known physics; the next step away from "reality" towards complete computer mastery of the problem is to consider the case of direct calculation from empirical physics. Examples of this are the determination of accelerator shielding using the Moyer model, the use of empirical formulae for skyshine problems involving the calculation of population doses or the use of simplified range-energy relations in muon transport problems or the evaluation of proton radiators in neutron personal dosimeters. In these problems the integrals encountered are generally not analytically integrable and computer oriented quadrature techniques are "de rigeur". This type of problem has often even to be expressed in a way that makes computer integration possible. It is only a short step from numerical quadrature to the iterative convergence techniques which are essential in the solution of the first-order Fredholm equations of gamma and neutron spectrometry. The last step away from the deterministic towards the statistical philosophy of the universe is to accept that any problem can be solved by using a random number generator and Monte Carlo techniques. This leaves the user with the embarrassing and often difficult choice of which computer method to use to solve his problem. For example why use a code with very complicated particle splitting and weighting techniques to determine energy deposition in the core of a hadron cascade when a more simple code with no Monte Carlo frills and where each event produced has a one-to-one counterpart in reality is more than adequate; why use a Monte Carlo technique at all in a muon transport problem in slab geometry with no magnetic fields when very elegant solutions of the Boltzmann transport equation exist; why use a generalized minimization routine to determine the proportions of different known radio-isotopes in a radioactive gas mixture with a defined decay curve when a linear least squares analysis is an order of magnitude faster and equally accurate? The educational aspect of the following lectures is for me the most important aspect of the course so that I am more able to make the optimum choice of program with which to solve my problems.

THE OBJECTIVE

In the preceding section the difficulty of choosing the code best suited to the problem was introduced; it is useful to probe deeper into this question. The criteria for an acceptable solution to a problem may not be based on physical correctness but on the cost of computer time. I am reminded of what I should like to call the Awschalom Criterion: "When the cost of the shield is less than the cost of running the transport code, calculation is inefficient". This implies that for a given problem there should be a range of codes available where, if absolutely necessary, precision and physical correctness may be balanced against the cost of obtaining a solution. I am not advocating that cost is the only criterion or that fastest is best, but I am trying to point out that there is a place for the fast, approximate calculation when one is at the start of a problem and is trying to fix the important variables in the same way that there is a place for the very detailed, exact calculation when one is designing a shield to the last centimetre or when one is aiming to compare the calculation with physical measurements. This puts another responsibility on the authors, to delimit clearly and explain the objectives of their programs.

THE INTERFACE

Most errors in computer methods come from the inability to enter the required data correctly or to understand the output. In spite of the present abundance of terminal facilities, such as WYLBUR or ELECTRIC for the IBM computers, or INTERCOM for the CDC series, it is my opinion that the data input for a problem should still be designed around cards as the input medium. The reason for this is simple -- there are still many computers with no terminal facilities, and if terminals do exist it is trivial to set up the input on a card format. A more serious comment is directed towards the elegance with which data is entered -- this is especially difficult to design efficiently when defining the input geometry to a problem. It should not be necessary to enter the same variable more than once in the input data; I know of one code at least which was not written in this way and where it was necessary to repeat the same coordinate many times. In order to check on the input data, all entered data appear on the output stream, preferably in one block and not spread throughout the whole of the output. How many times has one inadvertently mixed the boundaries of different media in a transport program? Referring to the output, I wish to put in a plea this time for the terminal users -- all output should now be made suitable for examination on a CRT. This is more difficult to design adequately than the normal output intended for a printer. Finally the output should be understandable without the program manual. Simple instructions take up next to no memory space and are very easy to include in the output. Outputs should be logical and correct:

COMPUTER METHODS IN RADIATION PROTECTION

for example, there are electromagnetic cascade programs which refer to star densities in their output, there are graphs where it is impossible to determine whether the output is linear or logarithmic, there is often a confusion between cm and g/cm^2, etc. These only take a small amount of effort on the part of the authors to eliminate, but they cause infinite amounts of pain to the users.

THE SOLUTION

The final comments that I wish to make concern the reliability of the output from the computer program. There are numerous effects which can cause the output to be in error, e.g. false input data, the wrong algorithm giving rise to round-off errors in the case under consideration, an insufficient number of iterations in a recursive algorithm, insufficient convergence criteria in an integration subroutine or an insufficient number of events/bin in a Monte Carlo program. The author of the program should always provide indications on the output data whenever any of these latter errors occur in a program; the one thing that cannot be guarded against are errors in the input data -- this is entirely the responsibility of the user. My last words concern the correctness of the solution. This must be checked by repeated tests of the program against detailed physical measurements. It is only when the solution is an interpolation between existing measurements that one can have any real confidence in the correctness of the solution.

LECTURE 2:   THE PHYSICS OF RADIATION TRANSPORT

Keran O'Brien

Environmental Measurements Laboratory
U.S. Department of Energy
New York, New York 10014
USA

INTRODUCTION

The Boltzmann equation is an integrodifferential equation describing the behavior of a dilute assemblage of corpuscles. It was derived by Ludwig Boltzmann in 1872 to study the properties of gases. It applies equally to the description of the behavior of "radiation" which, for the purpose of this paper, will comprise nuclei, leptons, mesons, baryons, and energetic photons.

Boltzmann's equation is a continuity equation in phase space which is made up of the three space coordinates of euclidian geometry and the kinetic energy, and the direction of motion, $\vec{\Omega}$, of the particle (Fig. 1).

The vector $\hat{\Omega}$ is a unit vector in the particle's direction of motion. Its three components, in cartesian form, in terms of the polar angle θ and the azimuthal angle ψ are

$$\Omega_x = \cos \theta \; ,$$

$$\Omega_y = \sin\theta \cos\psi \, ,$$

$$\Omega_z = \sin\theta \sin\psi \, .$$

Boltzmann's equation is in terms of the variable $\phi_i(\vec{x}, E, \vec{\Omega}, t)$, called the angular flux. The angular flux is the number of particles of a given type (nuclei, leptons, mesons, etc.) in the volume element dx dy dz about $\vec{x}$ in the kinetic energy element dE about E with a direction of motion $d\vec{\Omega}$ about $\vec{\Omega}$, multiplied by the speed of these particles. The speed is in terms of the scalar $v = pc/W$, where p is the particle momentum $(E^2 + 2 Em_i)^{1/2}$, c is the speed of light, W is the total particle energy $E + m_i$, and m_i is the particle mass. (in c=1 units)

The angular flux gives the number of particles per cm^2 per MeV per steradian per second of a given kind at a given location at a

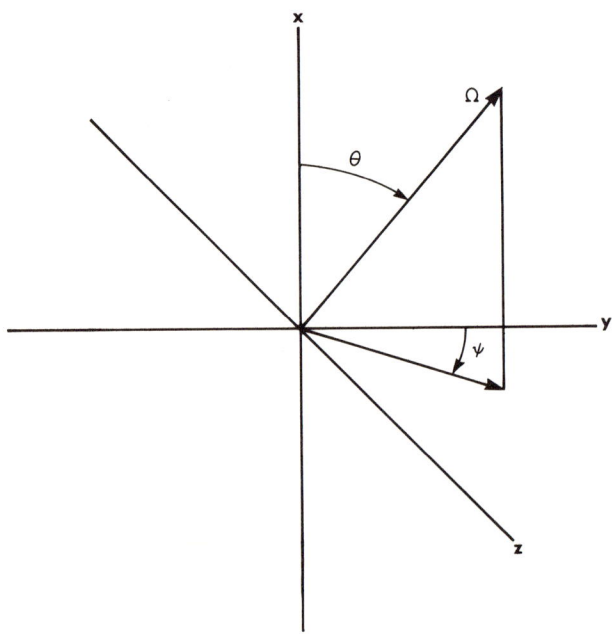

Fig. 1. The coordinate system of the Boltzmann equation.

given time. It is related to the scalar flux, or flux density by

$$\phi_i(\vec{x}, t) = \int_{4\pi} d\vec{\Omega} \int dE \, \phi_i(\vec{x}, E, \vec{\Omega}, t) \quad,$$

to the fluence by

$$\Phi_i(\vec{x}) = \int_{4\pi} d\vec{\Omega} \int dE \int dt \, \phi_i(\vec{x}, E, \vec{\Omega}, t) \quad,$$

to the current (in the x-direction, say) by [assuming symmetry]

$$\vec{J}_i(\vec{x}, t) = 2\pi \int_0^\pi \sin\theta d\theta \int dE \times$$

$$\times \cos\theta \phi_i(\vec{x}, E, \vec{\Omega}, t) \quad,$$

to the density of the radiation by

$$n(\vec{x}, t) = \int_{4\pi} d\vec{\Omega} \int dE \, \phi_i(\vec{x}, E, \vec{\Omega}, t)/v \quad,$$

and to the energy spectrum by

$$\phi_i(\vec{x}, E, t) = \int_{4\pi} d\vec{\Omega} \, \phi_i(\vec{x}, E, \vec{\Omega}, t) \quad.$$

THE BOLTZMANN EQUATION

The density of radiation in a volume of phase space may change in five ways:

1. Uniform translation, where the spatial coordinates change, but the energy-angle coordinates remain unchanged;
2. Collisions, as a result of which the energy-angle coordinates change, but the spatial coordinates remain unchanged;
3. Continuous slowing down, in which process uniform translation is combined with continuous energy loss;
4. Decay, where particles are changed through radioactive transmutation into particles of another kind;
5. Introduction of a source, the direct emission of a particle into the volume of phase space of interest: electrons or photons from radioactive sources, or neutrons from an α-n emitter, for instance.

The first process results in a current that leads to a reduction of the phase-space density, or

$$\frac{1}{v_i}\left[\frac{\partial \phi_i(\vec{x}, E, \vec{\Omega}, t)}{\partial t}\right]_1 = - \text{div}\, [\vec{\Omega} \phi_i(\vec{x}, E, \vec{\Omega} t)] \quad,$$

which becomes

$$\frac{1}{v_i}\left(\frac{\partial \phi_i}{\partial t}\right)_1 = - \vec{\Omega} \cdot \text{grad}\, \phi_i \quad.$$

The second process involves changes in energy, angle, and particle type as a result of collisions between radiations and nuclei:

$$\frac{1}{v_i}\left[\frac{\partial \phi_i(\vec{x}, E, \vec{\Omega}, t)}{\partial t}\right]_2 = \sum_j \left[\int d\vec{\Omega}'\, dE_B\, \sigma_{ij}(\vec{x}, E_B \to E, \vec{\Omega}' \to \vec{\Omega}) \times \right.$$

$$\left. \times \phi_j(\vec{x}, E_B, \vec{\Omega}', t) - \int d\vec{\Omega}'\, dE_B \times \right.$$

$$\times \sigma_{ij}(\vec{x}, E \to E_B, \vec{\Omega} \to \vec{\Omega}',)\phi_j(\vec{x}, E,\vec{\Omega}, t)\Big]$$

$$- \sigma_i(\vec{x}, E)\phi_i(\vec{x}, E, \vec{\Omega}, t) \ .$$

The cross section $\sigma_{ij}(\vec{x}, E_B \to E, \vec{\Omega}' \to \vec{\Omega})$ represents the probability of producing an i-type particle with phase space coordinates $(\vec{x}, E, \vec{\Omega}, t)$ as a result of a collision with a j-type particle with phase-space coordinates $(\vec{x}, E_B, \vec{\Omega}', t)$. The energy E_B is higher than the energy E, and the first term in the square braces is sometimes called the "down-scattering" integral.

The cross section $\sigma_{ij}(\vec{x}, E \to E_B, \vec{\Omega} \to \vec{\Omega}')$ represents the probability of the production of a higher energy particle at E_B from a lower energy particle. This "up-scattering" integral is significant in treating neutrons with energies low enough so that they may gain energy in colliding with the atoms of a medium - the so-called thermal neutrons, or where a reaction is so exothermic that the energy gain of the outgoing particles is too large to be neglected. All cross sections may depend on the position coordinate in phase space.

The third process describes particles losing energy continuously at a rate S per unit path length (S is called the "stopping power"). The density of these particles at an energy E_B is

$$\phi_i(\vec{x}, E_B, \vec{\Omega}', t)S(\vec{x}, E_B),$$

and on slowing down to E, the density becomes

$$\phi_i(\vec{x}, E, \vec{\Omega}', t)S(\vec{x}, E);$$

hence,

$$\frac{1}{v_i}\left(\frac{\partial \phi_i}{\partial t}\right)_3 = \frac{\partial(\phi_i S)}{\partial E} \quad .$$

Note that energy loss is associated with motion through the medium. Hence, the gradient term should appear here too as it does in (1). I have, for simplicity, left it out. In this case, we are simply discussing the projection of the density change on the energy axis.

The decay process acts as a simple removal; therefore

$$\frac{1}{v_i}\left(\frac{\partial \phi_i}{\partial t}\right)_4 = \left(-\frac{1}{\lambda_i}\right)\phi_i \quad ,$$

where the decay probability, λ_i, per unit path-length is given by

$$\lambda_i = \tau_i c \frac{\beta_i}{(1-\beta_i)^{1/2}} \quad ,$$

where

τ_i is the mean lifetime in the rest frame of the particle,

c is the velocity of light, and

β is the particle velocity relative to the velocity of light.

In the last process, particles are supplied from a "source," such as neutrons from a α-n source, photons or electrons from a radioactive material, or as a result of collisions from high-energy particles. In this case,

$$\frac{1}{v_i}\left(\frac{\partial \phi_i}{\partial t}\right)_5 = Y(\vec{x}, E, \vec{\Omega}, t) \quad .$$

Combining all these processes gives us the Boltzmann equation:

$$\frac{1}{v_i}\left(\frac{\partial \phi_i}{\partial t}\right) = -\vec{\Omega} \cdot \text{grad } \phi_i + \sum_j \left[\int d\vec{\Omega}' \, dE_B \, \sigma_{ij}(\vec{x}, E_B \to E, \vec{\Omega}' \to \vec{\Omega}) \times \right.$$

$$\times \phi_j(\vec{x}, E_B, \vec{\Omega}', t) - \int d\vec{\Omega}' \, dE_B \times$$

$$\left. \times \sigma_{ij}(\vec{x}, E \to E_B, \vec{\Omega} \to \vec{\Omega}')\phi_j(\vec{x}, E, \vec{\Omega}, t) \right] -$$

$$- \sigma_i(\vec{x}, E)\phi_i(\vec{x}, E, \vec{\Omega}, t) + (\partial/\partial E)(\phi_i S) -$$

$$- (1/\lambda_i)\phi_i + Y_i(\vec{x}, E, \vec{\Omega}, t) \quad .$$

In the discussion to follow, only the stationary form of the Boltzmann equation will be considered, so that $\partial \phi_i/\partial t = 0$. Pre-arranging terms and abbreviating somewhat:

$$\vec{\Omega} \cdot \text{grad } \phi + \sigma\phi + (\phi/\lambda) - [\partial(\phi S)/\partial E] = Q + Y \quad ,$$

where

Q represents the sum over the up- and down-scattering integrals in the square brackets,

and subscripts and arguments have been dropped.

This form of the Boltzmann equation is sometimes given in operator notation as

$$B\phi = Q + Y \quad ,$$

where

$$B = \vec{\Omega} \cdot \text{grad} + \sigma + \lambda^{-1} - (\partial S/\partial E) \ .$$

This equation is quite difficult to solve, in general, and a number of approximations and special techniques have been devised to yield useful results.

THE STRAIGHT-AHEAD APPROXIMATION

In many ways, the simplest approximation of the Boltzmann equation arises from the neglect of changes in particle direction that occur as a result of collisions, the second kind of process noted above:

$$\frac{1}{v_i}\left(\frac{\partial \phi_i}{\partial t}\right)_2 = \sum_j \int d\vec{\Omega}' \ dE_B \ \sigma_{ij} \ \phi_j - \phi_i \ \sigma_i \ ,$$

$$= \sim \sum_j \int dE_B \ \sigma_{ij} \ \phi_j - \phi_i \ \sigma_i \ .$$

Since no processes will result in a change in direction, we may without loss in generality, consider a plane wave of radiations traveling in the x-direction of Fig. 1. The Boltzmann equation becomes

$$(\partial \phi_i/\partial x) + (\phi_i/\lambda_i) + \sigma_i \phi_i - (\partial \phi_i S_i/\partial E) = Q_{ij} + Y_i \ ,$$

which is an ordinary differential equation.

Classical electron slowing down can be obtained from the

THE PHYSICS OF RADIATION TRANSPORT

straight-ahead equation. This is done by neglecting all processes except electron-atom collisions in which electron-electron collisions (or electron "stopping") and bremsstrahlung are represented as a continuous slowing down process with no changes in direction. Under these assumptions, the Boltzmann equation in the absence of sources becomes

$$(d\phi/dx) - (\partial\phi S/\partial E) = 0 \quad .$$

Noting that

$$dx = -(1/S) \, dE \quad ,$$

one obtains by integration ,

$$\phi(E, x) = \phi(E_B, 0) S(E_B, 0)/S(E, x)$$

where x and E are related by

$$x(E_B, E) = \int_E^{E_B} [dE'/S(E')] \quad .$$

Since the stopping power is a slowly varying function of energy, the assumption is often made that $S(E)$ can be replaced with its average over the electron range, R,

$$S(E) = \sim E_0/R = \bar{S} \quad .$$

When this assumption is made, the expression for $x(E_B, E)$ reduces to an approximation due to Harder[1],

$$E = E_B(1 - x/R) \quad .$$

This solution to the transport equation yields a plane wave of electrons traveling in the x-direction with an intensity proportional to $1/S(E, x)$. If the assumption is made that $S(E)$ is a constant, as is made to derive Harder's expression, the plane wave has a constant intensity. The electrons comprising the wave all have the same energy which is a monotonically decreasing function of x, as given by the equation for $x(E_B, E)$.

As we will see in a later lecture in this course,[2] this approach to electron transport is a poor one and yields inaccurate results in the 10-30 MeV range. This is due primarily to the representation of bremsstrahlung energy losses as continuous.

Muon propagation can be treated in a similar way. Since muons are radioactive with a mean life of 2.2 μs, an additional term to account for decay is necessary:

$$(d\phi/dx) + (\phi/\lambda) - (\partial\phi S/\partial E) = Y .$$

The source term, Y, is included since muons are the result, in many interesting cases, of pion decay extending over a significant portion of the volume of interest.

For a plane wave incident at $x = 0$ and traveling above x, the approach is identical to the earlier one.

Once again the relation ,

$$dx = - [1/S(E)] \, dE ,$$

is applied, yielding in the absence of sources,

$$\phi(E, x) = \phi(E_B, 0) \frac{S(E_B, 0)}{S(E, x)} \exp\left(- \int_E^{E_B} \frac{dE'}{\lambda(E')S(E')} \right) ,$$

$$x(E_B, E) = \int_E^{E_B} \frac{dE'}{S(E')} \ .$$

As in the electron case, this solution yields a plane wave moving along x, all the muons having an energy E which depends on x according to the equation for $x(E_B, E)$. The intensity of the wave is proportional to $1/S(E)$, but also declines exponentially with increasing x due to the radioactivity of the muon.

This approach has been applied to cosmic-ray muons traveling vertically below the earth's surface.[3] The angular muon flux (see Fig. 2), $\phi(x) = \int_0^\infty \phi(E, x)dE$, was calculated for depths from 0.1 to 1 tonne/cm^2 in the earth, and compared with experimental data (1 tonne/cm^2 = ~ 4km depth below the surface of the earth).[4-11] The agreement in this case is rather good.

The straight-ahead approach has also been applied successfully to high-energy hadron propagation. This is due to the fact that high-energy hadron-nucleus collisions yield secondary hadrons which travel in essentially the same direction as the incident colliding primary. The straight-ahead form for the Boltzmann equation in this case is just

$$B_i \phi_i = Q_{ij} + Y_i \ ,$$

$$B_i = (d/x) + \sigma_i + (1/\lambda) - (\partial S_i/\partial E) \ .$$

The solution for one species i without slowing down, which might apply to neutrons with energies between 25 and 500 MeV for a plane wave of particles incident at x = 0, is

$$\phi(E, x) = \phi_0(E, x) \times \left[e^{-\sigma x} + \int_0^x e^{-\sigma x'} Q(x') \, dx' \right]$$

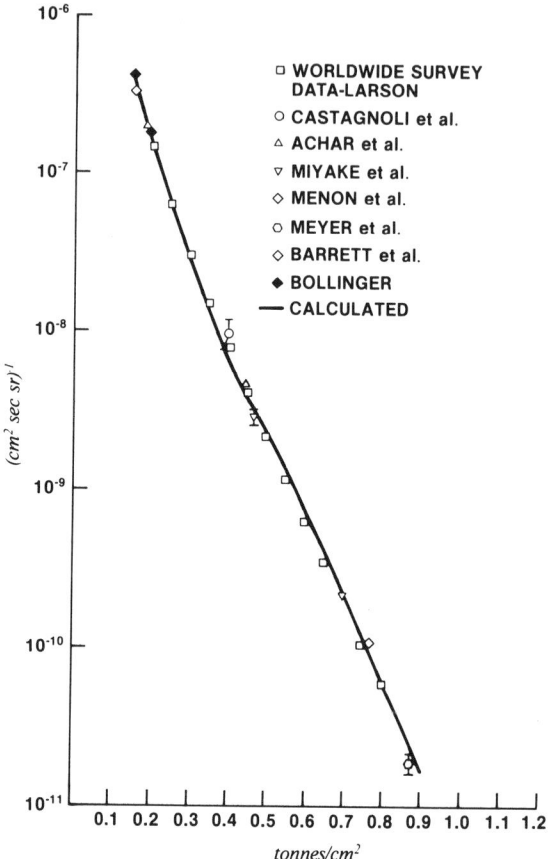

Fig. 2. The cosmic-ray angular muon flux underground.

This solution is no longer a simple plane wave. Neutrons of every energy can be found at every location along x.

Passow[12] has come up with a useful solution to the straight-ahead equations for hadron propagation. At energies sufficiently high that both charged particle stopping and decay processes may be neglected, the Boltzmann equation in the straight-ahead approximation becomes

$$B_i \phi_i = Q_{ij} + Y_i \ ,$$

$$B_i = (d/dx) + \sigma_i \ .$$

Consider a plane wave of hadrons of unit intensity normally incident at x = 0. If σ_i is independent of energy and the cross section σ_{ij} has the functional form,

$$\sigma_{ij} = (N_j/E)(E_B/E)^\ell \ ,$$

where

> N_j is an arbitrary constant depending on emitted particle type, and
>
> ℓ is a constant which is the same for all i, j, then

$$\phi_i(E, x) = N_j \ e^{-\sigma x} \ (\sigma/E)(E_B/E)^\ell [x/B(E_B, E)]^{1/2} \times$$

$$\times I_1 \left\{ 2[xB(E_B, E)]^{1/2} \right\} \ ,$$

$$B(E_B, E) = \sum_k N_k \ \ln(E_B/E) \ ,$$

where I_1 is the modified Bessel function of the first kind.

Passow's approximation has been applied to the calculation of cosmic rays in the earth's atmosphere. The constants, N_j and ℓ were determined from cross-section data in the literature[13], and the vertical angular neutron flux at sea level calculated and compared with experiment.[14,15] The results are shown in Fig. 3 and are clearly good.

Alsmiller[16] has written a very useful review of solutions to the Boltzmann equation in the straight-ahead form.

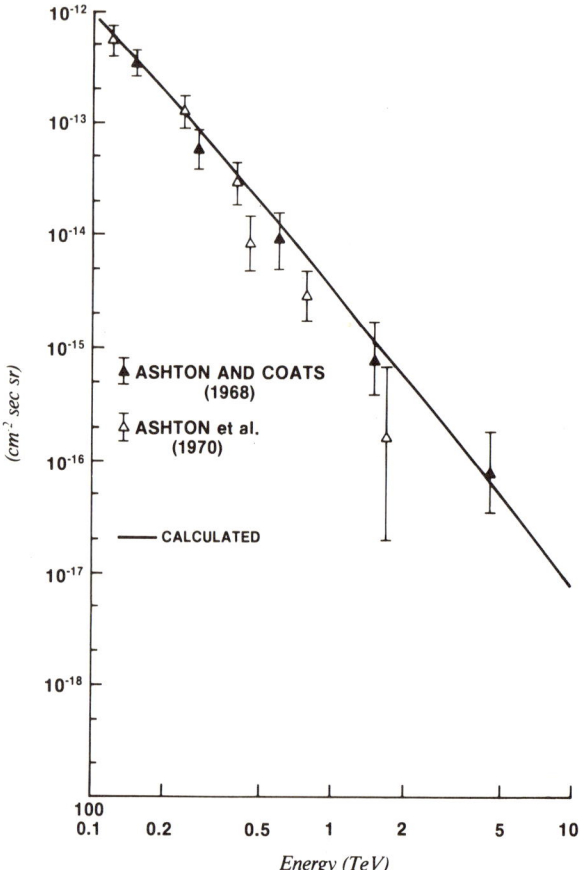

Fig. 3. The cosmic-ray angular neutron flux at sea level.

There are a great many important problems where the straight-ahead approach is inapplicable, as particles coming out of a collision may undergo big changes in direction. One such approach, widely applied to neutron problems, is the method of spherical harmonics.

THE METHOD OF SPHERICAL HARMONICS

While the method of spherical harmonics is applicable to all geometries, this discussion will be restricted to slab geometries,

THE PHYSICS OF RADIATION TRANSPORT

and to one particle type. Under these restrictions, the Boltzmann equation becomes

$$(\mu d\phi/dx) + \sigma\phi + (\phi/\lambda) - (\partial\phi S/\partial E) = Q + Y ,$$

$$\mu = \Omega_x = \cos\theta .$$

It is convenient to lump particle decay in with other absorption processes, so we write $\sigma\phi$ in place of $\sigma\phi + (\phi/\lambda)$, leaving just

$$B = (\mu d/dx) + \sigma - (\partial S/\partial E)$$

for the Boltzmann operator.

We can, without loss of generality, carry out the expansions,

$$\sigma(E_B \to E, \mu_0) = \sum_{L'=0}^{\infty} \left(\frac{2L'+1}{2}\right) q_{L'} P_{L'}(\mu_0) ,$$

$$q_{L'} = \int_{-1}^{+1} d\mu_0 \int_{E}^{\infty} dE_B \, \sigma(E_B \to E, \mu_0) ,$$

$$\phi(\mu) = \sum_{L'=0}^{\infty} \left(\frac{2L'+1}{2}\right) F_{L'} P_{L'}(\mu) ,$$

$$F_{L'} = \int_{-1}^{+1} d\mu \, P_{L'}(\mu) \phi(\mu) ,$$

$$Y = \sum_{L'=0}^{\infty} \left(\frac{2L'+1}{2}\right) U_{L'} P_{L'}(\mu) ,$$

$$Y = \sum_{L'=0}^{\infty} \left(\frac{2L'+1}{2}\right) U_{L'} P_{L'}(\mu) ,$$

$$U_{L'} = \int_{-1}^{+1} d\mu \, P_{L'}(\mu) Y(\mu) .$$

μ_0 is the cosine of the collision angle, i.e., the angle at which a secondary particle comes out of a collision, with respect to the direction of the colliding particle; the functions P_L are the Legendre polynomials. This leads to

$$Q = \int_{-1}^{+1} d\mu \left[\sum_{L'} \left(\frac{2L'+1}{2}\right) F_{L'} P_{L'}(\mu) \right] \times$$

$$\times \left[\sum_{L'} \left(\frac{2L'+1}{2}\right) q_{L'} P_{L'}(\mu_0) \right] \int_E^{\infty} dE_B ,$$

$$= \int_{-1}^{+1} d\mu \left[\sum_{L'} \left(\frac{2L'+1}{2}\right)^{1/2} F_{L'} Y_{L'}^0(\mu) \times \right.$$

$$\left. \times \left[\sum_{L'} q_{L'} \sum_{m=-L'}^{+L'} Y_{L'}^m(\mu) Y_{L'}^m(\mu') \right] \int_E^{\infty} dE_B .$$

Using the spherical harmonics addition theorem,[17] where the functions Y_L^m are the normalized zonal harmonics such that

$$\left(\frac{2L+1}{2}\right)^{1/2} Y_L^0(\mu) = \frac{2L+1}{2} P_L(\mu) ,$$

THE PHYSICS OF RADIATION TRANSPORT

and making use of the orthogonal properties of the zonal harmonics,

$$\int_{-1}^{+1} \left[Y_L^m(\mu) \right]^2 d\mu = \frac{2}{2L+1}$$

after carrying out the integration over μ, we get

$$Q = \int_E^\infty dE_B \sum_{L'} \left(\frac{2L'+1}{2} \right) F_{L'} \, q_{L'} \, P(\mu) \ .$$

We now make the assumption that only a finite number of terms is necessary to represent the flux, that for some L sufficiently large,

$$\sum_{L'=0}^{L} \left(\frac{2L'+1}{2} \right) F_{L'} \, P(\mu) = \sim \sum_{L'=0}^{L} \left(\frac{2L'+1}{2} \right) F_{L'} \, P(\mu) = \phi,$$

to sufficient accuracy for some purpose. This means setting $F_{L'} = 0$ for all $L' > L$. Note that in the expression for Q, the values of $q_{L'}$ for all $L' > L$ do not contribute to the sum. This means that the number of moments necessary to represent ϕ accurately determines the value of L, not the accuracy of the representation of $\sigma(E_B \to E, \mu_0)$.

To digress briefly, a result important to the S_n theory to be discussed later is gotten if we put the polynomial representation of Q in the Boltzmann equation,

$$\frac{\mu_i d\phi}{dx} + \sigma\phi - \frac{\partial \phi S}{\partial E} = \sum_{L'=0}^{L} \left(\frac{2L+1}{2} \right) F_{L'} \, q_{L'} \, P(\mu_i) + Y \qquad (*)$$

where the μ_i are no longer continuous variables but are values chosen according to some prescription, and

$$F_{L'} = \sum_{i=1}^{N} w_i \phi(\mu_i) P_{L'}(\mu_i) \,,$$

i.e., a numerical quadrature using the nodes μ_i and the weights w_i are used to obtain the moments $F_{L'}$. Equation (*) will be the starting point for the discussion of the S_n method.

The method of spherical harmonics proceeds by substituting a suitably truncated Legendre polynomial expansion into the Boltzmann equation, and by replacing the integrals for q_L, by numerical quadrature formulae. This discretization of the energy variable is referred to as imposing a "group structure", and the number of groups is the number of discrete energies into which the energy range is divided. As a consequence, Q becomes an L + 1 by g matrix, where g is the number of groups.

A word about the continuous slowing-down operator should be inserted at this point, as it may not be evident how it can be discretized, and because it is sometimes used in neutron transport to describe elastic neutron scattering with elements heavier than hydrogen.

We have previously observed in deriving the operator that particles slowing down from E_B to E have a density $\phi(E_B) S(E_B)$ at E_B, and a density $\phi(E) S(E)$ at E, so we write

$$\frac{1}{v} \left(\frac{\partial \phi}{\partial t} \right) = \frac{\partial \phi S}{\partial E} = \sim \frac{\phi_{g'+1} S_{g'+1}}{(E_{g'+1} - E_{g'})} - \frac{\phi_{g'} S_{g'}}{(E_{g'+1} - E_{g'})} \,,$$

THE PHYSICS OF RADIATION TRANSPORT

and the Boltzmann operator becomes

$$(\mu d/dx) + \sigma + (S/\Delta E) ,$$

and the second term corresponding to g'+1 is inserted into the matrix Q. In the theory of neutron slowing-down, the stopping power S corresponds to $\sigma_s \xi E$ where ξ is the so-called logarithmic decrement, and σ_s is the elastic scattering cross section. The slowing-down in the operator is usually included in the absorption cross section.

The substitution of a truncated polynomial expansion series into the Boltzmann equation yields a set of coupled differential equations which are solved through the application of suitable boundary conditions. It is easy to show that the interface boundary conditions most suitable for these differential equations, between regions, for L odd, is to impose continuity on the $F_{L'}$ across the boundary. This is accomplished by writing the Boltzmann equation for a thin region between two other regions and allowing the thin region to go to zero, when it becomes an interface boundary. In this limit, all the $F_{L'}$ on the left side of the boundary are equal to the F_L' on the right side of the boundary.

Odd order approximations are generally preferred over even-order approximations, yielding better results. One problem with even-order approximations has to do with the fact that the $F_{L'}$ cannot all be made continuous across an interface.

External boundary conditions are more difficult to apply. Consider the case in Fig. 1 where the half-space defined by x < 0 is a vacuum. The physics of the situation demands that if there is no flux incident,

$$\phi(0, \mu) = 0 \quad \text{for } \mu < 0 \ .$$

However, this can only be approximated by a finite polynomial expansion in μ over the interval $[-1, +1]$. If, instead of Legendre polynomials over $[-1, +1]$, we had used two sets of polynomials, one over $[-1, 0]$ the other over $[0, +1]$, the condition can be satisfied exactly. The use of such half-range polynomials is the basis of Yvon's method.

One approach is to require

$$\phi(0, \mu_i) = \sum_{L'=0}^{L} \left(\frac{2L'+1}{2}\right) F_{L'} P_{L'}(\mu_i) ,$$

where the μ_i are the roots of $P_{L+1}(\mu)$ that are less than zero, $\phi(0)$ is the incident flux. This yields the lower boundary conditions, which, when combined with the upper boundary conditions, yields the solution. This is known as Mark's boundary condition and, physically, is the equivalent to treating the lower half-space as a perfect absorber.

Another approach, known as Marshak's boundary condition, is to set the odd moments of the flux to zero--this, of course, causes the current returning from the vacuum to be zero. In this case,

$$\int_{-1}^{0} \mu^j \phi(0, \mu) \, d\mu = \sum_{L'} (-1)^{L'} a_{L'j} F_{L'} \quad (j \text{ odd}) ,$$

$$a_{L'j} = \frac{2L'+1}{2} \int_{0}^{1} \mu^j P_{L'}(\mu) \, d\mu ,$$

where $\phi(0)$ is the incident flux. Experience has shown that Marshak's conditions are generally superior to Mark's.

Other truncated polynomial sets can be used in the same fashion as the Legendre functions given here, such as the half-range polynomials, or the Tschebyscheff polynomials.[18] The half-range or Yvon's method offers distinct advantages near boundaries. There is no addition theorem to assist in cutting off the influence of slowly converging cross-section expansions that would be orthogonal to Legendre coefficents of the flux in the spherical harmonics approach.

The product of the substitution of a truncated polynomial series into the Boltzmann equation, the replacement of numerical integrals by angular quadratures and by energy groups, the establishment of constraints by the use of outer and interface boundary conditions, is a coupled set of inhomogeneous ordinary differential equations which can be solved by a variety of conventional means.

THE DIFFUSION EQUATION

The diffusion equation is an important tool in the evaluation of the behavior of dilute fluids in a variety of situations outside, as well as inside, the study of radiation transport. Here, it will be derived as the lowest order spherical harmonics approximation.

We represent the radiation flux by the first two terms of the Legendre polynomial expansion:

$$\phi(x, \mu) = \frac{1}{2} F_0(x) P_0(\mu) + \frac{3}{2} F_1(x) P_1(x),$$

and

$$Q = \frac{1}{2} q(x), \quad Y = \frac{1}{2} U(x).$$

Making the polynomial substitution leads to

$$F_1 = -\frac{D d F_0}{dx} ,$$

$$-\frac{D d^2 F_0}{dx^2} = q + u ,$$

which is the diffusion theory result, with $D = 1/3$ equal to the diffusion coefficient. The top equation can be interpreted to give the familiar statement that current is proportional to the gradient of the concentration.

THE METHOD OF DISCRETE ORDINATES

We return to Eq. (*) above, replacing the integral over energy with a numerical integral as before, so that Q is now an L + 1 by g matrix. The transport equation is now made spatially discrete as well by introducing a set of cells (Fig. 4) with discrete values of the s, y, z coordinates. The intervals are chosen in such a way as to be consistent with interface and outer boundaries, and with the magnitude of the gradient of the flux expected.

Eq. (*) is then written ,

$$\frac{\mu(\phi_{k+1} - \phi_k)}{\Delta x} + \sigma \phi_{k+1/2} = Q + Y ,$$

where $\Delta x = x_{k+1} - x_k$ is the cell width, ϕ_k is the flux at x_k, ϕ_{k+1} is the mean flux in the range k to k+1, and the slowing-down term is assumed to be included, as before, in the matrix Q and in σ since the energy variable has been discretized. The discrete ordinate equations are solved using finite difference methods, iteratively

THE PHYSICS OF RADIATION TRANSPORT

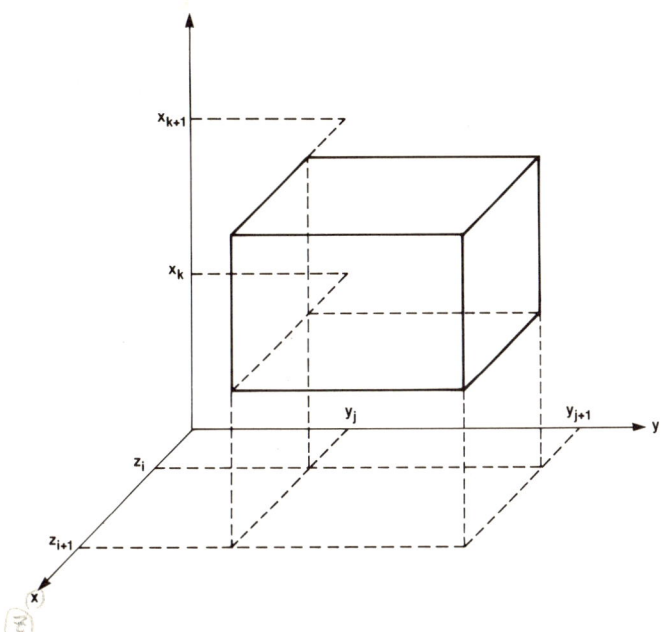

Fig. 4. The cells used in a three-dimensional rectangular coordinate geometry discrete ordinates calculation. [From Carlson and Lathrop.[19]]

recalculating terms in Q that depend on the fluxes being estimated.[19]

THE MONTE CARLO METHOD

The Monte Carlo method is based on the use of random sampling to obtain the solution of the Boltzmann equation. The calculation proceeds by constructing a series of trajectories, each segment of which is chosen at random from a distribution of applicable processes.

If $p(x)\, dx$ is the probability of an occurrence at x in the interval $[a, b]$,

$$\int_a^b p(x')\, dx' = 1 \quad ,$$

then

$$P(x) = \int_a^x p(x') \, dx'$$

is the probability that the event will occur in the interval [a, x], and is monotonically increasing, satisfying $P(a) = 0$, $P(b) = 1$. It is assumed that a random number ξ can be chosen, uniform on the interval [0, 1], from a computer routine. The equation

$$\xi = P(x)$$

amounts to a random choice of the value of x, where the event $p(x)$ can be inverted, as

$$x = P^{-1}(\xi) .$$

To show this, let $p(x) = \delta(x - x_0)$. Then

$$\xi = P(x) = H(x - x_0) ,$$

where δ is Dirac's improper function and H is the Heaviside function [$H(Y < 0) = 0$, $H(Y \geq 0) = 1$]. Performing the inversion,

$$x_0 = \xi$$

for all values of ξ.

Two δ-functions with amplitudes $a_1 + a_2 = 1$ at x_1 and x_2 can easily be seen to yield x_1 with a frequency a_1 and x_2 with a frequency a_2. Proceeding to the limit of a continuous distribution yields the desired result.

THE PHYSICS OF RADIATION TRANSPORT

In Fig. 5 is shown how the distribution function $P(x)$ is constructed from the differential probability (the density of probabilities $p(x)$).

An example of the use of this kind of sampling is in the processes contained in the Boltzmann operator. A particle entering a region travels a certain distance before undergoing a reaction. In this case, the particle satisfies

$$B\phi = 0 ,$$

or

$$(\mu d\phi/dx) + \sigma\phi = 0 ,$$

neglecting slowing-down for the nonce.

The solution to this equation is

$$\phi = \phi_0 \, e^{-\sigma x/\mu} .$$

We write $\sigma x/\mu$ as r, the number of mean-free-paths the radiation travels in the medium. The differential probability per unit mean-free-path for an interaction is given by

$$p(r) = e^{-r} .$$

Hence,

$$\xi = P(r) = 1 - e^{-r} ,$$

and

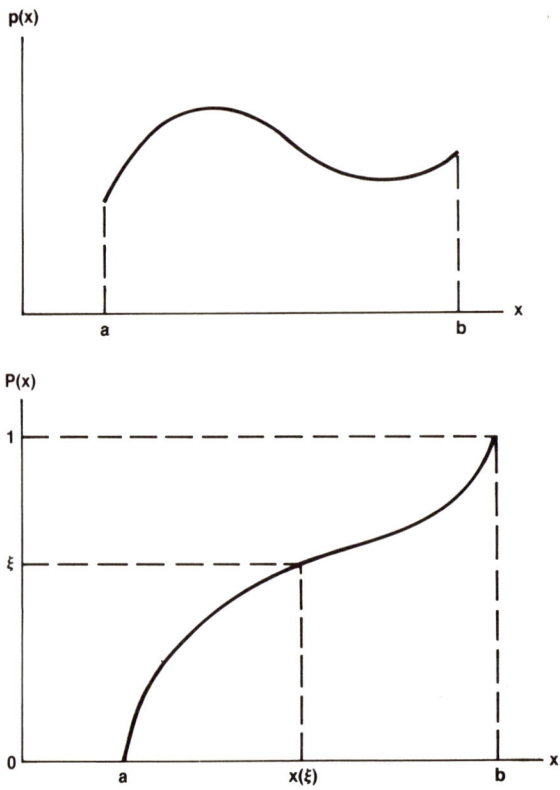

Fig. 5. The density function and the probability distribution constructed from it. [From Carter and Cashwell.[20]]

$$r = -\ln(1 - \xi) ,$$

or, since $1 - \xi$ is just as probable as ξ ,

$$r = -\ln \xi .$$

By taking into account charged particle slowing down during passage along r, the correct energy-dependent cross section can be chosen.

Another common sampling technique is known as the rejection

THE PHYSICS OF RADIATION TRANSPORT   43

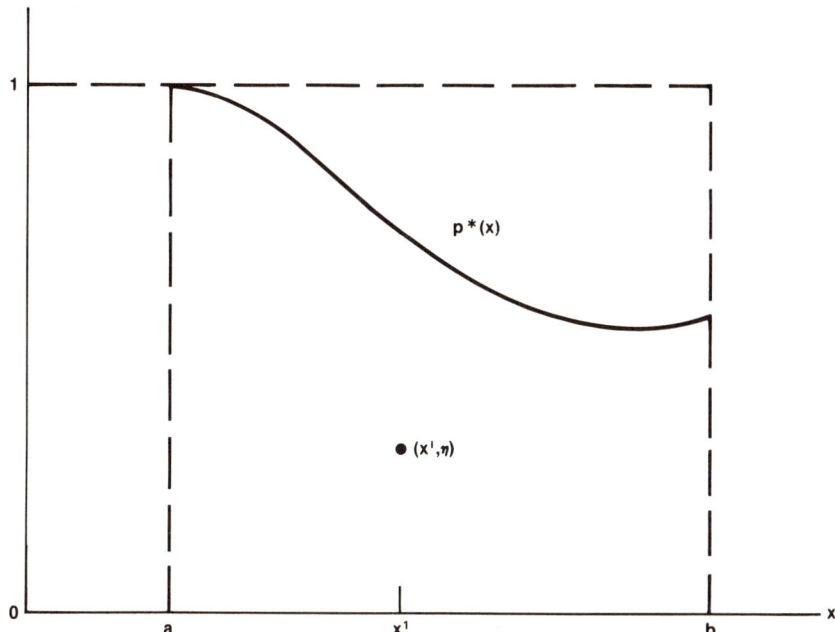

Fig. 6. A diagram of the rejection method. [From Carter and Cashwell.[20]]

technique. This is indicated graphically in Fig. 6. The technique is to redefine p(x) so that

$$p^*(x) = \frac{p(x)}{\text{Max } p(x)} .$$

A pair of random numbers (ξ_1, ξ_2) is chosen. Define $x' = a + \xi_1 (b - a)$. This moves the origin over till it lies under the maximum point of P(x) on the graph. If $p^*(x') < \xi_2$, x' is accepted as our sample.

Clearly, the set of coordinates made up of a column at x', which is dx wide with a height given by P*(x) will have an area proportional to p(x), as can be seen from the figure.

In the simplest and most widely used form of Monte Carlo, a history is obtained by calculating travel distances between collisions, then sampling from distributions made up from the cross sections,

$$\sigma_{ij}(E_B \to E, \Omega' \to \Omega) ,$$

in energy and in angle.

The result of the interaction may be a number of particles of varying types, energies, and directions each of which will be followed in turn. The results of many histories will be processed which, typically, may mean some sort of mean and standard deviation will be calculated.

It is by no means clear that the distributions obtained using the Monte Carlo method will be normally distributed, so that a statistical test of the adequacy of the mean and standard deviation may be required.[21]

The last three methods of calculating radiation transport we have discussed, the spherical harmonics method (in the P_3 approximation),[22] the method of discrete ordinates[23] and the Monte Carlo method[23], have been applied to neutron transport in a concrete slab irradiated with 400 MeV neutrons, normally incident, and the results are shown in Fig. 7. In the Monte Carlo case, the calculation was carried out using a two-component transport equation where proton production from neutron collisions and the reverse were permitted, and for a one-component transport equation containing only neutrons.

The results are presented in terms of the scalar flux with energies higher than thermal, 0.1 and 20 MeV. They are usefully indicative of the sort of agreement one may expect when one does

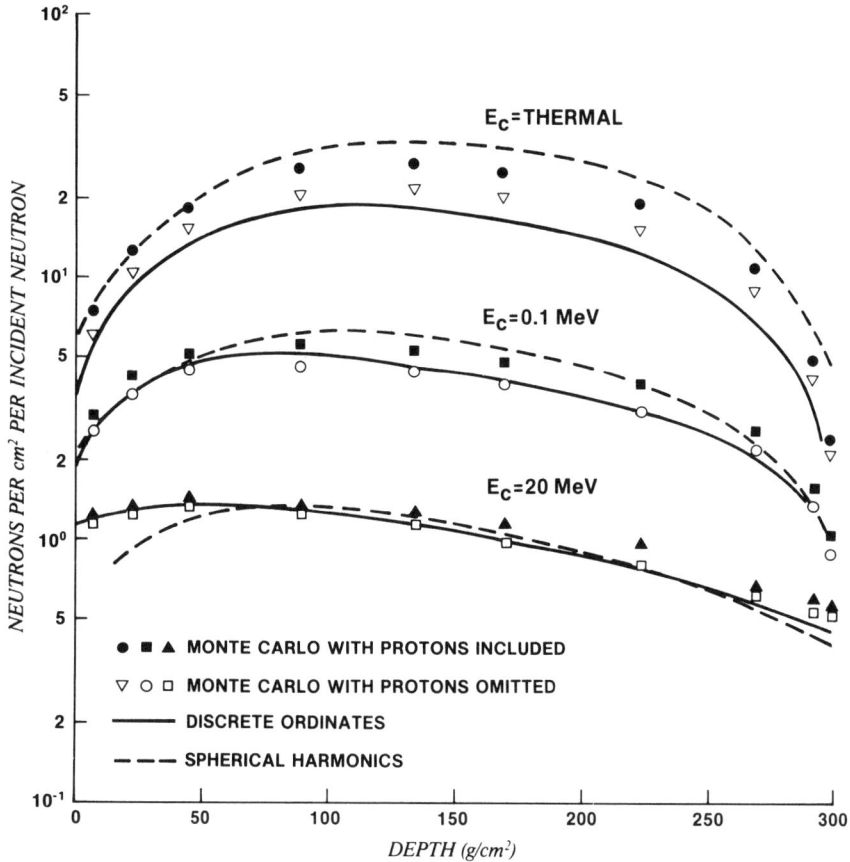

Fig. 7. Neutron fluxes with energies above E_c in a 300-g/cm^2 concrete slab calculated by four methods.

calculations of this sort, and are also indicative by inference of the general validity of all three of these very different approaches.

IMPORTANCE

In the calculation of angular particle fluxes or some integral of the angular flux such as current, scalar flux density, or detector response, some domains of the problem are more "important" than others in that they contribute more of the particle trajectories that make up the desired result. In the calculation of the dose absorbed by a dosimeter in a medium, for instance, those electron trajectories intersecting with the detector region are more important than others. Since the hadrons emitted in essentially the same direction as the incident particle in a hadron--nucleus collision are much more energetic than particles emitted at large angles, the forward parts of the cross section are the most important, and it is now obvious that the straight-ahead approximation, when it is applicable, gains its ease and power by importance-weighting the collision angle of the emerging secondary particles.

This concept is utilized in the Monte Carlo method, usually to reduce the number of histories necessary to obtain some result by reducing the variance. It is accomplished by sampling from fictitious density functions, and correcting for the resultant error using a weight function. This is a common approach to importance in Monte Carlo, and this example,[20] following Carter and Cashwell, may help illuminate it.

Let

$$\bar{f} = \int_a^b f(x)p(x)\,dx\ ,$$

where $p(x)$ is the probability density function discussed in the section on Monte Carlo, and $\bar{f}$ is the mean of the function $f(x)$ defined on the interval $[a, b]$, obtained by random sampling from $p(x)$.

Let us define a different density function $q(x)$, and sample from it. For each value of x_i chosen from the sampling procedure, a weight $w(x_i)$ is assigned, where $w(w_i) \equiv p(x_i)/q(x_i)$.

The mean is obtained by taking the average of the product $w(x_i)f(x_i)$. This mean is the mean of a new function

THE PHYSICS OF RADIATION TRANSPORT

$$g(x) = f(x)w(x) = f(x)p(x)/q(x) \ ,$$

and substituting in the equation for $\bar{f}$,

$$\bar{f} = \int_a^b g(x)q(x) \ dx = \int_a^b f(x)p(x) \ dx \ ,$$

shows that the correct mean is obtained.

The variance, however, is different as we have noted above.

Let

$$\overline{f^2} = \int_a^b f^2(x)p(x) \ dx \ ,$$

$$\overline{g^2} = \int_a^b g^2(x)q(x) \ dx \ ,$$

$$= \int_a^b f^2(x) \frac{p^2(x)}{q^2(x)} q(x) \ dx \ ,$$

$$= \int_a^b \frac{p(x)}{q(x)} f^2(x)p(x) \ dx \ .$$

The variance is given by

$$v^2 = \overline{(g^2)} - (\bar{g})^2 \ ,$$

or

$$v^2 = \overline{g^2} - (\bar{f})^2 \ ,$$

since, as we have seen, $\bar{f} = \bar{g}$, and therefore

$$v^2 = \int_a^b \left[\frac{p(x)}{q(x)}\right] f^2(x) p(x) \, dx - (\bar{f})^2 \, .$$

If $q(x)$ is chosen judiciously, v^2 can be reduced without limit. For instance, if $q(x)$ is chosen so that

$$f(x) p(x) / q(x) = \text{constant} \, ,$$

then

$$\bar{f} = \int_a^b g(x) q(x) \, dx \, ,$$

$$= \int_a^b \frac{f(x) p(x)}{q(x)} q(x) \, dx \, ,$$

$$= \int_a^b \text{constant } q(x) \, dx \, ,$$

$$= \text{constant} \, ,$$

which choice leads to

$$q(x) = \frac{f(x) p(x)}{\bar{f}} \, .$$

Hence, the variance becomes

$$v^2 = (\bar{f})^2 - (\bar{f})^2 = 0 \, .$$

In a practical problem, neither $f(x)$ nor $\bar{f}$ is likely to be known well enough so that such a choice can be made. A good estimate may be possible, however, and $f(x)p(x)/q(x)$ may be made approximately constant, which will reduce the variance. We note that $q(x)$ may define many more sample choices in a given interval than $p(x)$, and if $q(x)$ is chosen so as to reduce the variance, then this is the interval of greatest importance to the determination of $\bar{f}$.

One of the most important methods of calculating particle trajectories of importance to some computational goal is the solution of the adjoint transport equation.

THE ADJOINT FORM OF THE BOLTZMANN EQUATION

Let us abbreviate our coordinate representation by

$$\vec{P} = (x, E, \vec{\Omega}) \quad,$$

and introduce the following notation

$$R = (f, f^+) = \int d\vec{P}\, f(\vec{P}) f^+(\vec{P}) \quad,$$

where R is called the scalar product of f and f^+.

Given a linear operator L, the adjoint L^+ to the operator is defined by[24]

$$(f^+, Lf) = (f, L^+ f^+) \quad.$$

If we replace f, f^+, L and L^+ with the elements of the Boltzmann equation,

$$(\phi^+, B\phi) = (\phi, B^+ \phi^+) \quad,$$

we define the adjoint operator and the adjoint flux, B^+ and ϕ^+. By forming the scalar product indicated, one can construct[25] the adjoint transport equation,

$$B^+\phi^+ = Q^+ + Y^+ ,$$

where

$$B^+ = -\vec{\Omega}\,\text{grad} + \sigma + 1/\lambda - (\partial s/\partial E) ,$$

$$Q^+ = \int dE_B\, d\vec{\Omega}'\, \sigma(E \to E_B, \vec{\Omega} \to \vec{\Omega}') .$$

The adjoint source will frequently be chosen to have the units of a cross section, yielding ϕ^+ in units of a reaction rate[26], such as number of captures or dose.

The adjoint to the Boltzmann equation can be recognized as an "anti-Boltzmann equation" describing particles starting at a low energy from a target, gaining energy with each collision, and emerging at the source.

Naturally, the boundary conditions must be formulated appropriately. For instance, the proper outer vacuum boundary condition for the forward flux requires that no radiation re-enter the medium. The corresponding adjoint boundary condition requires that the adjoint radiation flux does not leave the medium.

Hansen and Sandmeier[27] have shown that, when the adjoint source is chosen to be the macroscopic capture or collision cross section for some effect or process, the adjoint flux on the outer surface of a finite medium in a vacuum is the reaction rate of the system as a function of energy; in other words, the energy response. Hence, given the incident angular flux $\phi(S, E, \vec{\Omega})$, where S is the surface, the integrated reaction rate is

$$R = \int_S ds\, dE\, d\vec{\Omega}\, \phi(s, E, \vec{\Omega}) \phi^+(s, E, \vec{\Omega}) \vec{\Omega} \cdot \vec{n},$$

where $\vec{n}$ is the inward normal at s.

Also we have

$$R = (\phi, Y^+) = (\phi^+, Y)$$

from the definition of the adjoint operator, substituting Y^+ for $B^+\phi^+$, and Y for $B\phi$.

We have seen that the solution of the adjoint form of the Boltzmann equation will yield the energy response of a system directly. In addition, it will yield Monte Carlo estimates of the flux at a point. The adjoint particles can be started at the point desired and followed to the source, whereas the set of forward particles intersecting any given point will have measure zero.

Finally, the relation to importance theory: the adjoint source can be a region of low importance to an unbiased Monte Carlo calculation. The adjoint flux has a high importance in relation to this region as it originates there. Of course, if the source and detector are widely separated, there may be many adjoint trajectories of low importance to the source, and the adjoint calculation may be of minimal value.

The solution of the adjoint to the multicomponent Boltzmann equation is the sum of the solutions for each adjoint source, integrated over the incoming forward flux. Thus, the reaction rate due to an incident forward flux ϕ_j is

$$R_j = \sum_{i=1}^{n} (\phi_j Y_i^+) .$$

Thus for n components, n adjoint solutions are required.

CONTRIBUTONS

Gerstl[27] and Williams and Engle[28] have independently proposed the formation of a fictitious flux,

$$\psi(\vec{x}, E, \vec{\Omega}) = \phi(\vec{x}, E, \vec{\Omega})\psi^+(\vec{x}, E, \vec{\Omega}) ,$$

to be called a "contributon."[28] The product is the flux weighted with its importance. Contributon theory is in its infancy, but a few results are of interest.

The contributon flux can be calculated by first calculating ϕ, then ϕ^+, and then forming the product. This approach has demonstrated its usefulness in shield analysis.[28]

Attempts have been made to formulate a contributon transport equation,[27,29] and to solve for ψ directly, since it would concentrate on just that part of the radiation flux that results in a specified reaction rate. As we have seen, the product (ϕ, ϕ^+) on a surface surrounding a detector (or alternatively, a source) yields the reaction rate.

Painter[29] has obtained the contributon transport equation in slab geometry for the energy-independent case (all particles and pseudoparticles having the same energy, or alternatively, energy-independent cross sections)

$$\mu \frac{\partial \psi(x, \mu)}{\partial x} + \bar{\sigma}_s(x,\mu)\psi(x, \mu) = 2\pi \int_{-1}^{+1} \bar{\sigma}_s(x, \mu' \to \mu)\psi(x, \mu') \times$$

$$\times d\mu' + \phi^+(x, \mu)Y - \phi(x, \mu)Y^+,$$

where the contributon scattering cross section is given by

$$\bar{\sigma}_s(x, \mu' \to \mu) = \sigma_s(x, \mu' \to \mu) \frac{\phi^+(x, \mu)}{\phi^+(x, \mu')},$$

and

$$\bar{\sigma}_s(x, \mu) = 2\pi \int_{-1}^{+1} \sigma_s(x, \mu \to \mu') d\mu'.$$

The principal difficulty with this approach is the formulation of the scattering cross section.

Another approach by Dubi et al.[30] uses a forward Monte Carlo calculation and calculates the adjoint flux in a prespecified volume. A source particle is emitted and its path followed to a collision point in the volume. Each collision is used as an estimate of the collision rate and the flux at these points. At each collision point, a secondary particle is emitted and followed to obtain an estimate of its importance in entering the detector, yielding the adjoint flux. The product of the adjoint and direct fluxes integrated over the volume can be converted into a reaction rate by dividing the volume into surfaces and calculating R, the reaction rate, on the surface as we did while discussing the adjoint to the Boltzmann equation.

The efficiency of the method hinges on the fact that the surface can be located deep in the medium and close to the detector, so that the computation of the adjoint flux can be made reasonably efficiently.

This method does not track contributons directly, but does gather information on detector response with enhanced efficiency by using the formal relations derived from contributon theory.

REFERENCES

1. "Radiation Dosimetry: Electrons with Initial Energies Between 1 and 50 MeV," Report 21, ICRU Publications, Washington, D.C. (1972).
2. K. O'Brien. "Electron Dosimetry Using Monte Carlo Techniques," this course.
3. K. O'Brien. Phys. Rev. D 5, 597 (1972).
4. M. O. Larson, Ph.D. Thesis, University of Utah, 1968 (unpublished).
5. C. Castagnoli, A. DeMarco, A. Longhetto, and P. Penengo. Nuovo Cimento 35, 969 (1965).
6. C. V. Achar, V. S. Narasimhan, P. V. Romana Murthy, D. R. Creed, J. L. Osborne, and A. W. Wolfendale. Proc. Phys. Soc. (London) 86, 1305 (1965).
7. S. Miyake, V. S. Narasimhan, and P. V. Romana Murthy. Nuovo Cimento 32, 1505 (1964).
8. M. G. K. Menon, S. Naranan, V. S. Narasimhan, K. Hinotani, N. Ito, S. Miyake, D. R. Creed, J.L. Osborne, and A. W. Wolfendale. Can J. Phys. 46, 5344 (1968).
9. B. S. Meyer, J. P. F. Sellschop, M. F. Crouch, W. R. Kropp, H. W. Sobel, H. S. Gurr, J. Lathrop, and F. Reines. Phys. Rev. D 1, 2229 (1970.
10. P. H. Barrett, L. M. Bollinger, G. Cocconi, Y. Eisenberg, and K. Greisen. Rev. Mod. Phys. 24, 133 (1952).

11. L. M. Bollinger, Ph.D. Thesis, Cornell University, 1951 (unpublished).
12. C. Passow. "Phenomenologische Theorie zur Berechnung einer Kaskade aus schweren Teilchen (Nukleonenkaskade) in der Materie," Deutches Elektronen Synchorotron Report DESY Notiz A 2.85 (1962).
13. K. O'Brien. Nuovo Cimento 3A, 521 (1971).
14. F. Ashton and R. B. Coats. J. Phys. A1, 169 (1968).
15. F. Ashton, N. I. Smith, K. King, and E. A. Mamidzian. Acta Physica Acad. Sci. Hung. 29, Suppl. 3, 25 (1970).
16. F. S. Alsmiller. "A General Category of Soluble Nucleon-Meson Cascade Equations, Oak Ridge Report ORNL-3746 (1965).
17. E. Jahnke and F. Emde. Tables of Functions with Formulae and Curves (Dover Publications, 1954).
18. B. G. Bennett and H. L. Beck. "Legendre, Tschebyscheff, and Half-Range Legendre Polynomial Solutions of the Gamma Ray Transport Equation in Infinite Homogeneous and Two Media Geometry," USAEC Report HASL-185 (1967).
19. B. G. Carlson and K. D. Lathrop. "Transport Theory the Method of Discrete Ordinates," in Computing Methods in Reactor Physics, H. Greenspan, C. N. Kelber, and D. Okrent, eds. (Gordon and Breach, 1968).
20. L. L. Carter and E. D. Cashwell. Particle-Transport Simulation with the Monte Carlo Method (Technical Information Center, USERDA, 1975).
21. G. L. Burrows and D. B. MacMillan. Nucl. Sci. and Eng. 22, 384 (1965).
22. K. O'Brien. "Shielding Calculations for Broad Neutron Beams Normally Incident on Slabs of Concrete," USAEC Report HASL-221 (1970).
23. R. G. Alsmiller, Jr., F. R. Mynatt, J. Barish, and W. W. Engle, Jr. Nucl. Sci. and Eng. 36, 251 (1969).
24. M. Hammermesh. Group Theory and Its Application to Physical Problems (Addison Wesley Pub. Co., 1962).

25. H. S. Isbin. *Introductory Nuclear Reactor Theory* (Reinhold Pub. Co., 1963).
26. G. E. Hansen and H. A. Sandmeier. *Nucl. Sci. and Eng.* 22, 315 (1965.
27. S. A. W. Gerstl. "A New Concept for Deep Penetration Transport Calculations and Two New Forms of the Neutron Transport Equation," Los Alamos Report LA-6628-MS (1976).
28. M. L. Williams and W. W. Engle, Jr. *Nucl. Sci. and Eng.* 62, 92 (1977).
29. J. W. Painter. "An Alternative Approach to the Contributon Problem," Los Alamos Report LA-7131-PR (1978).
30. A. Dubi, S. A. W. Gerstl, D. J. Dudziak. "Monte Carlo Aspects of Contributons," Los Alamos Report LA-UR (1977).

LOW ENERGY NEUTRON AND GAMMA-RAY PROGRAMS
AND THEIR APPLICATIONS

LECTURE 3: THE METHODS AND APPLICATIONS OF DISCRETE ORDINATES IN
            LOW ENERGY NEUTRON-PHOTON TRANSPORT (ANISN, DOT).
            PART I: METHODS

W. W. Engle, Jr.

Oak Ridge National Laboratory
Oak Ridge, TN 37830
USA

INTRODUCTION

Lecture 2 contained a brief introduction to the differential form of the Boltzman transport equation which is solved by the discrete ordinates codes. A rather complete description of the derivation of the finite difference form of the transport equation can be found in Reference 1; therefore that derivation will not be discussed here. Attention will be focused on the additional equations required to solve the transport equation which are often referred to as flux models and on the iteration process and efforts to accelerate the convergence of the iteration process. All equations discussed here will be limited to the one-dimensional, time-independent case, but they may be extended in a straightforward manner to multidimensional, time-dependent geometries.

THE FINITE DIFFERENCE EQUATION

The finite difference form of the transport equation can be derived by considering a flow balance across a typical spatial interval as pictured in Fig. 1.

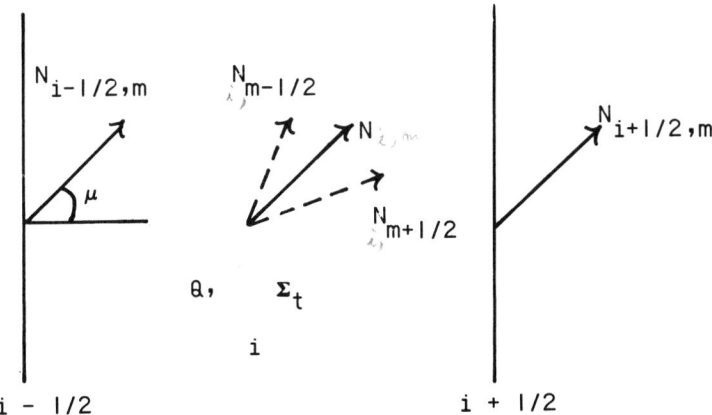

Fig. 1.

Here we consider the flow of particles in a solid angle about the direction defined by the cosine μ. The fraction of a unit sphere occupied by the solid angle about μ is defined as w, often called the angular weight. The subscript i refers to the ith spatial interval and the subscript m refers to the mth angular interval. The subscripts i ± 1/2 and m ± 1/2 refer to the boundaries of the spatial and angular intervals respectively. A third subscript, g, is implied since in general the discrete ordinates equations are solved in multigroup form; however, for the purposes of this discussion, consideration of a single group is sufficient since the groups are interrelated through the source term, Q. Q is assumed to be a uniform source across the interval and may consist of scattering from all groups, particles produced by fission in a multiplying medium, and user-defined constant sources. Σ_t is the total cross section of the material in the interval. In general in the discussion which follows, centered subscripts are omitted; therefore $N_{i,m,g}$ is written simply as N and refers to the average number of particles per unit time per unit area per unit solid angle in the mth angular interval in the ith spatial interval in group g.

The net convective flow of particles through interval i in direction m is given by the number entering the i-1/2 boundary minus the number leaving the i+1/2 boundary and is written as

$$w \mu A_{i-1/2} N_{i-1/2} - w \mu A_{i+1/2} N_{i+1/2}$$

where A refers to the area of the interval interface. The source and removal of particles in the i,m interval are written as wQV and $w\Sigma_t NV$ respectively where V is the volume of interval i. In curved geometries the angular convection shown in Fig. 2 must be considered. As a particle traverses the ith curved interval, its direction cosine relative to the origin changes from μ to μ' and it is clear that an angular interval may gain and lose particles as they traverse a spatial interval. Note, however, that particles always change angles such that the cosine μ' is greater than μ even when μ is negative. In a manner analogous to the spatial convection terms, the angular convection may be written as

$$\alpha_{m-1/2} N_{m-1/2} - \alpha_{m+1/2} N_{m+1/2}$$

where the coefficients, α, are to be determined.

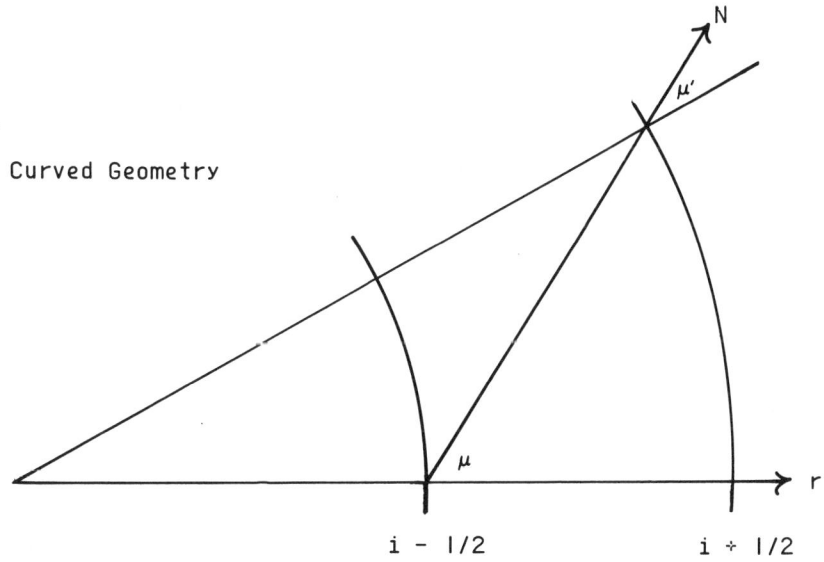

Fig. 2.

Equating the source terms and the loss terms and dividing by the angular weight w the basic finite difference form of the transport equation is obtained:

$$\mu A_{i+1/2} N_{i+1/2} - \mu A_{i-1/2} N_{i-1/2} + \frac{\alpha_{m+1/2}}{w} N_{m+1/2} \quad (1)$$

$$- \frac{\alpha_{m-1/2}}{w} N_{m-1/2} + \Sigma_t VN = QV .$$

In order to determine the form of the α's, consider an infinite homogeneous medium with an everywhere uniform isotropic source of single velocity particles. It is obvious that this situation results in an everywhere equal isotropic flux, $N_{i+1/2} = N_{i-1/2} = N_{m+1/2} = N_{m-1/2}$, and total removal equal to the source $\Sigma_t VN = QV$. If the transport equation is applicable to this situation — and it surely must be — a recursive relationship for the α's is obtained:

$$\text{and } \alpha_{m+1/2} = \alpha_{m-1/2} - w\mu (A_{i+1/2} - A_{i-1/2}) . \quad (2)$$

There are two angular interfaces in spherical geometry for which $\alpha = 0$, namely $\mu = \pm 1.0$. Since, as previously mentioned, the angular streaming is always toward a larger μ, the calculation in a discrete ordinates program always proceeds from $\mu = -1.0$ to $\mu = +1.0$ and $\alpha_{1/2}$ and $\alpha_{M+1/2}$ are both zero (M is the number of angular intervals.) Note that Equation 2 then defines all $\alpha = 0.0$ in slab geometry as it should.

LINEAR AND STEP FLUX MODELS

Having determined a relationship to calculate the α coefficients, consider the solution of Equation 1. There are five unknowns; therefore some additional relationships must be defined in order to solve the equation. The oldest and most often used flux model is the linear model or diamond difference model defined by the following equations:

$$\begin{aligned} N_{i+1/2} &= 2N - N_{i-1/2} \\ N_{m+1/2} &= 2N - N_{m-1/2} \end{aligned} \quad (3)$$

for $\mu > 0$.

Substituting 3 in 1 and solving for N yields:

$$N = \frac{\mu(A_{i+1/2} + A_{i-1/2})N_{i-1/2}}{2\mu A_{i+1/2} + 2\frac{\alpha_{m+1/2}}{w} + \Sigma_t V} \genfrac{}{}{0pt}{}{+}{-}$$

$$\genfrac{}{}{0pt}{}{+}{-} \frac{\left(\frac{\alpha_{m+1/2}}{w} + \frac{\alpha_{m-1/2}}{w}\right)N_{m-1/2} + QV}{} \qquad (4)$$

$N_{i-1/2}$ and $N_{m-1/2}$ are determined from boundary conditions or calculations in the previous spatial or angular interval.

Considering the calculational speed on a computer, the following relationships are defined:

$$\alpha = 1/2(\alpha_{m+1/2} + \alpha_{m-1/2}) \qquad (5)$$
$$A = 1/2(A_{i+1/2} + A_{i-1/2}) . \qquad (6)$$

Combining Equations 5 and 6 with the recursive relationship for the α's and substituting in 4 yields:

$$N = \frac{2\mu A N_{i-1/2} + 2\alpha/w\, N_{m-1/2} + QV}{2\mu A + 2\alpha/w + \Sigma_t V} \qquad (7)$$

which is a more efficient form of the solution if the coefficients of the N's are precomputed.

If $\mu < 0$ the following relationship is substituted:

$$N_{i-1/2} = 2N - N_{i+1/2}$$
$$N_{m+1/2} = 2N - N_{m-1/2} \qquad (8)$$

and the resulting solution for N is:

$$N = \frac{2|\mu|A N_{i+1/2} + 2\alpha/w\, N_{m-1/2} + QV}{2|\mu|A + 2\alpha/w + \Sigma_t V} . \qquad (9)$$

After solving for N, the flux models are used to determine the unknown boundary fluxes and the calculation proceeds to the next interval. Note, however, that the linear flux model can lead to negative flux values if the average flux is less than one-half of the known boundary flux. For this reason most discrete ordinates codes which use the linear model also utilize one or more "fixup" flux models when negatives occur. One of the first fixup models was the "step" model defined by

$$N_{i+1/2} = N_{m+1/2} = N \tag{10}$$

when $\mu>0$.

Solving for N as before gives:

$$N = \frac{\mu A_{i-1/2}\, N_{i-1/2} + \frac{\alpha_{m-1/2}}{w} N_{m-1/2} + QV}{\mu A_{i+1/2} + \frac{\alpha_{m+1/2}}{w} + \Sigma_t V} \tag{11}$$

Using the α recursive relationship, it can be shown that the sum of the first two terms in the denominator of 11 is equal to the sum of the coefficients of $N_{i-1/2}$ and $N_{m-1/2}$ and a form of 11 analogous to 7 may be obtained. Similarly for $\mu<0$, interchanging $N_{i-1/2}$ and $N_{i+1/2}$ in the step model yields:

$$N = \frac{|\mu|\, A_{i+1/2}\, N_{i+1/2} + \frac{\alpha_{m-1/2}}{w} N_{m-1/2} + QV}{|\mu|\, A_{i-1/2} + \frac{\alpha_{m+1/2}}{w} + \Sigma_t V} \tag{12}$$

BOUNDARY CONDITIONS

As stated previously some assumption must be made about the incoming boundary flux in the spatial intervals at the boundaries of the model. Five boundary conditions are described in this section for the right boundary of a one-dimensional geometry. These may be extended to other boundaries in multidimensional geometries.

Boundary conditions at the right boundary are applied to all incoming directions m, where $\mu_m<0$. A vacuum boundary simply implies that all $N_{I+1/2,m} = 0$ where I is the number of spatial intervals in the model. The reflected boundary condition implies that $N_{I+1/2,m} = N_{I+1/2,m'}$ where $\mu_m = -\mu_{m'}$. These two boundary conditions are the most commonly used. The periodic boundary condition is used when modeling repeating segments and is defined by $N_{I+1/2,m} = N_{1/2,m}$.

The albedo boundary condition does not rigorously describe any physical situation, but it is useful in some approximations and is defined by

$$N_{I+1/2,m} = \frac{R \sum_{m'} N_{I+1/2,m'} \mu_{m'} w_{m'}}{\sum_{m'} \mu_{m'} w_{m'}} \qquad (13)$$

where $\mu_{m'} > 0$

$\mu_m < 0$

R = albedo or reflection coefficient.

The last boundary condition is often referred to as a boundary source but, in fact, is a complete specification of the incoming angle dependent flux at a boundary.

INITIAL DIRECTIONS

Examination of Equations 9 and 12 reveals that while the coefficient of the $N_{1/2}$ term in the numerator is zero for the step model ($\alpha_{1/2} = 0$), the coefficient is non-zero for the linear model. It is required, therefore, to calculate the flux in a singular direction, which is a boundary condition for the first angular interval in each spatial interval. In general the assumptions for the linear model initial direction are:

$w = 0$

$\mu < 0$

$\alpha_{m-1/2} = 0$

$N_{i-1/2} = 2N - N_{i+1/2}$.

$N_{m+1/2} = N$

$\frac{\alpha_{m+1/2}}{w} = -\mu(A_{i+1/2} - A_{i-1/2})$

Substituting the above into Equation 1 yields

$$N = \frac{2|\mu|A\, N_{i+1/2} + QV}{2|\mu|A_{i+1/2} + \Sigma_t V} \quad . \qquad (14)$$

Since the linear model assumed for the initial direction calculation can also generate negatives, a step initial direction model used as a fixup may be provided. Changing the spatial model to

$N_{i-1/2} = N$ gives the step model form of the initial direction flux:

$$N = \frac{|\mu|A_{i+1/2} N_{i+1/2} + QV}{|\mu|A_{i+1/2} + \Sigma_t V} . \tag{15}$$

FLUX MODEL COMPARISONS

The preceding descriptions of the linear and step model are useful not only because the linear model with a step fixup is probably the oldest flux model combination, but because all other flux models widely used in discrete ordinates codes lie somewhere between the linear model and the step model. Historically, all flux models have been a compromise between simplicity (speed), accuracy and positivity. (Here the concern is to generate positive fluxes when the source term is positive. Because the scattering cross sections in most discrete ordinates codes are represented as truncated expansions of Legendre polynomials, negative scattering sources are quite often generated.) In general the linear model is very simple, accurate if the mesh interval size is reasonable, but not always positive. Of course positivity can be assured (assuming positive sources) if the spatial and angular mesh intervals are sufficiently small, but as a practical matter this can seldom be accomplished in real problems, especially in multidimensions. The step model is simple and positive but distressingly inaccurate. Another model used as a fixup is referred to as the "zero" model since the assumption is made that the offending flux is zero and then Equation 1 is resolved for the average flux. This model is simple and non-negative, but it also lacks accuracy particularly in deep penetration calculations. The linear model with zero model fixup does appear to be the best fixup model compromise for criticality calculations since it does not suffer as severely as the linear model with step fixup from model switching oscillations. Recently much attention has been focused on the weighted difference model which is used both as the primary model and in some codes as the fixup for the linear model. As will be shown in the next section, the weighted difference model can be both positive and accurate, but it is not simple.

WEIGHTED DIFFERENCE MODEL

Consider the following flux model equations:

$$N = bN_{i+1/2} + (1 - b) N_{i-1/2}$$
$$N = cN_{m+1/2} + (1 - c) N_{m-1/2} \tag{16}$$

If $b = c = 1/2$, Equations 16 are identical to the linear model. If $b = c = 1$, they are identical to the step model equations for $\mu > 0$. Combining Equations 16 with Equation 1 and solving for $N_{i+1/2}$ and $N_{m+1/2}$ gives the following:

$$N_{i+1/2} = \frac{QV + \frac{\alpha_{m-1/2}}{w} + (1/c - 1) \frac{\alpha_{m+1/2}}{w} N_{m-1/2}}{b(\mu A_{i+1/2}/b + \frac{\alpha_{m+1/2}}{wc} + \Sigma_t V)} \;\underline{+}$$

$$\underline{+}\; \frac{\mu A_{i-1/2} - (1 - b) \frac{\alpha_{m+1/2}}{wc} + \Sigma_t V \quad N_{i-1/2}}{} \tag{17}$$

$$N_{m+1/2} = \frac{QV + \mu A_{i-1/2} + (1/b - 1) A_{i+1/2} \quad N_{i-1/2}}{c(\mu A_{i+1/2}/b + \frac{\alpha_{m+1/2}}{wc} + \Sigma_t V)} \;\underline{+}$$

$$\underline{+}\; \frac{\frac{\alpha_{m-1/2}}{w} - (1 - c) \frac{\mu A_{i+1/2}}{b} + \Sigma_t V \quad N_{m-1/2}}{} \;. \tag{18}$$

It is clear by inspection that except for the bracketed terms, all terms in Equations 17 and 18 are always positive (again, assuming positive sources). Several options are available which lead to a positive solution for $N_{i+1/2}$ and $N_{m+1/2}$. The first option is to choose b and c such that the bracketed terms are positive:

$$\mu A_{i-1/2} > (1 - b) \left(\frac{\alpha_{m+1/2}}{wc} + \Sigma_t V\right) \tag{19}$$

$$\frac{\alpha_{m-1/2}}{w} > (1-c)\left(\frac{\mu A_{i+1/2}}{b} + \Sigma_t V\right) \tag{20}$$

An approximate and conservative solution to the above inequalities is to assume c = 1/2 when solving 19 for b and assume b = 1/2 when solving 20 for c.

$$1-b = \mu\, A_{i-1/2} / \left(\frac{2\alpha_{m+1/2}}{w} + \Sigma_t V\right) \tag{21}$$

$$1-c = \frac{\alpha_{m-1/2}}{w} / (2\mu\, A_{i+1/2} + \Sigma_t V) \tag{22}$$

It is clear that this solution yields the maximum value for both b and c consistent with the assumption that all models must lie between the linear model (most accurate) and the step model (always positive). In this case and in subsequent cases, if b or c is calculated less than 1/2, it is set equal to 1/2, and if b or c is calculated greater than 1, it is set equal to 1. Setting a smaller value to 1/2 still assures positivity since larger values approach the step model which is always positive. Unfortunately this option is equivalent to using the step model or something near the step model over the entire mesh for most practical problems and is therefore not accurate.

A second approach is to calculate b and c such that the entire numerators of Equations 17 and 18 are positive. This perhaps seems reasonable since the desired results are positive solutions for $N_{i+1/2}$ and $N_{m+1/2}$. However, this approach has been observed to produce $N_{i+1/2}$ and $N_{m+1/2}$ that are positive but negligibly small giving results similar to the zero model. A third option is to consider the bracketed term and other selected terms (such as the source term) but not the entire numerator. This technique was used for a time in ANISN and DOT at ORNL and works reasonably well in deep penetration calculations. However, in criticality calculations, consideration of selected terms often ignores the dominant particle source or flow in some intervals resulting in use of step or near step models when not required for positivity.

A compromise solution is to consider the bracketed terms plus some fraction, β, of all the remaining terms in the numerator. This leads to the following inequalities:

$$\frac{\mu A_{i-1/2} + \beta\left[QV + \left(\frac{\alpha_{m-1/2}}{w} + (1/c - 1)\frac{\alpha_{m+1/2}}{w}\right)N_{m-1/2}\right]}{N_{i-1/2}} >$$

$$(1 - b)\left(\frac{\alpha_{m+1/2}}{wc} + \Sigma_t V\right) \qquad (23)$$

$$\frac{\frac{\alpha_{m-1/2}}{w} + \beta\left[QV + \mu\left(A_{i-1/2} + (1/b - 1)A_{i+1/2}\right)N_{i-1/2}\right]}{N_{m-1/2}} >$$

$$(1 - c)\left(\frac{\mu A_{i+1/2}}{b} + \Sigma_t V\right). \qquad (24)$$

Again with conservative assumptions for b and c, approximate solutions for b and c are:

$$1 - b = \frac{\left(\mu A_{i-1/2} + (QV + \frac{\alpha_{m-1/2}}{w}N_{m-1/2})\frac{\beta}{N_{i-1/2}}\right)}{\left(\frac{2\alpha_{m+1/2}}{w} + \Sigma_t V\right)} \qquad (25)$$

$$1 - c = \frac{\left(\frac{\alpha_{m-1/2}}{w} + (QV + \mu A_{i-1/2} N_{i-1/2})\frac{\beta}{N_{m-1/2}}\right)}{(2\mu A_{i+1/2} + \Sigma_t V)} \qquad (26)$$

Substituting Equations 16 in Equation 1 and solving for N gives:

$$N = \frac{QV + \mu\left[A_{i-1/2} + (1/b - 1)A_{i+1/2}\right]N_{i-1/2}}{\Sigma_t V + \mu A_{i+1/2}/b + \alpha_{m+1/2}/wc} +$$

$$+ \frac{1/w\left[\alpha_{m-1/2} + (1/c - 1)\alpha_{m+1/2}\right]N_{m-1/2}}{} \qquad (27)$$

Equation 27 is solved by calculating b and c from Equations 25 and 26. The unknown boundary fluxes are calculated using Equations 16. Again b and c are limited to the range 1/2 to 1. The solution does not seem sensitive to the value of β and 0.9 has been used successfully. In practice the solution is not as complex as it appears since many of the terms in Equations 25 and 26 have been precomputed. All weighted difference equations discussed thus far have been for $\mu > 0$. Similar equations are obtained for $\mu < 0$ by reversing the i+1/2 and i-1/2 subscripts in Equation 16 and repeating the process. Equations for initial directions using weighted difference are obtained as before, and the results are:

$$N = \frac{QV + |\mu|\ (A_{i+1/2} + (1/b-1)A_{i-1/2})\ N_{i+1/2}}{\Sigma_t V + |\mu|\ [A_{i+1/2} + (1/b - 1)\ A_{i-1/2}]} \quad (28)$$

with b determined from the relation:

$$1-b = \left(|\mu|\ A_{i+1/2} + QV \frac{B}{N_{i+1/2}}\right) \bigg/ \left[\Sigma_t V + |\mu|\ (A_{i+1/2} - A_{i-1/2})\right] \quad (29)$$

If $\Sigma_t = 0$ in a slab, inspection of the equation for $N_{i-1/2}$ shows that b = 1/2 will not cause negatives to be calculated. As before b is limited to the range from 1/2 to 1.

This form of weighted difference is available in both ANISN and DOT-IV and has been used successfully in criticality problems where other forms of weighted difference failed.

THE ITERATION PROCEDURE

Having chosen an appropriate spatial mesh, angular mesh, flux model, fixup model if required, set of boundary conditions and cross section library, the problem of interest can be defined for the discrete ordinates code and the iteration process can begin. The inner or flux iteration begins with one sweep of the spatial mesh in order to calculate the initial direction flux in the first energy group. Additional sweeps of the spatial mesh are completed for each direction in the angular mesh. The calculation always moves through the spatial mesh in the general direction of the particle flow. For example in one-dimensional geometries, if $\mu>0$ the calculation moves from right to left and when $\mu<0$ the calculation moves from left to right. One complete sweep through all spatial intervals for all angular intervals in one energy group is defined as one inner iteration. Note that the within-group scattering cross section is a part of the removal term on the left side of Equation 1 and a part of the scattering source term on the right side of Equation 1. The removal term is associated with the flux in the current iteration, and the source term must use the flux calculated in the previous iteration. Additional inner iterations are required to resolve this unbalance in the within-group scattering term. The iterations continue until the fluxes in successive iterations satisfy some convergence criterion or until a specified maximum number of iterations is completed. A complete set of inner iterations for each group is defined as an outer iteration. The fission process and thermal upscattering cause an outer iteration unbalance and when either of these processes are present, the outer iteration process is also repeated.

INNER ITERATION CONVERGENCE ACCELERATION

Equation 30 is a slightly expanded form of Equation 1 to illustrate the within-group scattering unbalance.

$$\mu A_{i+1/2} N_{i+1/2} - \mu A_{i-1/2} N_{i-1/2} + \frac{\alpha_{m+1/2}}{w} N_{m+1/2} - \frac{\alpha_{m-1/2}}{w} N_{m-1/2} + \Sigma_r NV + \Sigma_{g \to g} NV = \Sigma_{g \to g} \phi^P V + Q' \qquad (30)$$

Here Σ_t has been replaced by the sum of the removal cross section Σ_r and the within-group scattering cross section $\Sigma_{g \to g}$ and the source Q has been replaced by the sum of the within-group scattering source and Q' which includes scattering from all other groups plus fission and fixed sources. $\phi = \Sigma N_m w_m$ is the scalar flux; the superscript p refers to the previous inner iteration. For simplicity isotropic scattering has been assumed. If Equation 30 is summed over all angular intervals and all spatial intervals, a system balance equation is obtained:

$$L + R + SS = SS^p + Q' \qquad (31)$$

where L is the net leakage, R is the removal by absorption and scattering to other groups and SS is the within-group scattering term. It is possible to define a flux scale factor f such that Equation 32 is satisfied.

$$f(L + R + SS) = fSS + Q' \qquad (32)$$

Equation 32 is the desired balance equation when the inner iteration process converges. Combining Equations 31 and 32 yields

$$f = \frac{Q'}{Q' + SS^p - SS} \qquad (33)$$

For many years the scale factor f was calculated at the end of each inner interation and all flux quantities were multiplied by f. As discrete ordinates codes began to be used for shielding calculations it became obvious that a single scale factor for the entire spatial mesh was hopelessly inadequate. The single scale factor is determined by the total self-scattering difference and relatively large differences in the self-scattering term in regions far from the source were insignificant in the total difference. Problems containing cross sections in particular groups with a dominance ratio ($\Sigma_{g \to g}/\Sigma_t$) near 1.0 require many iterations to achieve convergence.

As a result space-dependent scaling, sometimes called space-dependent rebalance was developed. If Equation 30 is summed over all angular intervals for the ith spatial interval, a balance equation for the ith interval is obtained.

$$L_i + R_i + SS_i = SS_i^p + Q'_i + LL_i + LR_i \qquad (34)$$

where L_i is the leakage loss, R_i is the removal, SS_i is the within-group scattering, LL_i is the leakage gain across the left boundary and LR_i is the leakage gain across the right boundary. Now a scale factor f_i is defined for each spatial interval such that:

$$f_i(L_i + R_i + SS_i) = f_i\, SS_i + Q'_i + f_{i-1}\, LL_i + f_{i+1}\, LR_i \qquad (35)$$

Combining Equations 34 and 35 gives

$$-LL_i f_{i-1} + (Q'_i + LL_i + LR_i + SS_i^p - SS_i)f_i - LR_i f_{i+1} = Q'_i \qquad (36)$$

The system of equations represented by Equation 36 can be solved directly in one-dimensional geometries but requires iteration in multidimensional geometries. This form of space-dependent rebalance was used for a few years but sometimes resulted in oscillations in convergence patterns and occasional divergence. Reed suggested a solution to these problems which effectively consisted of a method of damping the calculation of the rebalance factors f_i without destroying the balance equation. A single term is added to both partial leakage terms at each spatial boundary; thus the net leakage at each boundary remains unchanged. Reed[2] showed that the proper term for stability is $\phi_i \Sigma_{g \to g} V/4$. The method suggested by Reed was an improvement but still failed in some problems which had very fine spatial meshes. Additional experimentation with ANISN[3] resulted in the development of a modified method which has been very reliable. The technique in ANISN utilizes Reed's method on an automatically calculated coarse spatial mesh for each energy group. Each coarse mesh boundary must coincide with a fine mesh boundary and the coarse mesh is the smallest subset of the fine mesh in which each coarse mesh interval is at least one mean free path in width. Table 1 displays the results of each acceleration technique for one

Table 1.

Group	Dominance Ratio	Inner Iterations		
		Standard Iteration	Space-Dependent Scaling	Coarse Mesh Damped SDS
1	0.850	254	40	8
2	0.856	241	15	7
3	0.887	294	16	7
4	0.862	140	11	7
5	0.890	266	13	7
6	0.922	357	20	7
7	0.922	328	17	7
8	0.909	256	17	8
9	0.932	270	32	8
10	0.897	146	34	7
11	0.707	29	9	7
12	0.671	36	15	6
13	0.425	13	9	8
14	0.473	18	9	9
15	0.423	13	8	8
16	0.399	14	9	8
17	0.429	17	9	8
18	0.393	15	7	8
19	0.332	9	7	7
20	0.344	10	8	7
21	0.416	16	9	9
22	0.428	17	9	9
23	0.417	16	9	9
24	0.319	13	7	8
25	0.320	14	7	8
26	0.380	17	8	8
Total		2819	354	200

particularly difficult problem, a neutron adjoint calculation through 3000 m of air. The standard iteration with the single scale factor is very sensitive to the dominance ratio, the original space-dependent scaling is still sensitive to the dominance ratio but the coarse mesh-damped scaling requires about the same number of inner iterations for each group.

Beacause it is impossible, in general, to define a one mean free path spatial mesh in multidimensional geometries using a rectangular grid, the DOT-IV[4] program calculates space-dependent rebalance factors on a user-defined coarse mesh.

FUTURE DEVELOPMENT

Future development of the discrete ordinates technique at ORNL will include the continued study of flux models, convergence acceleration techniques and possibly a three-dimensional, time-independent, θ-R-Z geometry code. The diffusion acceleration technique appears very promising for inner iteration convergence acceleration, and a suitable method of outer iteration convergence acceleration should be implemented in DOT IV.

REFERENCES

1. F. R. Mynatt, F. J. Muckenthaler, and P. N. Stevens, "Development of Two-Dimensional Discrete Ordinates Transport Theory for Radiation Shielding," CTC-INF-952 (1969).
2. W. H. Reed, "The Effectiveness of Acceleration Techniques for Iterative Methods in Transport Theory," Nucl. Sci. Eng., 45:245-254 (1971).
3. W. W. Engle, Jr., "Users Manual for ANISN," K-1693 (1967).
4. W. A. Rhoades, D. B. Simpson, R. L. Childs, and W. W. Engle, Jr., "The DOT IV Two-Dimensional Discrete Ordinates Transport Code with Space-Dependent Mesh and Quadrature," ORNL/TM-6529 (August 1978).

LECTURE 4:   THE METHODS AND APPLICATIONS OF MONTE CARLO IN LOW
             ENERGY ($\lesssim$ 20 MeV) NEUTRON-PHOTON TRANSPORT (MORSE).
             PART I: METHODS

T. A. Gabriel

Oak Ridge National Laboratory
Oak Ridge, TN 37830
USA

INTRODUCTION

   The Monte Carlo solution of the Boltzmann transport equation differs considerably from other standard numerical solutions such as discrete ordinates.[1,2] The discrete ordinate solution usually provides a rather complete description of the particle fluxes in all of phase space (spatial coordinates, x, y, z; angular direction, u_x, v_y, w_z; time, t; energy, E; etc.) in contrast to the Monte Carlo solution which usually only provides information about certain regions of the phase space. The Monte Carlo method utilizes random sampling from the physical probability distributions which describe what happens to the individual particles. For example, by sampling from $e^{-\Sigma_T x}$ where Σ_T is the total interaction cross section, one knows how far the particle travels before interacting, and from the ratio Σ_i / Σ_T, where $\sum_{j=1}^{n} \Sigma_j = \Sigma_T$, what is the probability that the ith interaction will occur. The analysis of what happens to these particles constitutes the solution, that is, do the particles escape from the system, are they captured, how many particles cross a given boundary, etc.

MORSE[3] is a multipurpose neutron and gamma ray transport code which utilizes Monte Carlo techniques for the solution of the Boltzmann transport equations. Some general features of the code are:

1) uses multigroup cross section sets,
2) anisotropic scattering can be utilized,
3) includes time dependence,
4) includes a generalized three-dimensional geometry package,
5) solution of a given problem may be obtained in either the forward or adjoint mode,
6) much of the drudgery associated with flux estimation, biasing, and data handling has been reduced,
7) it is not an analog Monte Carlo program, that is, weighting and averaging are used to describe many of the physical phenomena.

Multigroup cross sections[4] represent cross sections which have been averaged over a given energy range and all particles within that energy range are required to behave according to that average cross section. A group averaged cross section can be defined as follows

$$\sigma_j = \int_{E_j}^{E_{j+1}} \sigma(E)\phi(E)\,dE \bigg/ \int_{E_j}^{E_{j+1}} \phi(E)\,dE \qquad (1)$$

where $\phi(E)$ is an appropriate flux weighting factor. When particles have collisions, they scatter from group to group rather than from a specific energy to a specific energy.

A simplistic view of the way MORSE, using Monte Carlo techniques, solves the transport equation is as follows:

1) pick a source particle,
2) obtain cross section and select a travel distance to the collision site,
3) if the travel distance is in the system and no medium

change has occurred, transport the particle to the new
position; obtain new energy, direction, and weight of
particle; store any produced particles; and go to step 2),
4) if the travel distance is out of the system, transport the
particle to the boundary and obtain a previously produced
particle, if any, and go to step 2) or go to step 1),
5) if the total travel distance places the particle in another
cross section media, transport the particle to the boundary
separating the media and go to step 2).

The basic divisions of the MORSE code are as follows

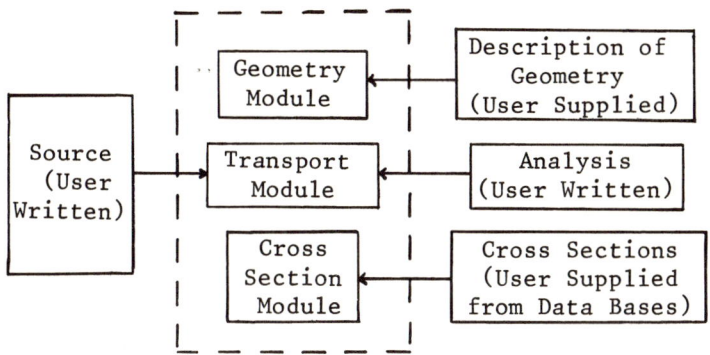

In the sections to follow, several of the above modules of the
MORSE code will be briefly discussed. The main emphasis will be
to present some concepts which a novice user of MORSE will need
during his break-in period with the code. Even though much of the
information is directly oriented to MORSE, it is still applicable
to many other Monte Carlo transport codes.

SOURCE ROUTINE AND SAMPLING METHODS

Source Routine

The information required for a complete source particle description is the group number of the particle (i.e., its energy),
the spatial coordinates (x, y, z), the direction of movement (u_x,
v_y, w_z), the statistical weight of the particle (wt) and the age
of the particle. In many applications, the source particle

description can be handled by normal input. These include point sources, isotropic or monodirectional sources, one energy group or energy spectrum sources, and one initial age for all source particles. If any of the above conditions cannot be met, subroutine SOURCE must be written by the user to describe his requirements. Since a few basic sampling techniques are needed if subroutine SOURCE need be written, these are described next.

Sampling Methods

There are several rather simple sampling methods which the users of Monte Carlo programs should be aware of. These include sampling directly from the probability distribution function (P.D.F.), rejection techniques, importance sampling, and sampling from a multidimensional distribution. Only a brief description is given here and the reader is referred to Refs. 1 and 2 for a more detailed description.

Basic Sampling

Let $f(x)$ define our probability distribution function which could be an energy spectrum, an angular distribution, etc. To obtain proper values of x from $f(x)$, the following equation needs to be solved for x

$$\xi = \int_{x_1}^{x} f(x')dx' \bigg/ \int_{x_1}^{x_2} f(x')dx' \qquad (2)$$

where $x_1 \leq x \leq x_2$ and ξ is a set of random numbers uniformly distributed between 0 and 1. As an example, assume that source particles are distributed in x according to the distribution $f(x) = x^2$, $0 \leq x \leq x_{max}$. Therefore,

$$\xi = \int_{0}^{x} x'^2 dx' \bigg/ \int_{0}^{x_{max}} x'^2 dx'$$

$$= \frac{x^3}{3} \bigg/ \frac{x_{max}^3}{3}, \qquad (3)$$

$$\text{or} \quad x = \sqrt[3]{\xi}\, x_{max},$$

and the proper selection of x is governed by the cube root of the random number.

In many instances the P.D.F. will be so complicated that closed form integration is impossible or the P.D.F. will not be a mathematical formula, but will be in terms of a histogram. The same basic equation applies, but numerical integration, table look up, and interpolation must be used in these cases. For example,

$$\xi = \frac{\int_{x_1}^{x} f(x')dx'}{\int_{x_1}^{x_2} f(x)dx} \approx \frac{\sum_{i=1}^{n} f(x_i)\Delta x_i}{\sum_{i=1}^{N} f(x_i)\Delta x_i} = \xi_n. \qquad (4)$$

Once the value of n has been determined relative to a random number ξ' by table lookup, the value of x on the interval between x_n and x_{n-1} can be obtained by linear interpolation,

$$x = x_n - \frac{\xi_n - \xi'}{\xi_n - \xi_{n-1}}(x_n - x_{n-1}). \qquad (5)$$

Rejection Technique

Let f_{max} represent a value which is larger than the P.D.F. ($f(x)$) in the interval of interest (from x_1 to x_2). Select a pair of random numbers (ξ,η) and define $x' = x_1 + \xi(x_2-x_1)$. x' will be acceptable as a fair sample if $\eta \leq \frac{f(x')}{f_{max}}$, or all x's can be accepted if the sample is given a weight of $f(x')/f_{max}$. This technique is very useful if computer storage is at a premium. However, if the rejection technique rather than weights is used, unacceptable computer time may result due to the efficiency of this technique. This efficiency can be calculated by

$$\int_{x_1}^{x_2} f(x)dx \bigg/ f_{max}(x_2-x_1), \qquad (6)$$

which represents the ratio of the areas.

Therefore, if f_{max} is chosen too large or $f(x)$ is very sharply peaked about a given value of x, this may not be the best technique to use.

Importance Sampling

The objective in importance sampling is to concentrate the distribution of the samples to those portions of the P.D.F. that are the most important.

Consider

$$\int f(x)\,dx = \int \frac{f(x)}{g(x)} g(x)\,dx \quad . \tag{7}$$

The technique is to sample from $g(x)$ which is the importance function, and give the sample a weight $f(x)/g(x)$ (note that the rejection technique is a form of importance sampling). The hardest part in importance sampling is the selection of the importance function. An importance function usually can be selected by using knowledge of the problem. For example, in a deep penetration problem which involves a fission source, the higher energy neutrons will usually be the more important ones. Therefore, an importance function which samples more from the higher energy neutrons should be selected. The adjoint flux obtained for the problem represents an importance function. However, if the adjoint solution is known the problem is solved (see last section)! In many very complicated three-dimensional systems, approximate one- or two-dimensional adjoint fluxes may be obtained which will yield reasonable importance functions.

Multidimensional Sampling

The sampling from a multidimensional P.D.F. ($f(x_1, x_2, \ldots, x_n)$) is just an extension of the sampling in a one-dimensional space. Consider a three-dimensional P.D.F., $f(x_1, x_2, x_3)$. Define the marginal density function $g(x_1)$, which is obtained by integrating over the remaining space,

$$g(x_1) = \iint f(x_1, x_2, x_3) dx_2 dx_3 . \tag{8}$$

By using techniques already described, one can sample from $g(x_1)$ to obtain x_1'. Now define the conditional density function

$$h(x_2|x_1') = \int f(x_1', x_2, x_3) dx_3 . \tag{9}$$

Sample from $h(x_2|x_1')$ to obtain x_2'. Finally, sample from $I(x_3|x_1', x_2')$ to obtain x_3'. This type of sampling is used extensively when the source is distributed over a volume in space.

BASIC INFORMATION ABOUT THE TRANSPORT AND CROSS SECTION MODULES

As has been previously stated, MORSE is not an analog transport code. Neglecting the production of fission neutrons and secondary gamma rays, all remaining reactions are accounted for at each collision site by considering what happens on the average. To account for the probability of all reactions, the weight of the incoming particle is modified. For neutrons, the weight of the outgoing neutron is given by

$$wt \cdot \left(\frac{\Sigma(n,n) + 2\Sigma(n,2n) + 3\Sigma(n,3n) + \text{etc.}}{\Sigma_T} \right), \tag{10}$$

where wt is the weight of the neutron producing the collision, the Σ's are the partial cross sections (where at least one or more neutrons are returned to the system) and Σ_T is the total.

The energy and direction of the outgoing particle is sampled from an averaged distribution which represents all possible outgoing neutrons. When fission is allowed to occur (the user can control this), the energy of the neutron is sampled from an appropriate distribution (which is inputed by the user) and the neutron is stored for later transport. The weight assigned to fission neutrons is the weight of the neutron producing the collision times

$$\nu_f \Sigma_f / \Sigma_T \tag{11}$$

divided by the user assigned fission probability, where ν_f is the neutron multiplicity for fission and Σ_f is the fission cross section. Produced gammas are treated in the same fashion as fission neutrons.

Presently in MORSE, fission neutrons and secondary gammas are assumed to be produced isotropically.

Presented in Tables 1-3 are very useful cross section data and summary sheets that the MORSE cross section module prints out. The data is for Fe and represents 11 neutron and 5 gamma groups.* The first column in Table 1 represents the transport cross sections, and column 4 represents the nonabsorption probability as discussed previously about neutron weight modification following collisions. Columns 5 and 6 give the average number of gammas and fission neutrons produced per collision. The remaining data represents the downscatter matrix (i.e., 58.58% of the time a neutron in group 1 will stay in group 1 after undergoing a collision). Table 2 shows the energy distribution probability of the secondary gammas corresponding to the producing neutron group. For example, at each collision site for neutrons in group 2, 0.40395 gammas will be produced, of which 10.26% will be in the 3rd gamma group. Given in Table 3 are the angular distributions of the particles following a collision. The cross section set that was used in MORSE for this example was a P_3 Legendre expansion. For a P_3 expansion of the cross section data, MORSE calculates only 2 angles of scatter. This may present problems in many applications of MORSE, and higher orders of expansions must be used. The number of angles is calculated as the order of the expansion plus one divided by two. The -1.0's given in the tables under probability indicate an isotropic emission. For these, MORSE automatically samples from an isotropic distribution and all angles are possible.

---

*This is a very crude group structure and is used only to demonstrate features of the code.

Table 1. Cross Sections for Fe

Grp	SIGT	SIGST	PNUP	PNABS	GAMGEN	NU*FIS	Downscatter Probability
1	1.445E-01	1.425E-01	0.0	0.9863	1.0099	0.0	0.5858 0.2492 0.0743 0.0458 0.0369 0.0060 0.0015 0.0005 0.0000 0.0000
2	1.312E-01	1.309E-01	0.0	0.9976	0.4039	0.0	0.7808 0.1220 0.0774 0.0182 0.0015 0.0001 0.0000 0.0000 0.0000 0.0000
3	1.078E-01	1.077E-01	0.0	0.9991	0.1683	0.0	0.8063 0.0987 0.0854 0.0096 0.0000 0.0000 0.0 0.0
4	1.279E-01	1.277E-01	0.0	0.9984	0.0088	0.0	0.9521 0.0434 0.0009 0.0033 0.0003 0.0 0.0 0.0
5	1.459E-01	1.456E-01	0.0	0.9985	0.0041	0.0	0.9911 0.0089 0.0 0.0 0.0 0.0
6	4.218E-01	4.213E-01	0.0	0.9988	0.0000	0.0	0.9937 0.0063 0.0 0.0 0.0
7	2.373E-01	2.367E-01	0.0	0.9976	0.0052	0.0	0.9537 0.0463 0.0 0.0 0.0
8	3.746E-01	3.717E-01	0.0	0.9924	0.0164	0.0	0.9893 0.0107 0.0 0.0
9	4.590E-01	4.560E-01	0.0	0.9934	0.0149	0.0	0.9845 0.0155 0.0
10	4.630E-01	4.560E-01	0.0	0.9848	0.0334	0.0	0.9699 0.0301
11	5.238E-01	4.560E-01	0.0	0.8706	0.2930	0.0	1.0000
12	1.111E-01	1.755E-01	0.0	1.5798	0.0	0.0	0.0192 0.0599 0.0597 0.0419 0.8193
13	1.106E-01	1.545E-01	0.0	1.3970	0.0	0.0	0.0675 0.1282 0.0782 0.7262
14	1.300E-01	1.478E-01	0.0	1.1362	0.0	0.0	0.1918 0.2089 0.5993
15	1.829E-01	1.838E-01	0.0	1.0049	0.0	0.0	0.2682 0.7318
16	6.355E-00	3.301E-01	0.0	0.0519	0.0	0.0	1.0000

Table 2. Neutron to Gamma Transfers for Fe

Neut. Group	GAMGEN	Transfer Probabilities				
1	1.0099E-00	0.0000	0.0171	0.2787	0.2784	0.4258
2	4.0395E-01	0.0	0.0006	0.1026	0.1632	0.7336
3	1.6828E-01	0.0	0.0040	0.0043	0.0025	0.9892
4	8.7923E-03	0.0	0.1262	0.1361	0.0397	0.6980
5	4.1052E-03	0.0	0.2590	0.2745	0.0794	0.3871
6	2.3119E-05	0.0	0.2642	0.2720	0.0779	0.3859
7	5.2124E-03	0.0	0.3695	0.2707	0.0781	0.2818
8	1.6371E-02	0.0	0.3760	0.2039	0.0906	0.3295
9	1.4887E-02	0.0	0.3623	0.1981	0.0985	0.3411
10	3.3447E-02	0.0	0.3623	0.1981	0.0985	0.3411
11	2.9303E-01	0.0	0.3605	0.1979	0.0991	0.3426

In the analog case, capture of particles is not allowed and a method must be used to dispose of particles with small statistical weights. If this is not done, computer time can be wasted by tracing particles with little statistical significance. Russian Roulette is the procedure by which a particle's weight (which is less than some predetermined weight, W_{low}) is set equal to an average weight (also predetermined, W_{ave}), or the particle is killed by setting its weight to zero.

One Russian Roulette procedure is: if $R \leq \frac{\text{Weight of Particle}}{W_{ave}}$, where R is a random number between 0 and 1, then the particle survives; otherwise the particle is killed. If the particle survives, its weight becomes $W = W_{ave}$. It must be remembered that particles need to be conserved, and placing a too stringent requirement on the killing of particles may bias the results.

Splitting is performed if a particle's weight is too large ($W > W_{high}$). Splitting consists of replacing an original particle of weight W by n particles of weight W/n. This procedure is not unique and there are many ways to split particles.

Table 3. Scattering Probabilities and Angles for Fe

Gp	to Gp	PROB	ANGLE	PROB	ANGLE
1	1	0.8951	0.8905	1.0000	-0.4449
1	2	0.5003	-0.5767	1.0000	0.5892
1	3	-1.0000	0.0	0.0	0.0
1	4	-1.0000	0.0	0.0	0.0
1	5	-1.0000	0.0	0.0	0.0
1	6	-1.0000	0.0	0.0	0.0
1	7	-1.0000	0.0	0.0	0.0
1	8	-1.0000	0.0	0.0	0.0
1	9	-1.0000	0.0	0.0	0.0
1	10	-1.0000	0.0	0.0	0.0
1	11	-1.0000	0.0	0.0	0.0
2	2	0.6948	0.7976	1.0000	-0.5723
2	3	0.5342	-0.6299	1.0000	0.5556
2	4	-1.0000	0.0	0.0	0.0
2	5	-1.0000	0.0	0.0	0.0
2	6	-1.0000	0.0	0.0	0.0
2	7	-1.0000	0.0	0.0	0.0
2	8	-1.0000	0.0	0.0	0.0
2	9	-1.0000	0.0	0.0	0.0
2	10	-1.0000	0.0	0.0	0.0
2	11	-1.0000	0.0	0.0	0.0
3	3	0.6547	0.7456	1.0000	-0.6280
3	4	0.5025	-0.6085	1.0000	0.5602
3	5	-1.0000	0.0	0.0	0.0
3	6	-1.0000	0.0	0.0	0.0
-	-	-	-	-	-
-	-	-	-	-	-
14	14	0.8601	0.9697	1.0000	0.8682
14	15	0.5455	0.8913	1.0000	0.7158
14	16	0.5514	0.4744	1.0000	-0.5744
15	15	0.7700	0.9572	1.0000	0.8189
15	15	0.6310	0.6157	1.0000	-0.5332
16	16	0.6251	0.7139	1.0000	-0.5545

There are other concepts which the new user of MORSE should be aware of but are not usually needed in relatively simple problems. These include: 1) the exponential transform which involves the biasing of the flight path of the particles, that is, forcing the particles to traverse greater (important to deep penetration problems) or lesser distances before collisions; 2) secondary energy spectra biasing; that is, modifying the downscatter matrix probability distributions given in Table 1; and 3) albedo boundaries; that is, boundaries within the system which reflect or return the particle back into the system (useful in labyrnths calculations).

MONTE CARLO ESTIMATION OF THE PARTICLE FLUX

In general, the particle flux is a function of spatial position, energy, time, direction, etc. However, in practice, one usually only needs the flux as a function of spatial position and energy. There are four methods which are usually used in practice to calculate the particle fluxes. These are referred to as tracklength estimation, collision density estimation, boundary crossing estimation, and statistical estimation.

Given an arbitrary volume in space at a location of interest, the energy dependent flux per unit source particle can be calculated by the tracklength estimation as

$$\phi(E,\vec{r}) = \frac{\sum_{i=1}^{N} \ell_i \cdot wt_i}{\Delta v \cdot \Delta E \cdot M} , \qquad (12)$$

where

ℓ_i = tracklength of the ith particle in the volume,
wt_i = statistical weight of the ith particle,
Δv = volume,
M = total number of source particles,
N = total number of tracklengths in the volume by particles of E about ΔE.

The volume can be of any size and shape and large volumes are sometimes needed to improve the statistics. However, if the volume is too large, the ability to determine the spatial variation of the flux becomes more difficult.

The collision density estimation of the particle flux is also a very useful method. However, in general, the collision density method will be inferior to the tracklength method because the number of collisions within a given volume only approximates the total tracklength, but this technique works very well if there are a large number of collisions in an area of interest. The collision density estimation of the particle flux can be calculated as

$$\phi(E,\vec{r}) = \sum_{i=1}^{N} \frac{wt_i}{\Sigma_t(E) \cdot \Delta v \cdot \Delta E \cdot M} , \qquad (13)$$

where

wt_i = statistical weight of the particle producing the collision,

$\Sigma_t(E)$ = total cross section in cm^{-1} units,

Δv = volume of interest,

M = total number of source particles,

N = total number of collisions in the volume by particles of energy E about ΔE.

The estimate of ϕ by collision density can be improved by the addition of a fictitious scattering material within Δv with cross section Σ^*.[5]

The method works as follows

1) choose an interaction distance not from $e^{-\Sigma_T x}$, but from $e^{-(\Sigma_t + \Sigma^*) x}$

2) accept the collision as real if a random number $R \leq \Sigma_t / (\Sigma_t + \Sigma^*)$, otherwise, call it a pseudo-collision,

3) calculate the particle flux as indicated before, except replace Σ_t by $(\Sigma_t + \Sigma^*)$.

It should be noted that pseudo-collisions will not alter the energy or direction of the particle.

Very closely related to the tracklength estimation is the boundary crossing estimation of the particle flux. Given a surface area in a position of interest, the energy and spatial dependent flux per unit source particle can be calculated as

$$\phi(E,r) = \sum_{i=1}^{N} \frac{wt_i \cdot (|\hat{n}\cdot\hat{\Omega}|^{-1})_i}{\Delta A \cdot \Delta E \cdot M}, \qquad (14)$$

where

wt_i = statistical weight of particles closing the surface area,

ΔA = surface area,

$\hat{n}\cdot\hat{\Omega}$ = cosine of the angle between the direction of the particle and the normal vector $\hat{n}$ of the surface area at the point the particle crosses the surface,

N = total number of particles crossing the surface area with energy E about ΔE,

M = total number of source particles.

One problem with this estimation occurs when $\hat{n}\cdot\hat{\Omega} \sim 0$. The solution of this is to control the magnitude of the cosine of the angle near $\pi/2$. F. H. Clark[6] has shown that an acceptable solution would be to assign an average value of 0.025 to all cosines having an absolute value of less than some predetermined value, such as 0.05.

A very powerful method for obtaining the particle flux is the statistical estimation method. This method allows for the estimation of the flux at a point in space and is extremely powerful for deep penetration problems. The method can result in an increased computing time, but this is almost always offset by improved statistics. The statistical estimation of the flux is composed of two contributions: 1) an uncollided response representing source and produced particles, and 2) a collided response from all scattered particles leaving a collision site. The uncollided term, assuming all source and produced particles have an isotropic distribution, can be written

$$\phi(E,\vec{r})_{uncollided} = \sum_{i=1}^{N} \frac{e^{-\Sigma_t(E)|\vec{r}_i-\vec{r}|}}{4\pi|\vec{r}_i-\vec{r}|^2} \cdot \frac{wt_i}{M}, \quad (15)$$

where

wt_i = weight of the source or produced particles,

$\vec{r}_i$ = the coordinates of the source or produced particles,

$\vec{r}$ = the coordinates of the detector,

$\Sigma_t(E)$ = the total cross section in cm^{-1} or approximate units,

N = the sum over all source and produced particles of energy E about ΔE,

M = total number of source particles.

The collided flux term can be written

$$\phi(E,\vec{r})_{collided} = \sum_{i=1}^{N} \frac{g(E'\rightarrow E, \Omega'\rightarrow\Omega) e^{-\Sigma_t(E)|\vec{r}_i-\vec{r}|}}{|\vec{r}_i-\vec{r}|^2} \cdot \frac{wt_i}{M} \quad (16)$$

where

$g(E'\rightarrow E, \Omega'\rightarrow\Omega)$ = probability that a particle of energy E' and direction Ω' will scatter to energy E and direction Ω which is toward the detector located at $\vec{r}$,

N = total number of particle collisions,

wt_i = weight of particle producing the collision.

The remaining variables are as for the uncollided term.

All of the above methods for calculating the particle flux have been incorporated into the MORSE code and can be readily implemented.

As in all Monte Carlo transport codes, communication with the transport module is essential so that proper analysis of the particle cascade can be accomplished. In MORSE, all analyses of the transport are controlled by subroutine BANKR(NCOL) in conjunction with COMMON/NUTRON/. The argument list corresponding to NCOL values is given in Table 4. To be able to do collision density estimation of the flux, every time BANKR is called with a NCOL value of 5, the proper information must be stored (for boundary crossing estimation, NCOL is 7 or 8). The variables of COMMON/NUTRON/ are given in Table 5. All necessary information for a complete analysis is given here; that is, where the particle is located, x, y, z, where it came from, xOLD, yOLD, zOLD, which direction it is going, u, v, w, etc.

STATISTICAL ERROR

In all Monte Carlo calculations there will be errors associated with an answer since this type of calculation only produces an estimate, not an exact result. MORSE operates in a batch mode so several answers are obtained during one run. Each batch run should contain enough source particles so that a "reasonable" estimate can be made. The average of these batch results represents the answer, that is

$$\bar{x} = \left(\sum_{i=1}^{N} x_i \right) / N \quad , \tag{17}$$

where x_i is the ith batch answer, and N is the number of batches.[7] The above assumes that the same number of source particles are used to start each batch. The fluctuation of the answer (batch average answer) can be calculated as

$$\sigma_{\bar{x}} = \sqrt{\frac{\sum_{i=1}^{N} (x_i - \bar{x})^2}{N(N-1)}} \quad . \tag{18}$$

The results are usually reported as $\bar{x} \pm \sigma_{\bar{x}}$ (or as $\bar{x} \pm \sigma'$, where σ' (%) = 100 $\sigma_{\bar{x}} / \bar{x}$).

Table 4. BANKR Arguments

BANKR Argument	Location of call in walk
-1	After call to INPUT - to set parameters for new problem
-2	At the beginning of each batch of NSTRT particles
-3	At the end of each batch of NSTRT particles
-4	At the end of each set of NITS batches - a new problem is about to begin
1	At a source event
2	After a splitting has occurred
3	After a fission has occurred
4	After a secondary particle has been generated
5	After a real collision has occurred - post-collision parameters are available
6	After an albedo collision has occurred - post-collision parameters are available
7	After a boundary crossing occurs (the track has encountered a new geometry medium other than the albedo or void media)
8	After an escape occurs (the geometry has encountered medium zero)
9	After the post-collision energy group exceeds the maximum desired
10	After the maximum chronological age has been exceeded
11	After a Russian Roulette kill occurs
12	After a Russian Roulette survival occurs
13	After a secondary particle has been generated but no room in the bank is available

Table 5. Definition of Variables in COMMON/NUTRON

Variable	Definition
NAME	Particle's first name
NAMEX	Particle's family name (Note that particles do not marry)
IG	Current energy group index
IGO	Previous energy group index
NMED	Medium number at current location
MEDOLD	Medium number at previous location
NREG	Region number at current location
U, V, W	Current direction cosine
uOLD, vOLD, wOLD	Previous direction cosines
X, Y, Z	Current location
xOLD, yOLD, zOLD	Previous location
WATE	Current weight
OLDWT	Previous weight (Equal to WTBC if no path length stretching)
WTBC	Weight just before current collision
IBLZN	Current zone number
IBLZO	Previous zone number
AGE	Current age
OLDAGE	Previous age

There are no absolute standards as to how small σ' should be. The following are only guidelines

$\sigma' > 50\%$ - usually meaningless for design, and often misleading,

$20 < \sigma' < 50\%$ - some significance but are not regarded as good design data,

$\sigma' \lesssim 10\%$ - a good calculation.

ADJOINT MONTE CARLO

If a code has the capability of solving the adjoint transport equation[8], as does MORSE, it gives the user a much broader flexibility than exists with codes which just produce a forward solution. For example, if a response is needed for several types of source spectra, then only one adjoint solution is needed since the answer of interest can be represented by

$$A = \int \phi^* \, S dE \, , \tag{19}$$

where ϕ^* is the adjoint flux at the source S. The basic differences between the forward and adjoint solutions of the Boltzmann Transport equation with respect to the MORSE code are 1) in the adjoint case the response becomes the source term and the source term becomes the response, and 2) the cross sections become inverted (i.e., particles now scatter "up" in energy and gammas produce neutrons). The changes required in the input to the MORSE code to modify the problem from the forward case to the adjoint case are minimal. The methods for calculating forward fluxes still apply for adjoint fluxes and the biasing techniques for the forward case are also still applicable for the adjoint case.

REFERENCES

1. L. L. Carter and E. D. Cashwell, "Particle-Transport Simulation with the Monte Carlo Method," (ERDA Critical Review Series) TID-26607, National Technical Information Service.
2. H. Kahn, "Applications of Monte Carlo," RM-1237-AEC (April, 1954).
3. M. B. Emmett, "The MORSE Monte Carlo Radiation Transport Code System," ORNL-4972, Oak Ridge National Laboratory (1975).
4. N. M. Greene, J. L. Lucius, L. M. Petrie, W. E. Ford, III, J. E. White, and R. Q. Wright, "AMPX: A Modular Code System for Generating Coupled Multigroup Neutron-Gamma Libraries

from ENDF/B," ORNL/TM-3706, Oak Ridge National Laboratory (1976).

5. S. N. Cramer, "Application of the Fictitious Scattering Radiation Transport Model for Deep-Penetration Monte Carlo Calculations," ORNL/TM-4880, Oak Ridge National Laboratory (December, 1977).

6. F. H. Clark, Nucl. Sci. Eng. 27:235 (1967).

7. N. Barash-Schmidt, Phys. Letters 75B:No. 1 (1978).

8. "A Review of the Monte Carlo Method for Radiation Transport Calculations," edited by B. F. Maskewitz and V. Jacobs, ORNL-RSIC-29, Oak Ridge National Laboratory (1971).

LECTURE 5: THE METHODS AND APPLICATIONS OF DISCRETE ORDINATES IN LOW ENERGY NEUTRON-PHOTON TRANSPORT (ANISN, DOT). PART II: APPLICATIONS

W. W. Engle, Jr.

Oak Ridge National Laboratory
Oak Ridge, TN 37830
USA

SUMMARY

The applications discussed in this lecture have been published previously. The references listed below contain complete descriptions of these and other applications of the discrete ordinates method in low energy neutron and photon transport.

REFERENCES

1. L. S. Abbott and F. R. Mynatt, "Review of ORNL Radiation Shielding Analyses of the Fast Flux Test Facility Reactor," ORNL-5027 (July 1975).
2. W. W. Engle, Jr. and M. B. Emmett, "Review of ORNL Radiation Shielding Analyses of the Fast Flux Test Facility Reactor (1975-1976)," ORNL-5166 (June 1976).
3. W. W. Engle, Jr. et al., "Analyses of the Preliminary In-Vessel and Enclosure System Shield Designs for the Clinch River Breeder Reactor (July 1973-July 1975)," ORNL/TM-5338 (March 1976).
4. W. W. Engle, Jr., F. R. Mynatt, M. B. Emmett, and M. L. Williams, "A Summary of the ORNL Shield Design Supporting Analysis for the FFTF," Proceedings of the Fifth International Conference on Reactor Shielding (April 1977).
5. D. E. Bartine, C. O. Slater, and L. R. Williams, "Application of an Advanced Shielding Analysis System to Gas-Cooled Fast Reactor Designs," Proceedings of the Fifth International Conference on Reactor Shielding (April 1977).

6. R. E. Maerker and F. J. Muckenthaler, "Measurement and Calculations of Neutron Fluxes Through a Simulation of the CRBR Upper Axial Shielding," Proceedings of the Fifth International Conference on Reactor Shielding (April 1977).
7. C. O. Slater and M. B. Emmett, "Analysis of a Fuel Pin Neutron-Streaming Experiment to Test Methods for Calculating Neutron Damage to the GCFR Grid Plate," Proceedings of the Fifth International Conference on Reactor Shielding (April 1977).

LECTURE 6: THE METHODS AND APPLICATIONS OF MONTE CARLO IN LOW
ENERGY (≲ 20 MeV) NEUTRON-PHOTON TRANSPORT (MORSE)
PART II: APPLICATIONS

T. A. Gabriel

Oak Ridge National Laboratory
Oak Ridge, TN 37830
USA

INTRODUCTION

There are many problems of current interest in the area of radiation transport which involve the use of the neutron and gamma ray transport code MORSE.[1] Some of these are: 1) fusion reactor shielding[2], 2) accelerator breeder design[3] (electronuclear fuel production), 3) high energy nuclear instrumentation design[4] (uranium calorimeters), and 4) accelerator beam stop activation.[5] In the sections to follow, 1) and 2) will be discussed. The reader is referred to Refs. 4 and 5 for a discussion of 3) and 4).

FUSION REACTOR SHIELDING

A fusion reactor is a device which "burns" (combines) hydrogen isotopes to produce energy. The particular reaction under current consideration is

$$D + T \rightarrow \alpha + n + \sim 17 \text{ MeV}.$$

Of the ∼ 17 MeV in available energy, the neutron receives ∼ 14 MeV. There are other possible reactions, but this one offers the lowest threshold energy. Some methods for igniting the deuterium and tritium are: heating a plasma containing these elements (Tokamak reactors, mirror machines; in general, magnetic confinement devices)

and, heating a frozen pellet of D and T by laser. In any case, the
14-MeV neutrons must be contained for power generation and personnel safety.

The information which is usually needed for fusion development
is: tritium breeding (n + Li → T + χ), spatial energy deposition,
radiation damage (displacement per atom (DPA) and gas production[6]),
and biological dose (during operation and from activation). Most
of the above information can be obtained with ANISN or DOT, but
due to the 3-D characteristics of fusion devices, MORSE is also
needed to determine the effects of penetrations on the performance
of the reactor. The penetrations in Tokamak reactor blankets and
shields can lead to excessive heating and damage in cryopanels,
neutral beam injectors, and toroidal field (TF) coils. This section summarizes the results of calculations using MORSE in the
adjoint mode that were carried out to estimate the nuclear heating
and radiation damage increases at selected locations in the TF
coils adjacent to a rectangular neutral beam injector duct that
passes through the blanket and shield. The nuclear responses obtained using MORSE will be compared with similar data obtained
previously using the ANISN 1-D discrete ordinates code.[7] The 1-D
calculations are for the reactor model without penetrations and
without void spacing between the TF coils. A cylindrical representation of the experimental power reactor (EPR) was used as the
calculational model. This model is illustrated in Fig. 1.

The toroidal-shaped plasma region, blanket, and shield are
represented by coaxial right-circular cylinders 1415 cm long and
having radial dimensions corresponding to those in the Experimental
Power Reactor (EPR).[7] The composition of the blanket was taken to
be that of the tritium breeding module proposed for use in the EPR.
The dimensions and compositions of the components are given in
Table 1. The TF coils were treated as cylindrical segments having
approximately the same radial and lateral dimensions of the TF
coils that surround the assembly. The coil spacing along the

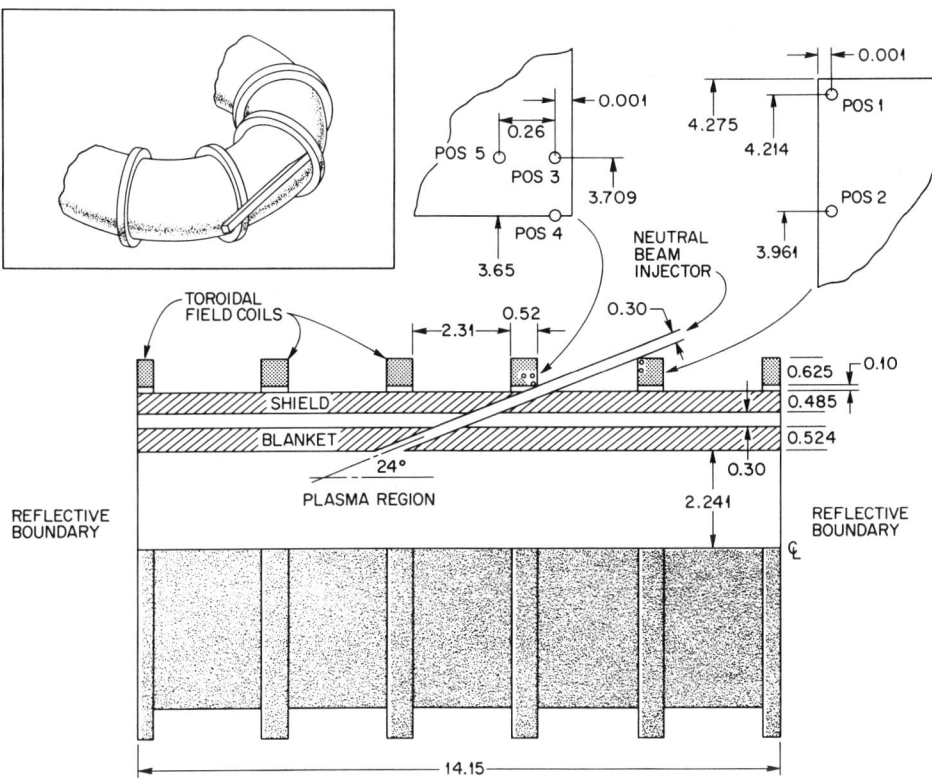

Fig. 1. The representation of the Tokamak reactor and injector duct used in the Monte Carlo calculations.

Table 1. Dimensions and Compositions of the Reactor Components

Region	Outer Radius (cm)	Composition
Plasma	224.1	Void
Blanket Assembly		
Graphite curtain	224.4	Impurity control
Water	225.0	
Type 316 stainless steel	226.1	0.636 Fe, 0.18 Cr, 0.13 Ni, 0.026 Mo, 0.028 Mn
Lithium	251.1	0.0742 ^{6}Li, 0.9258 ^{7}Li
Type 316 stainless steel	252.1	
Graphite reflector	262.1	
Type 316 stainless steel	263.4	
Lithium	265.9	
Type 316 stainless steel	268.4	
Gamma-ray shield	276.5	Alternating Type 316 stainless steel and lithium
Shield Assembly		
Type 316 stainless steel	311.5	
Shield	343.5	65 vol. % Type 316 stainless steel + 35 vol. % $H_2O \cdot B$
Lead liner	350.5	
Type 316 stainless steel	355.5	

cylinder corresponds to the spacing of the coils about the outer toroidal surface. For the purposes of calculation, the conductor material in the coil was a homogenized composition consisting of 52.5 vol% aluminum, 14.0 vol% copper, 4.9 vol% Nb_3Sn, 0.6 vol% tin, and 28 vol% helium.

The neutral beam injector duct passes through the blanket-shield assembly between two TF coils at a grazing angle of 24 deg with the axis of cylindrical symmetry. The duct has a cross-sectional area of 28 X 68 cm^2 and a Type 316 stainless steel wall thickness of 1 cm. Modeling the injector in this manner simulates tangential neutral beam injection. Also, the angle of injection was chosen so that little or no additional shielding could be placed adjacent to the TF coils. The nuclear heating rates and radiation damage at five locations in the TF coils adjacent to the injector were estimated using adjoint Monte Carlo radiation transport methods. Generally, the nuclear response is determined from the integral over all phase space $\bar{p}$ of the particle flux distribution, $\Phi(\bar{p})$, and the macroscopic response function, $\Sigma(E,\bar{r})$, given by

$$R = \int_{all\ phase\ space} \Phi(\bar{p}) \Sigma(E,\bar{r}) d\bar{p} , \qquad (1)$$

where $d\bar{p} = d\bar{r} dE d\bar{\Omega}$. In this equation, the particle flux distribution is determined from the solution of the forward transport equation with a particle source term $S(\bar{p})$. That is,

$$\hat{H}\Phi(\bar{p}) = S(\bar{p}) , \qquad (2)$$

where $\hat{H}$ is the forward Boltzmann operator. However, for this problem the large source volume, the locations of the detectors in the TF coils relative to the mouth of the duct, and the large attenuation factor ($\sim 10^5$) of the blanket and shield make forward Monte Carlo calculations problematic.[8] Strong source position and angular biasing are necessary, and preliminary forward calculations were performed, but the results were unsatisfactory from the point of view of running time and statistical uncertainty. Based on these considerations, the adjoint Monte Carlo approach was used. In adjoint Monte Carlo calculations, the nuclear response is obtained from the integral

$$R = \int_{\text{all phase space}} \Phi^*(\bar{p}) S(\bar{p}) d\bar{p} \quad , \tag{3}$$

where $\Phi^*(\bar{p})$ is the adjoint flux, which is the solution of the adjoint Boltzmann transport equation using the nuclear response function $\Sigma(E,\bar{r})$ as the source term. That is

$$\hat{H}^* \Phi^*(\bar{p}) = \Sigma(E,\bar{r}) \quad , \tag{4}$$

where $\hat{H}^*$ is the adjoint Boltzmann operator. A detailed derivation establishing the relationships between these equations and the identification of $\Sigma(E,\bar{r})$ as the adjoint source term are given by Hansen and Sandmeier.[9] In general, the disadvantage of using the adjoint radiation transport method is that a separate transport calculation must be carried out for each nuclear response and for each detector site. In multigroup adjoint Monte Carlo calculations, however, it is possible, as shown below, to obtain any number of nuclear responses at a given detector from one adjoint calculation, but a separate calculation is still required for each detector site. In multigroup adjoint Monte Carlo calculations, $\Sigma_j(\bar{r})$ is the adjoint source in energy group j defined by

$$\Sigma_j(\bar{r}) = \int_{\Delta E_j} \Sigma(E,\bar{r}) dE \quad ,$$

where the integral is evaluated at each energy group of width ΔE_j.

Since $\hat{H}^*$ is a linear operation and

$$\sum_{j=1}^{n} \Phi^*j(\bar{p}) = {}^*(\bar{p}) \quad ,$$

Eq. (4) can be broken up into n equations:

$$\hat{H}*\Phi*^1(\bar{p}) = \Sigma_1(\bar{r})$$

$$\hat{H}*\Phi*^2(\bar{p}) = \Sigma_2(\bar{r})$$

$$\vdots$$

$$\hat{H}*\Phi*^n(\bar{p}) = \Sigma_n(\bar{r}) \quad , \quad (5)$$

which are evaluated separately for each $\Sigma_j(\bar{r})$. The corresponding nuclear response for each of the above equations is

$$R_1 = \int \Phi*^1(\bar{p}) S(\bar{p}) d\bar{p}$$

$$R_2 = \int \Phi*^2(\bar{p}) S(\bar{p}) d\bar{p}$$

$$\vdots$$

$$R_n = \int \Phi*^n(\bar{p}) S(\bar{p}) d\bar{p} \quad , \quad (6)$$

and the total nuclear response is

$$R = \sum_{j=1}^{n} R_j \quad . \quad (7)$$

For a new response function $\Sigma'(E,\bar{r})$, the total nuclear response R' can be obtained from

$$R' = \sum_{j=1}^{n} R'_j \quad , \quad (8)$$

where

$$R'_j = R_j \left[\frac{\Sigma'_j(\bar{r})}{\Sigma_j(\bar{r})} \right] \quad . \quad (9)$$

To obtain a nuclear response due to a different response function $\Sigma'_j(\bar{r})$, one applies Eq. (9) every time a contribution to R_j is made.

To facilitate the calculations, adjoint source angular biasing was used to direct more particles toward the plasma region through the duct rather than transporting large numbers of particles through regions of the blanket and shield away from the penetration. Furthermore, path length stretching, nonabsorption weighting, Russian Roulette, and splitting (see Lecture 4) were employed.

Two estimators were used in the calculations. They are the next-flight track length estimator and the true track length estimator. For the true track length estimation, the tracks of adjoint particles inside the plasma volume were scored. For the next-flight track length estimation, adjoint particles that are directed toward the plasma were scored for each collision, that is, the probability that the particle will reach the plasma region times the track length produced upon reaching the plasma.

The radiation transport was accomplished using coupled 35-energy-group neutron, 21-energy-group gamma-ray cross sections obtained by collapsing the 100n-21γ DLC-37 cross-section library.[10,11] The nuclear heating rates were obtained using fluence-to-kerma response functions generated by the MACK code[11] and the SMUG code,[13] and radiation damage was estimated using atomic displacement and gas production cross sections generated by the RECOIL code.[6]

The calculated nuclear heating rates at the various locations in the TF coils adjacent to the neutral beam injector duct are summarized in Table 2. The detector position numbers correspond to those as shown in the inset of Fig. 1. The neutron, gamma-ray, and total (neutron plus gamma-ray) heating rates using both the next-flight and true track length estimation techniques are given at each detector location. In all cases, the kerma factor for the TF coil conductor was used as the source term in the adjoint calculations. The fractional values given below each entry are the calculated standard deviations for each response. The entries labeled "MORSE/ANISN Ratio" indicate the effect of penetration on

Table 2. Nuclear Heating Rates at Various Locations in the TF Coils

(Neutron Wall Loading = 1.0 MW/m^2)

Nuclear Heating Rate (W/cm^3)

Detector Position	1	2	3	4	5
Next-Flight Track Length Estimator					
Neutron	9.0×10^{-5}	1.2×10^{-4}	7.3×10^{-4}	2.1×10^{-3}	5.2×10^{-4}
	0.24	0.57	0.39	0.29	0.24
Gamma ray	6.8×10^{-5}	3.0×10^{-4}	1.7×10^{-3}	4.7×10^{-2}	9.9×10^{-3}
	0.40	0.90	0.17	0.58	0.70
Total	1.6×10^{-4}	4.7×10^{-4}	2.4×10^{-3}	4.9×10^{-2}	1.0×10^{-2}
	0.24	0.67	0.17	0.56	0.66
MORSE/ANISN Ratio					
Neutron	228	31	27	47	19
Gamma ray	21	12	13	234	77
Total	43	14	16	196	67
True Track Length Estimator					
Neutron	7.2×10^{-5}	4.4×10^{-5}	8.0×10^{-4}	1.9×10^{-3}	7.0×10^{-4}
	0.23	0.18	0.49	0.33	0.47
Gamma ray	5.2×10^{-5}	4.8×10^{-5}	1.1×10^{-3}	2.5×10^{-2}	1.5×10^{-2}
	0.40	0.41	0.24	0.45	0.85
Total	1.3×10^{-4}	9.1×10^{-5}	1.9×10^{-3}	2.7×10^{-2}	1.5×10^{-2}
	0.21	0.25	0.25	0.42	0.81
MORSE/ANISN Ratio					
Neutron	182	12	30	44	26
Gamma ray	16	2	9	125	115
Total	34	3	13	108	100
Number of histories	2×10^5	2×10^5	4×10^5	3.4×10^5	6×10^5

each response. The ANISN results[7] are for the reactor configuration without a penetration. All results have been normalized to a 14-MeV neutron wall loading of 1 MW/m^2 [4.43×10^{13} n/(cm^{2}s)] at continuous reactor operation.

The inclusion of the neutral beam injector leads to increases in the nuclear heating rates in the coils ranging from a factor of 3 to a factor of 196, depending on the location in the coils relative to the duct. The heating rates at each detector position, obtained using both track length estimators, agree with the statistical deviations albeit, in some cases, these uncertainties are quite large.

The neutron heating rate at detector position 1 is due, in large part, to neutrons streaming through the duct, as evidenced by the value ($\sim$ 200) in the MORSE/ANISN Ratio. This portion of the coil has a direct view of the plasma through the duct. The nuclear heating rates at detector position 2 are comparable to the values at location 1, but the effects of the injector are smaller. There is more shielding material between the plasma region and this portion of the coil. The uncertainties in the gamma-ray heating rates at detector positions 1 and 2 are large compared to the uncertainties in the neutron heating rates.

Detector positions 3 and 4 each lie along the same radius from the plasma, but are separated by 6 cm. The differences in the neutron and gamma ray heating rates at these locations are due to shielding of detector position 3 by the material in the coil. The gamma ray heating rates at detector position 4 are accompanied by large uncertainties, so the increase in the heating rate relative to the fully shielded case may be overestimated. The gamma-ray heating at position 4 may be attributed to photons produced by energetic neutron reactions near the outer layers of the shield. These photons are attenuated in the coil, resulting in lower heating rates at location 3. Similar increases in radiation damage, DPA and gas production, were also obtained and these are summarized in Table 3.

Table 3. Expected Increases in DPA and Gas Production
Due to Injector Port Streaming

DPA/yr	(2X-95X)
H (appm/yr)	(11X-5600X)
He (appm/yr)	(15X-7400X)

Unfortunately, in many situations, running MORSE for many design problems will be prohibitive in cost and time. In situations like these, comparisons between 3-D and 2-D calculations must be made to obtain insight into the problem so that 2-D (or 1-D) calculations can be used for the bulk of the design problems.[14]

Two- and three-dimensional radiation transport methods have been employed to estimate the nuclear performance of the neutral beam injectors being designed for the Tokamak Fusion Test Reactor (TFTR). The D-T fusion reactions will be initiated by the injection of neutral deuterium into a magnetically confined tritium plasma. Some of the neutrons formed in the D-T reaction will stream through the injection ducts as they did in the previous example and interact in the injector components, producing heat and other detrimental nuclear responses within the injector. The purpose of this study was to obtain the spatial dependences of the nuclear heating rates and the neutron and gamma-ray scalar flux distributions at various locations in the neutral beam injector. The motivation for using two radiation transport methods was to obtain a two-dimensional model for the neutral beam injector that allows for more cost efficient analyses of the nuclear heating rates in the injector and which estimates the nuclear responses consistently with the more detailed, but expensive, 3-D analysis. MORSE was used for the 3-D analysis and DOT[14] for the 2-D analysis. A detailed discussion of this calculation is given in Ref. 14.

ACCELERATOR BREEDER DESIGN

An accelerator breeder[3] is a device that is used to generate a particle cascade within a medium containing fertile nuclides (such as ^{232}Th or ^{238}U) to create an inventory of neutrons that can be captured by the fertile nuclides to produce fissile nuclides (such as ^{233}U or ^{239}P). The incident particles that are currently under consideration are protons and deuterons. The final accelerated energies of these particles will be ∿ 1 GeV since this energy offers efficient production of neutrons.[3] The high energy nucleon-meson transport code HETC[15] is used to calculate the transport of protons, charged pions (and muons) and neutrons with energies ≥ 15-20 MeV. MORSE is used to transport the neutrons which are created by HETC with energies below 15-20 MeV and to transport the fission and deexcitation gammas produced by the high energy collisions.

The type of calculated information that is needed is: spatial energy deposition, number of neutron captures, number of fissions, and number of escaping neutrons. In theory, the difficulties are minimal in using MORSE for a calculation like this: 1) there are no deep penetrations, 2) only integral quantities are needed, and 3) the source is distributed over the area of interest.

Results are present in Fig. 2* of an idealized case in which protons of various energies are incident on an infinite natural uranium system. The system is large enough so that no particle leakage is possible. Figure 2a indicates approximately 24 fissions/1 GeV of incident proton energy. Figure 2c shows the total energy deposition. This yield amounts to ∿ 5.4 GeV/1 GeV of incident proton energy. The fissions and captures were calculated by considering the probability at each collison site that a fission

---

*Figures 2d and 3d were used for calorimeter analysis and are not directly related to accelerator breeder calculations (see Ref. 4).

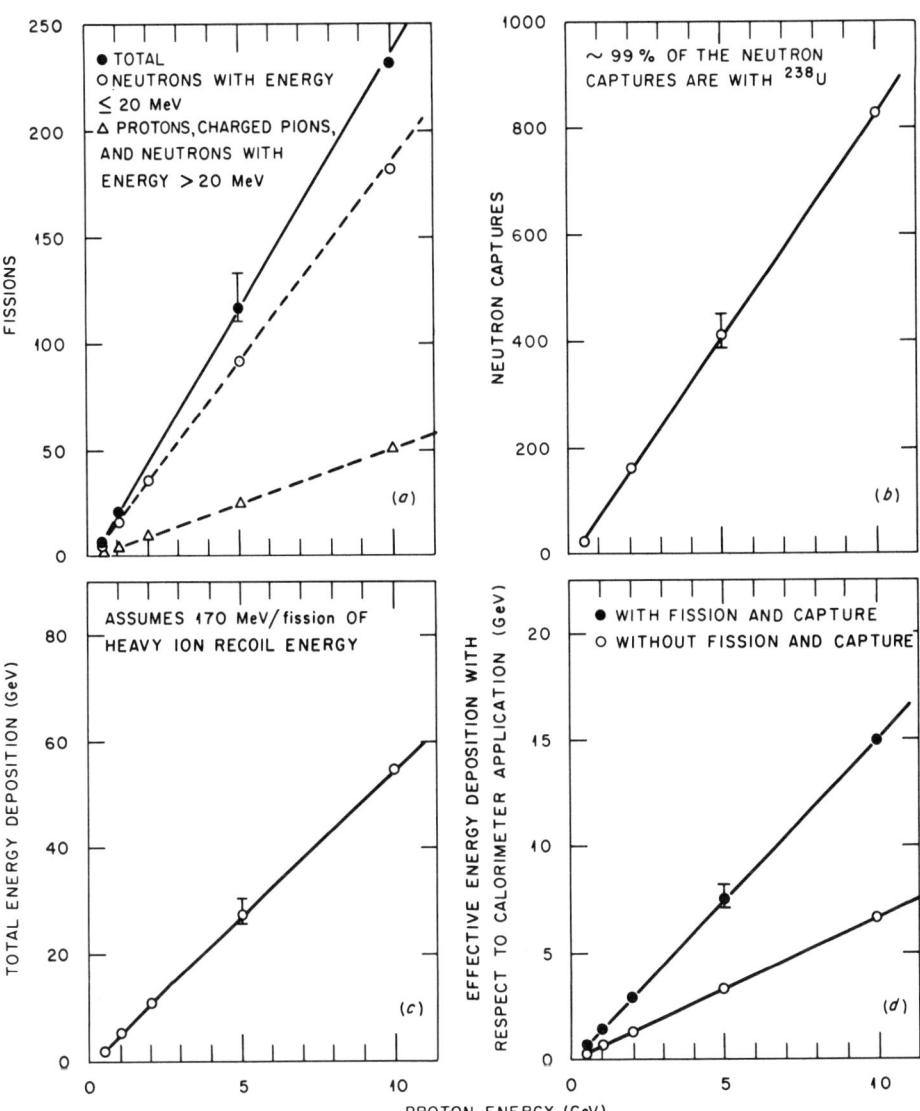

Fig. 2. Fissions, captures, energy deposition and effective energy deposition in an infinite uranium system produced by protons of various energies.

or capture occurred ($\Sigma_{fission}/\Sigma_{Total}$ or $\Sigma_{cap}/\Sigma_{Total}$).

Shown in Fig. 3* is the time scale associated with fission, capture and energy deposition. Since the high energy transport code does not have a timing scheme incorporated, it has been assumed that no time has passed for this phase of the particle cascade. Therefore, all energy deposited from this phase of the calculation is deposited at t = 0, and all low energy ($E \leq 20$ MeV) neutrons, high-energy fission and deexcitation gamma rays are also born at t = 0. In actuality, these particles are usually born and the energy deposited in a time window from 0 to $\sim$ 30 nsec. Therefore, the time scale shown must be considered an absolute lower limit for the processes indicated. Obviously, this type of system will bear no resemblence to an actual breeder blanket, but it does give a researcher a general estimate of what can be expected.

Shown in Fig. 4 is a more realistic system with respect to material composition, but not to geometric configuration. Since neutron containment is vital, this design will allow too much neutron leakage. However, this design does allow an easy means for comparing different material compositions. Given in Table 4 are the comparative results for $^{238}UO_2$ and $^{232}ThO_2$. The number of neutron captures in the $^{238}UO_2$ configuration is small compared to the infinite uranium system presented previously. The number of captures in ^{232}Th is less than that in ^{238}U, but not by a large amount. To a considerable extent, the small number of neutron captures in both configurations is due to the fact, as indicated before, that a large number of neutrons and energy leak from the systems. The larger energy deposition in the $^{238}UO_2$ configuration is a direct consequence of the larger number of fission in that system.

Since, in this example, the source for MORSE is based mostly on theoretical calculations, comparison with experimental data is a

---

*Figures 2d and 3d were used for calorimeter analysis and are not directly related to accelerator breeder calculations (see Ref. 4).

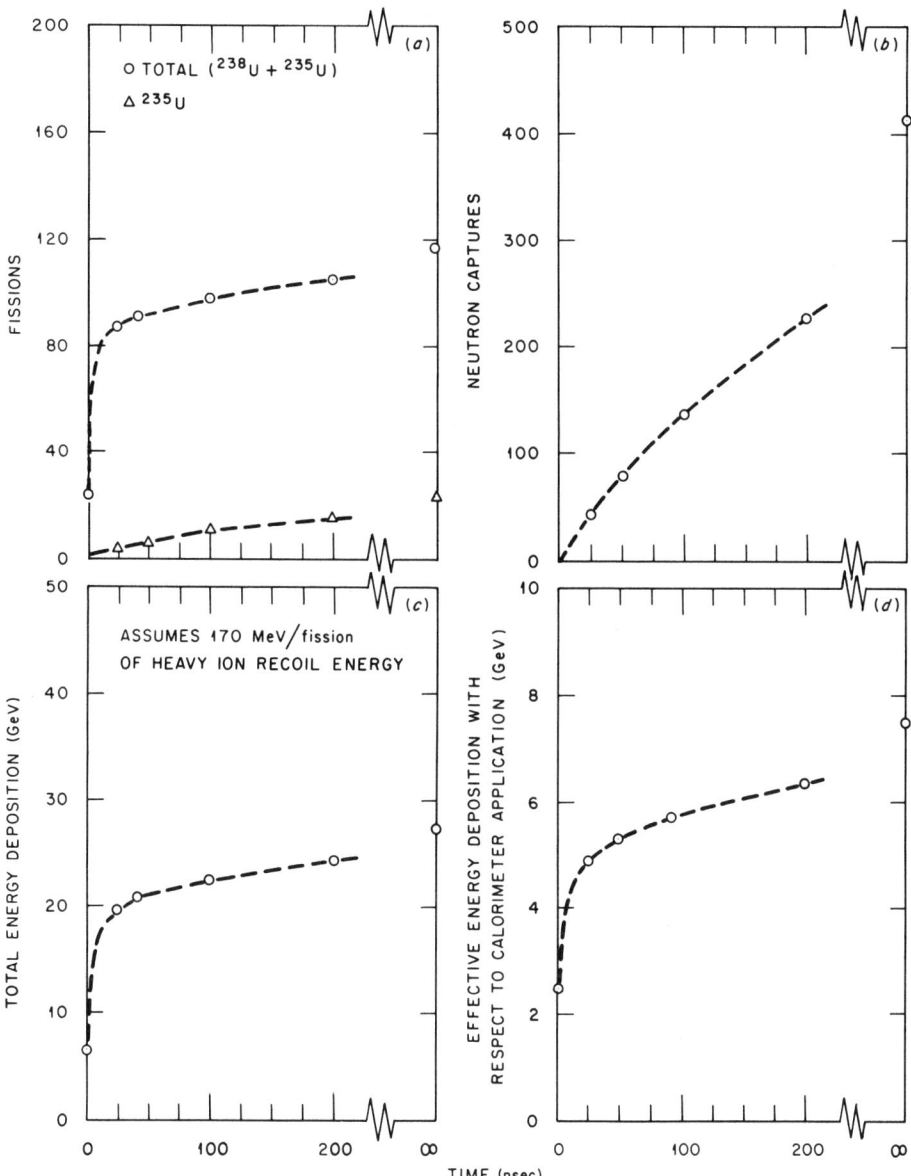

Fig. 3. Fissions, captures, energy deposition and effective energy deposition in an infinite uranium system as a function of time for incident 5-GeV protons.

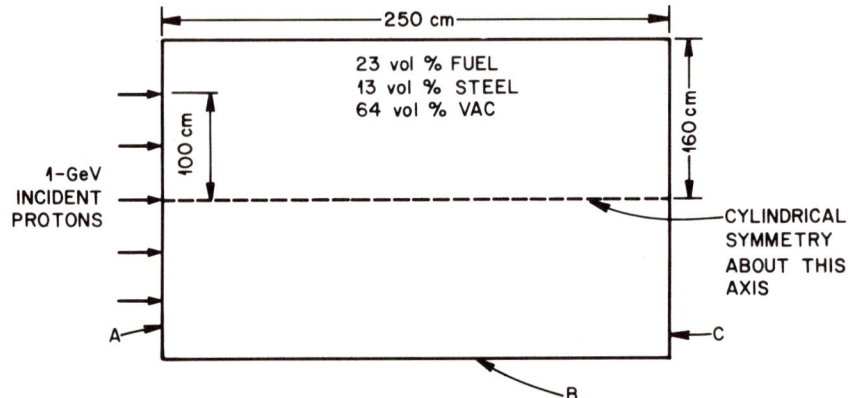

Fig. 4. Schematic diagram of configuration used in the calculations. Results have been obtained for $^{238}UO_2$ and $^{232}THO_2$ as fuel.

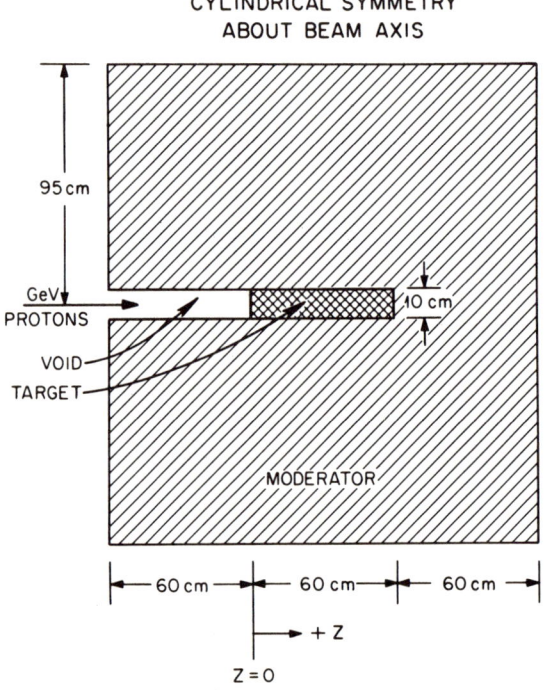

Fig. 5. Schematic diagram of the ^{238}U target surrounded by a H_2O moderator.

Table 4. Calculated Results for the $^{238}UO_2$ and the $^{232}ThO_2$ Configuration (All results are given per incident 1-GeV proton)

	Configuration with $^{238}UO_2$	Configuration with $^{232}ThO_2$
Neutron captures in fertile material	17.2	14.4
Neutron captures in all other elements	1.8	1.3
Number of fission	4.5	1.6
Total energy deposited (GeV)	1.7	1.0
Number of neutrons leaking out of the front face (A in Fig. 4)	8.7	7.5
Number of neutron leaking out of side and back (B and C in Fig. 4)	3.2	2.7

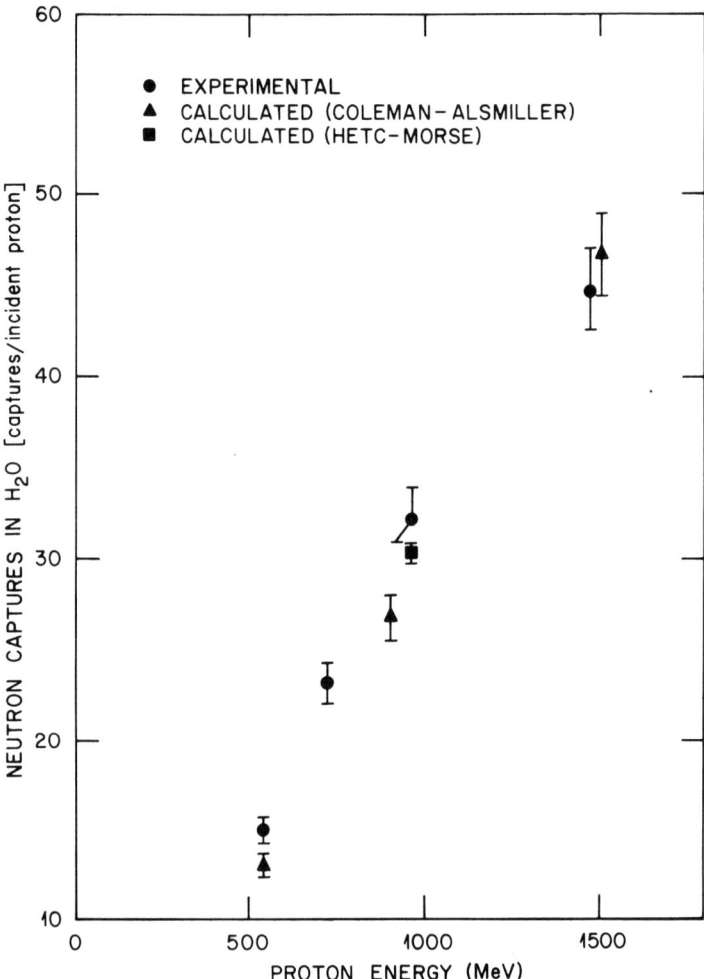

Fig. 6. Neutron captures in H$_2$O as a function of proton energy and comparison with experimental data.

necessity. Presented in Fig. 5 are the geometry and material composition of an experiment which was carried out to determine the number of neutron captures in water when high energy protons are incident on a high Z target, namely, ^{238}U. The results of the calculation are presented in Fig. 6. The agreement, though slightly lower than experimental data (see ref. 16), is very gratifying.

There are other comparisons with experimental data[17,18] which are currently being carried out. As before, the preliminary calculated results underestimates slightly the experimental data. The reason for this underestimation of the experimental data is not totally understood, but is under investigation.

ACKNOWLEDGEMENT

The author wishes to thank R. T. Santoro and R. G. Alsmiller, Jr. for their generosity in supplying data, text, and comments. Also, the author wishes to thank Mrs. C. Zeigler for her patience and understanding while typing these lectures.

REFERENCES

1. M. B. Emmett, "The MORSE Monte Carlo Radiation Transport Code System," ORNL-4972, Oak Ridge National Laboratory (1975).
2. R. T. Santoro, J. S. Tang, R. G. Alsmiller, Jr. and J. M. Barnes, "Monte Carlo Analysis of the Effects of a Blanket-Shield Penetration on the Performance of a Tokamak Fusion Reactor," ORNL/TM-5874, Oak Ridge National Laboratory (1977).
3. F. R. Mynatt, Editor, "Preliminary Report on the Promise of Accelerator Breeding and an Alternative Energy System" ORNL/TM-5750, Oak Ridge National Laboratory (1977).
4. T. A. Gabriel, "Uranium-Liquid Argon Calorimeters: A Calculational Investigation," ORNL/TM-5769, Oak Ridge National Laboratory (1977).
5. R. G. Alsmiller, Jr., T. A. Gabriel, and J. Barish, "Photon Dose Rate from Induced Activity in the Beam Stop of a 400-GeV

Proton Accelerator," ORNL/TM-6238, Oak Ridge National Laboratory (1978).

6. T. A. Gabriel and B. L. Bishop, "Sensitivity of Primary Knock-on Atom Spectra and Displacement per Atom Cross-Sections to Different Secondary Neutron Energy and Angular Distributions and "In-Group" Weighting Schemes," ORNL/TM-6108, Oak Ridge National Laboratory (1978).

7. R. T. Santoro, V. C. Baker, J. M. Barnes, "Neutronics and Photonics Calculations for the Tokamak Experimental Power Reactor," ORNL/TM-5466, Oak Ridge National Laboratory (1977).

8. M. A. Abdou, L. J. Milton, J. C. Jung and E. M. Gelbard, "Multi-dimensional Neutronic Analysis of Major Penetrations in Tokamaks," Proc. of the Second Topical Meeting on the Technology of Controlled Nuclear Fusion, Vol. II, p. 845, September 21-23, 1975, Richland, WA.

9. G. E. Hansen and H. A. Sandmeier, "Neutron Penetration Factors Obtained by Using Adjoint Transport Calculations," Nucl. Sci. Eng. 22:315 (1965).

10. D. M. Plaster, R. T. Santoro and W. E. Ford, III, "Coupled 100-Group Neutron and 21-Group Gamma-Ray Cross Sections for EPR Calculations," ORNL/TM-4872, Oak Ridge National Laboratory (1975).

11. W. E. Ford, III, R. T. Santoro, R. W. Roussin and D. M. Plaster, "Modification Number One to the Coupled 100n-21γ Cross-Section Library for EPR Calculations," ORNL/TM-5249, Oak Ridge National Laboratory (1976).

12. M. A. Abdou, C. W. Maynard and R. Q. Wright, "MACK - A Computer Program to Calculate Neutron Energy Release Parameters (Fluence-to-Kerma Factors) and Multigroup Neutron Reaction Cross Sections for Nuclear Data in ENDF Format," ORNL/TM-3994, Oak Ridge National Laboratory (1973).

13. N. M. Greene, J. L. Lucius, L. M. Petrie, W. E. Ford, III, J. E. White and R. Q. Wright, "AMPX: A Modular Code System

for Generating Coupled Multigroup Neutron-Gamma Libraries from ENDF/B," ORNL/TM-3706, Oak Ridge National Laboratory (1976).

14. R. T. Santoro, R. A. Lillie, R. G. Alsmiller, Jr., J. M. Barnes, "Two- and Three-Dimensional Neutronics Calculations for the TFTR Neutral Beam Injectors," ORNL/TM-6354, Oak Ridge National Laboratory (1978).

15. K. C. Chandler and T. W. Armstrong, "Operating Instructions for the High-Energy Nucleon-Meson Transport Code HETC," ORNL-4744, Oak Ridge National Laboratory (1972).

16. W. A. Coleman and R. G. Alsmiller, Jr., Nucl. Sci. Eng. 34:104 (1963).

17. R. G. Vasil'kov, V. I. Gol'Danskii, B. A. Pimenov, Yu. N. Potokilovskii and L. V. Chistyakov, Atomnaya Energiya, 44:329 (1978).

18. V. S. Baraschenkov, V. D. Toneev and S. E. Chigrunov, "On the Calculation of Electronuclear Method of Neutron Generation," JINR-R2-7694 (1974).

LECTURE 7:   THE EUROPEAN SHIELDING INFORMATION SERVICE - ESIS

C. Ponti

Joint Research Centre
Ispra (Varese)
Italy

ESIS is a service in the field of radiation shielding, intended for engineers engaged in design problems or research. It collects, analyzes and circulates information concerning the shielding of neutron and gamma radiation below 20 MeV, and in particular, the shielding of nuclear reactors.

ESIS was established in 1972. It has a staff of about ten members, all having experience in shielding design and experimentation. The main work of ESIS proceeds along the following lines:

- code assessment
- nuclear data libraries
- **shielding experiments**
- technical support
- training.

The assessment of a code is carried out by studying reference problems, or by comparing its results with those of other codes or with experiments, to determine the limits of applicability and the accuracy of the predictions. The main work performed in this domain concerns the codes SABINE, MERCURE, TRIPOLI, ANISN, DOT and MORSE.

The work in the field of nuclear data consists in studying cross section preparation and handling codes, and in developing and updating data libraries. For several years ESIS has promoted the development of a standard European cross section library. EURLIB, which is a 100 neutron, 20 gamma-ray, coupled cross section library, developed in collaboration with IKE-Stuttgart, is in fact the proposal for such a standard library, and many European laboratories have already implemented it.

ESIS was one of the promoters of the Common Benchmark Programme carried out by European and Japanese shielding groups (Cadarache, Casaccia, Karlsruhe, Saclay, Stuttgart, University of Tokyo, Wuerenlingen). As part of this common programme, ESIS performs benchmark experiments on a few important shielding materials at the EURACOS-II facility.

With the help of an information service such as ESIS, it is possible for a small group, without special competence, to perform shielding calculations with the same sophisticated tools (codes and nuclear data) as are used by highly qualified teams.

LECTURE 8: CROSS SECTION PROCESSING CODES AND DATA BASES (AMPX).

W. W. Engle, Jr.

Oak Ridge National Laboratory
Oak Ridge, TN 37830
USA

SUMMARY

The AMPX system, in continuing development at the Oak Ridge National Laboratory, is a collection of computer programs in a modular arrangement.[1] Starting with ENDF-formatted nuclear data files or pseudo-problem-independent multigroup cross-section data, the system includes a full range of features needed to produce problem-dependent neutron, gamma-ray production, and gamma-ray interaction cross-section data for use with neutronics transport codes such as ANISN, DOT, MORSE, KENO, CITATION, and VENTURE.

The capabilities of AMPX modules can be loosely categorized into the following areas: (1) basic cross-section processing of ENDF-formatted data into pseudo-problem-independent cross-section libraries, (2) resonance self shielding with the Nordheim integral treatment, the Bondarenko iterative treatment, or a treatment utilizing a one-dimensional integral transport solution, (3) spectral collapsing using either previously-prepared multigroup spectra to flux weight cross sections or one-dimensional discrete ordinates, one-dimensional diffusion, or infinite medium transport theories to calculate system-dependent spectra and flux weight cross sections,

(4) format conversion of cross-section libraries, (5) cross-section library service functions such as editing, punching, managing, modifying data, preparing macroscopic data, etc., (6) and miscellaneous cross-section functions such as first-order checking of data libraries, processing of ENDF-formatted elastic scattering, fission, capture, and total cross sections for resonance nuclides into point values, creating ENDF-type point string libraries from point cross-section libraries, etc.

The modularity of the AMPX system is particularly attractive since it allows the user to choose an arbitrary execution sequence from the approximately 40-50 modules available in the system. The modularity also allows selection from different cross-section treatments, permits efficient adjustment to technological change, and provides a means of utilizing features of other state-of-the-art multigroup cross-section processing systems with features of the AMPX system.

The AMPX system is distributed by the Radiation Shielding Information Center (RSIC) in a package identified as PSR-63/AMPX-II. RSIC's AMPX package includes the following: (1) a tape containing IBM source decks and assembler language decks for the various modules, data libraries, and input and output for sample problems, (2) the AMPX User's Guide, and (3) other documents which provide "hints" for the novice user of the system.

1. N. M. Greene, J. L. Lucius, L. M. Petrie, W. E. Ford III, J. E. White, and R. Q. Wright, "AMPX: A Modular Code System for Generating Coupled Multigroup Neutron-Gamma Libraries for ENDF/B," ORNL/TM-3706 (March, 1976).

LECTURE 9: RADIATION SHIELDING INFORMATION CENTER
            AND
            BIOMEDICAL COMPUTING TECHNOLOGY INFORMATION CENTER

T. A. Gabriel

Oak Ridge National Laboratory
Oak Ridge, TN 37830
USA

ABSTRACT

The scope of the Radiation Shielding Information Center (RSIC) located at the Oak Ridge National Laboratory (ORNL) includes the physics of interaction of radiation with matter; radiation production, protection and transport; radiation detectors and measurements; engineering design techniques; shielding materials properties; computer codes useful in research and design; and nuclear data compilations. The goals of RSIC are to function as a technical institute to provide bibliographic information, computer codes, and data upon request; collect, evaluate, enrich, distill, and repackage information to extend the state-of-the-art (this brings into the public domain technology more usable and more valuable than the sum of the input); and to initiate and effect research and development in areas of need that can be managed by RSIC personnel.

The Biomedical Computing Technology Information Center (BCTIC) was established at ORNL to collect, evaluate, and disseminate information in computing technology pertinent to biomedicine, in general, and nuclear medicine, in particular. The BCTIC functions were established to meet the full spectrum of needs of the biomedical community. These functions include a clearinghouse for computing technology; producing a newsletter; organizing meetings, conferences and proceedings; establishing standards; generating directories of clinical computing resources; and publishing state-of-the-art reviews.

Your are cordially invited to participate both as a contributor and consumer of information within the RSIC and BCTIC spheres.

ACKNOWLEDGEMENT

The author wishes to thank B. F. Maskewitz, Head of RSIC and BCTIC, for her support in preparing the ERICE lectures on RSIC and BCTIC.

LECTURE 10: APPROXIMATE METHODS IN REACTOR SHIELDING CALCULATIONS

Carlo Ponti

Joint Research Centre
Ispra (Varese)
Italy

The solution of any shielding problem requires the knowledge of the radiation flux and of its space-energy distribution. The determination of this flux by means of transport theory -- in one of its more or less rigorous forms -- is therefore always a necessary step. The methods of calculation that have been more widely applied for solving reactor shielding problems, are not the same as those applied for neutronic calculations inside the reactor core.

Inside the core, angular fluxes are nearly isotropic and their space distribution is rather uniform. Consequently the approximation provided by diffusion theory is generally suitable, except in the case of local dishomogeneities; greater accuracy is required for the eigenvalue, that is K_{eff}; fission and capture cross-sections must be well known, and their resonances also must be taken into account in detail. The calculation of the flux inside a shield is a non-homogeneous problem (it is a source problem): angular fluxes may be strongly anisotropic; they may vary by many orders of magnitude in space.

Under these conditions the approximation supplied by diffusion theory is not adequate. The most important cross-sections are the total, the differential elastic scattering, and the inelastic for high energies; special attention should be devoted to cross-section minima. Because of these different characteristics, special methods have been used in shielding calculations. As neutron transport models become more and more sophisticated, rigorous and general, the same method may be applied to both core and shield calculations. Thus, for instance, codes based on the discrete ordinates approximation or on the Monte Carlo approach may be applied to the solution either of core or of shielding problems. The discrete ordinates method and the Monte Carlo method are dealt with in other lectures in this course. This lecture will describe some methods of calculation that have been developed expressly for use in shielding problems. These are non-rigorous methods, which introduce some simplifying hypotheses and can give satisfactory results without expensive computing times. They are supported by comparison with experiments, and can be applied to all problems for which the basic assumptions of the methods in question are satisfied.

THE BUILD-UP FACTOR

The Boltzmann transport equation has its simplest solution under the hypothesis of a perfectly absorbing medium. In this case, the total flux coincides with the uncollided flux. For an isotropic point source located at P, emitting $S(E,P)$ particles/MeV·sec, there will be an uncollided flux at Q:

$$\phi(E,Q) = \frac{S(E,P) \, e^{-\tau}}{4\pi \overline{PQ}^2} \text{ particles/cm}^2 \cdot \text{MeV} \cdot \text{sec} \tag{1}$$

where
$$\tau = \int_P^Q \Sigma(s) \, ds,$$

τ being the distance from P to Q in mean free paths. Equation (1) expresses the well-known fact that the uncollided flux attenuates like a product of two factors: the first, a geometrical factor, varies with the inverse square of the distance; the second is the probability of non-collision and it decreases exponentially with the distance from the source. For an extended source emitting $S(E,P)$ particles/cm^3·MeV·sec, in a space of volume V, the total flux at Q is obtained by integrating Eq. (1) over V.

In the calculation of high-energy gamma-ray propagation, Eq. (1) can give results approximated within an order of magnitude, the scattering cross-section being small in comparison with the total cross-section. The total flux is obviously greater than the uncollided flux, since it includes all the components of the different scattering orders. The increase due to scattering contributions can be taken into account by means of a dimensionless build-up factor B (B ≥ 1):

$$\phi(Q) = \frac{S(E,P) \, e^{-\tau} \, B(\tau,E)}{4\pi \overline{PQ}^2} \, . \tag{2}$$

Here $\phi(Q)$ is the flux at Q of all the particles, of any energy, emitted with energy E at P. We now specialize in gamma-rays: the flux will be expressed in photons/cm^2·sec. The problem of determining the flux distribution is thus reduced to that of calculating the build-up factors.

Let us suppose that the point (isotropic) source is placed at the origin and that it radiates in an infinite homogeneous medium. Let us suppose, moreover, that the source is unitary and monoenergetic, with energy E. From Eq. (2) it follows that

$$\phi(r) = \frac{e^{-\tau} \, B(\tau,E)}{4\pi r^2} \, , \tag{3}$$

with $\tau = \mu r$, μ (cm^{-1}) being the total attenuation coefficient of the medium at the source energy.

If we know a suitable method for this extremely simplified problem which allows us to determine $\phi(r)$, then Eq. (3) will give us the required function $B(\tau,E)$. Once this is known we have the solution to the problem of an arbitrary source (provided that it is isotropic):

$$\phi(Q) = \int dP \int \frac{e^{-\tau} B(\tau,E) S(P,E) dE}{4\pi \overline{PQ}^2} \ . \tag{4}$$

The first method, by means of which the build-up factors have been computed and systematically tabulated, is the method of moments. The tables obtained by Goldstein and Wilkins are still widely used (for example, see Goldstein[1]). We point out that with this method we can only obtain an integral answer: Eq. (4) supplies the total flux at Q, integrated over the energy, and it does not give any information about the spectrum. If, on the other hand, we would like to know the equivalent dose, $D_e(Q)$, or the heat deposition H(Q), as well as the total flux, what should we do? It is not correct to simply obtain the product of the total flux times conversion coefficients, since the flux at Q is not monoenergetic and its spectrum is not known. It is therefore necessary to determine as many build-up factors as the number of answers sought. We can write the equations for the equivalent dose and the heat deposition in a way similar to Eq. (3):

$$D_e(r) = \frac{C(E) \ e^{-\tau} \ B_D(\tau,E)}{4\pi r^2} \ , \tag{5}$$

$$H(r) = \frac{K(E) \ e^{-\tau} \ B_E(\tau,E)}{4\pi r^2} \ ; \tag{6}$$

C(E) is the conversion factor from flux to dose equivalent rate; K(E) is the kerma factor; B_D and B_E are dimensionless factors, namely dose and energy deposition build-up. The same method that leads to the determination of B leads also to B_D and B_E. Actually, when the flux $\phi(E,r)$ produced by the monoenergetic point source is known, it is possible to determine not only the total flux (integrating over E), but also the dose and the heat deposition. This is done by multiplying by the corresponding conversion factors and integrating. Once $D_e(r)$ and H(r) have been determined, Eqs. (5) and (6) supply B_D and B_E.

There exist several tables of build-up factors[1-3], obtained by different methods and for different materials. They cover the interval of τ from zero to 30 mfp, and the energies from 0.5 MeV to 10 MeV. The build-up factors increase regularly with τ; they generally diminish with increasing energy, but this is not always true for heavy materials. Their trend is fairly regular; a rough approximation, valid in the interval from 1 to 3 MeV within a factor of about 2, is $B(X) = 1 + X$.

The attenuating properties of a material, with respect to gamma-rays in the energy interval we are considering here (i.e. below about 10 MeV), depend mainly on the number of electrons per cm^3 existing in the material, and consequently on its density. Something similar is true for the build-up factors; their variations according to material depend mainly on the density. This trend allows for simple interpolation to determine the build-up factors of a material, when those of materials with similar densities are known. It is sufficient to have tables for 5 or 6 materials of density in the range 1 to 11 g/cm^3 (water and lead are in general the lightest and heaviest material for which build-up factors are needed). For any other material it is then possible to interpolate.

Up to now we have limited our attention to homogeneous media. However, Eqs. (3), (5), and (6) are applicable even to non-homogeneous materials, provided that for the calculation of the answer at Q one applies the build-up factor of the material where Q is located. Attention should be paid to the fact that near interfaces, errors will arise because of the differences in the build-up factors of neighbouring materials. In fact, while the build-up factor strictly depends on all the materials and thicknesses passed through, as a first approximation it can be said that it depends essentially on the total thickness (expressed in mean free paths) and on the scattering properties of the last medium traversed.

It is actually this property which allows the application of the method of build-up factors to many problems of gamma-ray attenuation (and also the fact that the build-up factors have a regular trend versus energy and material density). When a better approximation is necessary in order to reduce the errors present near the discontinuities (especially between materials which exhibit strong differences in density), we can apply more complicated expressions which allow the smoothing away of differences in build-up factors of different materials[4].

The advantages of this method are the remarkable conceptual simplicity, the possibility of application in situations in which the source is extended and complicated, and a reasonable approximation.

Among the disadvantages we point out: the lack of differential information (the method gives the total flux, but it does not give the spectrum); the calculation must be repeated for all detector points Q. It should be checked that, around the optical path PQ connecting the source-point to the detector-point there are no preferential paths such as vacuums or materials lighter than those present along PQ. The method of build-up factors is applied exclusively to gamma-ray calculations for energies higher than 0.5 MeV. For lower energies, the build-up factors lose their regular behaviour, and the considerations we have just made are no longer valid.

Codes using this method are: MERCURE-3 [2], MERCURE-4 [5] and SABINE-3 [3,4].

DIFFUSION AND REMOVAL-DIFFUSION THEORIES

The Diffusion Approximation

Diffusion theory has, for many years, been the most widely used calculation method for neutronic calculations in reactor analysis. It has been applied (and still is) mainly to reactor core calculations. But it was of little use for shielding calculations until

APPROXIMATE METHODS IN REACTOR SHIELDING CALCULATIONS

particular adaptations and improvements were developed. From an analytical point of view, the assumptions necessary for the validity of the diffusion equation are the following:

a) The space dependence of the scalar flux can be represented by a Taylor, expansion cut-off at the second-order term; that is:

$$\phi(X) = \phi(0) + X\phi'(0) + \frac{X^2}{2} \phi''(0) .$$

b) The scattering is isotropic in the laboratory system.

If these two conditions are met, then Fick's equation for the current J,

$$J = -D \text{ grad} ,$$

as well as the diffusion equation,

$$D\nabla^2 \phi(X) - \Sigma_a \phi(X) + S(X) = 0 , \qquad (8)$$

can be shown to be valid. D is called the diffusion coefficient and depends on the properties of the material through the cross-sections:

$$D = \Sigma_s / 3 \Sigma_t^2 . \qquad (9)$$

The derivation of these equations can be found in reactor physics text-books[6,7]. We are interested here in the meaning and the consequences of the hypotheses which have been made. In this context it is interesting to remark how the same diffusion equation (8) can be derived starting from different hypotheses (see Weinberg and Wigner[7], p. 231); that is:

c) the angular flux is isotropic or linearly anisotropic (P_1 approximation).

Also, in this case, Eqs. (7) and (8) are valid, provided one sets:

$$D = 1/3 (\Sigma_t - \bar{\mu}_0 \Sigma_s), \qquad (10)$$

where $\bar{\mu}_0$ is the mean value of the cosine of the scattering angle (in the laboratory system).

Actually, the hypotheses (a)+(b) and (c) do not differ so much from each other, since we cannot have regularity in the angular flux if there is no regularity in the spatial distribution. One can see, for instance, from Eq. (7) (or, more directly, from P_1 equations) that the angular flux can be isotropic only if its spatial distribution is flat, and vice versa. The physical conditions which must be satisfied in order for these assumptions to be valid are:

- low absorption;

- away from boundaries.

These conditions allow a regular trend of both the spatial distribution and the angular flux. Therefore, $\Sigma_a \ll \Sigma_t$ (or $\Sigma_s \simeq \Sigma_t$), so that Eq. (9) can be written in the form:

$$D = 1/3 \Sigma_t .$$

This equation coincides with Eq. (10), provided that the scattering is isotropic ($\bar{\mu}_0 = 0$). Condition (c) and Eq. (10) are a little more general than (a) and (b) [and consequently Eq. (9)] because they enable us to take into account a possible scattering anisotropy.

The diffusion equation (8) states a balance between entering and emerging particles per unit volume, and it specifies that (in stationary conditions) the balance must be maintained. It is a monoenergetic equation because it assumes that all the neutrons have the same speed and the same cross-sections. The energy dependence is generally considered in the multigroup scheme: the energy range is divided into several intervals, within each of which neutrons are assumed to have constant (suitably averaged) cross-sections. In this scheme, let us indicate:

$i = 1, N$ the index of the energy group;

ϕ_i the flux of group i;

Σ_{ij} the transfer matrix from group to group;

$\phi_i \Sigma_{ij}$ the number of neutrons/cm^3·sec which, after scattering, are slowed down from group i to group j;

Σ_i the total cross-section of group i.

In this scheme the diffusion equation is written:

$$D_i \nabla^2 \phi_i - \Sigma_i \phi_i + \sum_{j=1}^{i} \phi_j \Sigma_{ji} + S_i = 0 , \qquad (11)$$

$$D_i = 1/3 \, (\Sigma_i - \bar{\mu}_i \Sigma_{ii}) .$$

Actually one might say that each author has his own way of defining the multigroup diffusion coefficients; the basic hypothesis for Eq. (11), which should always be remembered when applying the diffusion theory, implies for energy-dependent problems not only low absorption but also little slowing down.

The Penetration of Fast Neutrons

Let us now see what is really happening in a typical shield. Consider a hydrogenated infinite medium (water or concrete), containing an isotropic monoenergetic point source placed at the origin; let this be a fast source, of some MeV, the same order as the energies present in the fission spectrum. The typical trend of the total cross-section versus energy shows a constant decrease with increasing energy; this trend is specially relevant for hydrogenated materials, owing to the shape of the total cross-section of hydrogen.

We observe, moreover, that the possible events for a neutron of a few MeV (in a hydrogenated medium) are, in approximate order of likelihood:

- elastic scattering against light nuclei;

- inelastic scattering;

- elastic scattering against heavy nuclei;

- absorption.

All these events, except the third, either eliminate the neutron (absorption) or substantially change its speed and direction. The elastic scattering against heavy nuclei, at these energies, is anisotropic and forward; hence these collisions, contrary to the others, produce little variation in the energy or direction of the neutron. This allows us to divide the possible collisions into roughly two classes: those which produce strong variations in the characteristics of the uncollided neutron (removal collisions); and those which produce little variation (weak collisions). It is clear that the neutrons propagating far from the source are the ones which, not having suffered collisions of the first kind, maintain their direction and energy and consequently a mean free path of sufficient length for them to travel long distances. On the contrary, the neutrons undergoing removal collisions -- that is, collisions of the first type -- will change their direction and will be slowed down to an energy interval where the mean free path is shorter, and they will rapidly undergo further collisions and slowing down; consequently they will not have the chance to travel great distances.

For the reasons explained in the preceding section, these fast neutrons do not satisfy the hypothesis of the diffusion theory, because they move inside a medium where absorption (in this case removal) is not at all negligible but is in fact dominant. Assuming then that the "weak" collisions leave direction and energy unchanged (straight-ahead collisions), the "uncollided" flux (that is, not having suffered removal collisions) may be written in the form

$$\phi_0(r) = S \, e^{-\Sigma_r r}/4\pi r^2 , \qquad (12)$$

where Σ_r is the removal cross-section, which can be deduced according to what we said about the different possible collisions; that is, it can be assumed equal to the difference between the total cross-section and the elastic cross-section against heavy nuclei.

It is clear that all the considerations made so far are empirical and qualitative, and that the whole question must be verified through experience. Actual experience confirms this model because it shows the presence of an asymptotic equilibrium spectrum and of a trend of the kind seen in Eq. (12). That is, it confirms that penetration occurs mainly at energies near that of the source, and that the removed neutrons produce the slowing down spectrum which is typical of the material and, hence, independent of the position. The shape described by Eq. (12) therefore applies not only to the source energy but to any energy, including the thermal one. This circumstance, verified experimentally, proves the validity of the model, and at the same time supplies a method for measuring the removal cross-sections from the asymptotic trend of the thermal flux (which is more easily measured than the fast flux). Measurements have been performed of the spatial distributions of the neutron fluxes produced by a monoenergetic source in a water tank[8]. From these the values of the removal cross-section of water can be easily computed. Removal cross-sections for materials other than water may be obtained by placing between the source and the water tank a layer of the sample material, and measuring the amount by which the asymptotic thermal flux has been reduced[8].

The Removal-Diffusion Model

As previously stated, diffusion theory is not suitable for computing fast neutron fluxes. However, these may be computed by means of formulas similar to Eq. (12).

The removal neutrons, slowed down by inelastic collisions or by elastic collisions against hydrogen, have an energy below the inelastic scattering threshold. In this energy range the slowing down is poor, there is no absorption, and diffusion theory may be applied. A diffusion equation can be written where the source term accounts for the removal neutrons; that is, the product of ϕ_0 times the removal cross-section. This source is produced by inelastic collisions (which are isotropic) and by collisions against hydrogen (which are isotropic in the centre-of-mass system). Its angular distribution will not be far from isotropy, as required by diffusion theory.

The removal model and the diffusion model may be considered as complementary to each other. Each of them applies where the other fails, and together they provide a suitable approximation for the calculation of neutron transport in hydrogenous shields.

REFERENCES

1. H. Goldstein, "Fundamental Aspects of Reactor Shielding", Addison-Wesley Publ. Co., Reading, Mass. (1959).
2. C. Devillers, Programme MERCURE-3, Atténuation en ligne droite dans une géometrie à trois dimensions, CEA-R 3264 (1967).
3. C. Ponti and R. Van Heusden, SABINE-3, An Improved Version of the Shielding Code SABINE, EUR 5159 (1974).
4. C. Ponti, H. Preusch and H. Schubert, SABINE, A One-Dimensional Bulk Shielding Program, EUR 3636 (1967).
5. C. Devillers et C. Dupont, Manuel d'utilisation du programme MERCURE-4, Rapport SERMA/S 168.
6. S. Glasstone and M.C. Edlund, "The Elements of Nuclear Reactor Theory", D. Van Nostrand Co., Princeton, New Jersey (1952).
7. A.W. Weinberg and E.P. Wigner, "The Physical Theory of Neutron Chain Reactors", University of Chicago Press, Chicago (1958).
8. G. Perlini et al., Attenuation of Monoenergetic Source Neutrons in Different Shielding Materials, Nucl. Eng. Des. 13:377 (1970).

ELECTROMAGNETIC CASCADE SHOWER PROGRAMS
AND THEIR APPLICATIONS

LECTURE 11:   THE PHYSICS OF ELECTROMAGNETIC CASCADE

Keran O'Brien

Environmental Measurements Laboratory
U.S. Department of Energy
New York, New York 10014
USA

INTRODUCTION

When an electron with an energy of the order of one GeV or more is introduced in matter, it radiates photons of comparable energy. These photons produce negatron-positron pairs which radiate in their turn. Eventually, the energy of the original electron is divided in a "shower" or "cascade" (Fig. 1). When finally electrons are created, set in motion or have lost enough energy so that further energy losses will take place primarily through collision mechanisms rather than through radiation, they no longer contribute to the shower and the shower stops.

Radiation (or bremsstrahlung) and pair production are the processes that make the shower possible. Many other processes are involved, as we shall see, which modify the details of the distribution of the shower, but these two are dominant, and we shall see that they are of comparable strength.

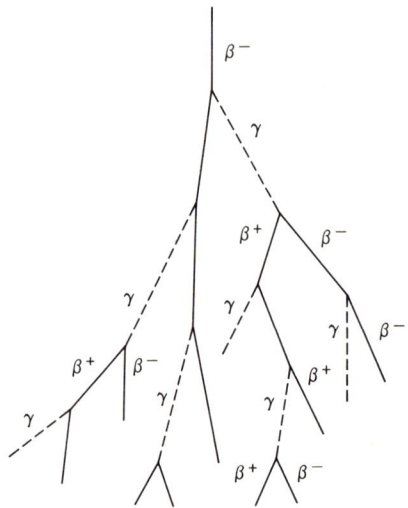

Fig. 1. Diagram of electromagnetic shower [taken from Fermi et al.[1]].

An approximate expression for the mean energy lost by an electron due to bremsstrahlung in a medium is[1]

$$-(dE/dx)_{rad} = 4\,Z^2(L/A)E(r_e^2/137)\,\ln(183/Z^{1/3}) = E/X_R$$

where

X_R is the radiation length,

$(dE/dx)_{rad}$ is the stopping power due to radiation,

Z is the nuclear charge of the atoms in the medium,

L is Avogadro's number,

A is the atomic weight of the medium,

E is the electron kinetic energy (in MeV), and

r_e is the classical electron radius (2.818 F).

The ratio of radiation to collision energy losses is given by[1]

THE PHYSICS OF ELECTROMAGNETIC CASCADE

$$\frac{(dE/dx)_{rad}}{(dE/dx)_{coll}} = \frac{ZE}{800}$$

Hence, the energy below which collision losses predominate and bring the shower to a stop, called the critical energy, is

$$E_{crit} = \frac{800}{Z} \ .$$

Since

$$-(dE/dx)_{rad} = (E/X_R) \ ,$$

we see that

$$E = E_0 \, e^{-x/X_R} \ .$$

Hence, the significance of the radiation length is that, given an electron of energy $E_0 \gg E_{crit}$, it will lose $1/e$ of its initial energy, on average, in passing through X_R grams per cm^2 of matter.

The cross section for pair production, when $E \gg E_{crit}$, is[2]

$$\sigma_{pp} = \frac{Z^2}{137} r_e^2 \frac{28}{9} \ln\left[\frac{183}{Z^{1/3}} - \frac{2}{27}\right] cm^2 \ .$$

A mean free path for pair production can be defined,

$$X_{pp} = \left[\frac{L}{A} \sigma_{pp}\right]^{-1} = \sim \left[\frac{L}{A} \frac{28}{9} \frac{Z^2}{137} r_e^2 \ln(\frac{183}{Z^{1/3}})\right]^{-1}$$

Given a beam of photons ϕ_0, the number not having undergone any interaction in a distance x g/cm^2 will be

$$\phi = \phi_0 \, e^{-x/X_{pp}}$$

and in this case, a beam of photons will lose 1/e of its total number, on average, in passing through X_{pp} g/cm^2 of matter.

Comparing X_R and X_{pp}, we see that

$$X_R = 0.8 \, X_{pp} \, ;$$

that is, the two mean free paths are nearly equal and, as was observed earlier, the two processes are of essentially equal importance to the propagation of the shower.

Analytic Forms of Cascade Theory

The full analytic treatment of the theory of cascade showers is prohibitively difficult. However, Rossi and Greisen[3,4] have introduced two levels of approximation. In Approximation A, energy loss through collision processes is neglected. Only radiation processes and pair production are taken into account. Further, the cross sections are somewhat simplified. As a consequence, the average behavior of particles in a shower under Approximation A is a function of two parameters only: the ratio E_0/E, where as above, E is the energy of the particle, and E_0 is the initiating particle energy, and of $t = x/X_R$, the depth of the particle in units of the radiation length. When expressed this way, Approximation A results are independent of material and particle energy.

In Approximation B, electron collisions loss is included as a constant term. Results are expressed in terms of the radiation length and the critical energy.

In Fig. 2 (taken from Hayakawa[5]) is shown the number of electrons for photon-initiated showers above an energy E for $E_0/E = 100$, where E = 10, 100, and 500 MeV, so that E_0 = 1, 10, and 50 GeV. The curves

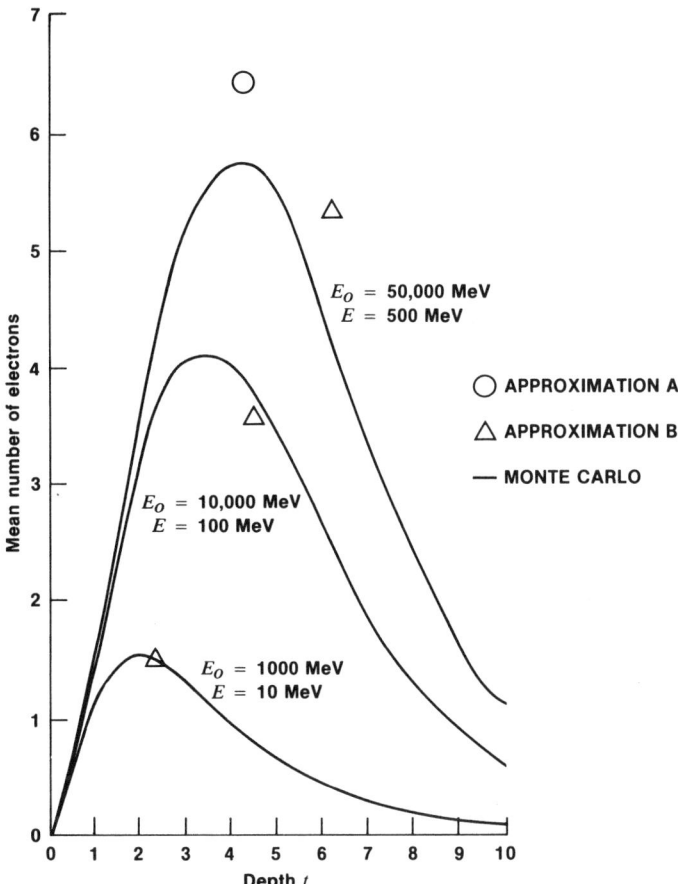

Fig. 2. Average number of electrons with energies above 10, 100, and 500 MeV for incident electrons with energies of 1, 10, and 50 GeV in air from Butcher and Messel.[6] The circle and triangles represent maxima in the shower curve calculated using Approximation A and B, respectively.

were calculated by Butcher and Messel[6] in air. The maxima range from 1.5 to 5.8 electrons, which are located at depths from 2 to 4.2 radiation lengths. Approximation A predicts a constant 6.5 electrons at 4.2 radiation lengths for all cases. Approximation B gives much better results, predicting maxima of 1.4, 3.6, and 5.4 electrons at

2.3, 4.6, and 6.2 radiation lengths.

Adequate as some analytic methods may be for some purposes in light media, they fail completely in heavy media, where cross section energy dependence and electron multiple scattering play an important role.

Nowadays, all serious shower calculations are carried out using the Monte Carlo technique, employing complete cross sections, accounting for all important processes, and permitting the solution of multimedium and three-dimensional problems. This development is a consequence of the availability of modern computers and computational techniques suited to machines. It is of interest to observe that the first Monte Carlo shower calculations were published in 1952.[7]

ELECTRON PROCESSES

Bremsstrahlung

An electron passing near a nucleus of atomic number Z is deflected due to the Coulomb forces between them (Fig. 3) and is accelerated, with the result that it radiates. The forces it undergoes depend on how close to the nucleus the electron passes. If this distance (the impact parameter) is smaller than the atomic radius, the field acting on the particles is just the force resulting from

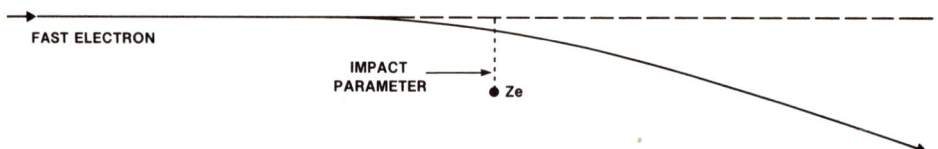

Fig. 3. Diagram of bremsstrahlung process [taken from Fermi et al.[1]]

THE PHYSICS OF ELECTROMAGNETIC CASCADE

the charge Ze at the nucleus. If the impact parameter is larger than the atomic radius, then this field will be screened to some extent by the electrons of the atom.

Of course, in quantum mechanics, the electron is represented by a plane wave and the preceding description of the bremsstrahlung process with a point electron deflected by the electrical fields of the atom is not exact. However, it is possible to define a quantum-mechanical impact parameter in terms of the distance at which the main contributions to the radiation process occur. This gives us[2]

$$b = \hbar c/p ,$$

where

- b is the impact parameter,
- $\hbar$ is Planck's constant,
- c is the velocity of light, and
- p is the momentum transferred to the nucleus by the radiation process.

The cross section for the radiation process has been calculated by Bethe and Heitler;[8]

$$\sigma_r(W, k)dk = 4Z^2 \alpha r_e^2 (dk/k) F(W, u), \qquad \alpha(e^2/\hbar c) = (137)^{-1},$$

where

- k is the phonton energy,
- W is the total incident electron energy, and
- u is the energy of the emitted photon divided by the incident electron energy (= k/W).

The function of $F(W, u)$ depends on the screening parameter ξ, which is

$$\xi = 100 \frac{\mu}{W} \frac{u}{1-u} Z^{-1/3},$$

where μ is the electron rest energy (= .511 MeV)

$$F(W, u) = [1 + (1 - u)^2 - \frac{2}{3}(1 - u)]\left[\ln \frac{2W}{\mu}\left(\frac{1 - u}{u}\right) - \frac{1}{2}\right]$$

for $\xi \gg 1$ (no screening),

$$F(W, u) = [1 + (1 - u)^2 - \frac{2}{3}(1 - u)]\left[\ln\left(\frac{2E}{\mu} \frac{1 - u}{u}\right) - \frac{1}{2} - c(\xi)^{-1/2}\right]$$

for $2 < \xi < 15$,

$$F(W, u) = [1 + (1 - u)^2]\left[\frac{f_1(\xi)}{4} - \frac{1}{3} \ln Z\right] - \frac{2}{3}(1 - u)$$
$$\left[\frac{f_2(\xi)}{4} - \frac{1}{3} \ln Z\right]$$

for $\xi < 2$,

$$F(W, u) = [1 + (1 - u)^2 - \frac{2}{3}(1 - u)] \ln(183Z^{-1/3}) + \frac{1}{9}(1 - u)$$

for $\xi \simeq 0$ (complete screening).

The functions $c(\xi)$, $f_1(\xi)$, and $f_2(\xi)$ are shown in Figs. 4 and 5.[4]

The function $F(W, u)$ is shown is Figs. 6 and 7 for air and lead. The distributions for various total electron energies in eV are quite flat and essentially material independent.[3]

A comparison of the radiation cross section with the formula given earlier for the radiation length X_R indicates that, roughly

$$\sigma \propto A/LX_R \propto Z^2$$

Coulomb Scattering

At low energies, the deflection shown in Fig. 3 may take place

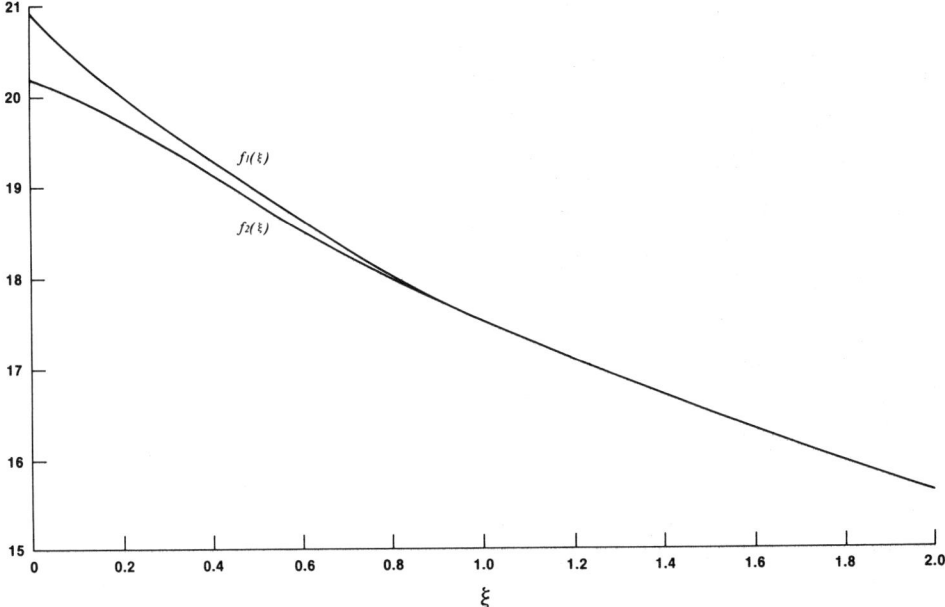

Fig. 4. The functions $f_1(\xi)$ and $f_2(\xi)$ [taken from Rossi[4]].

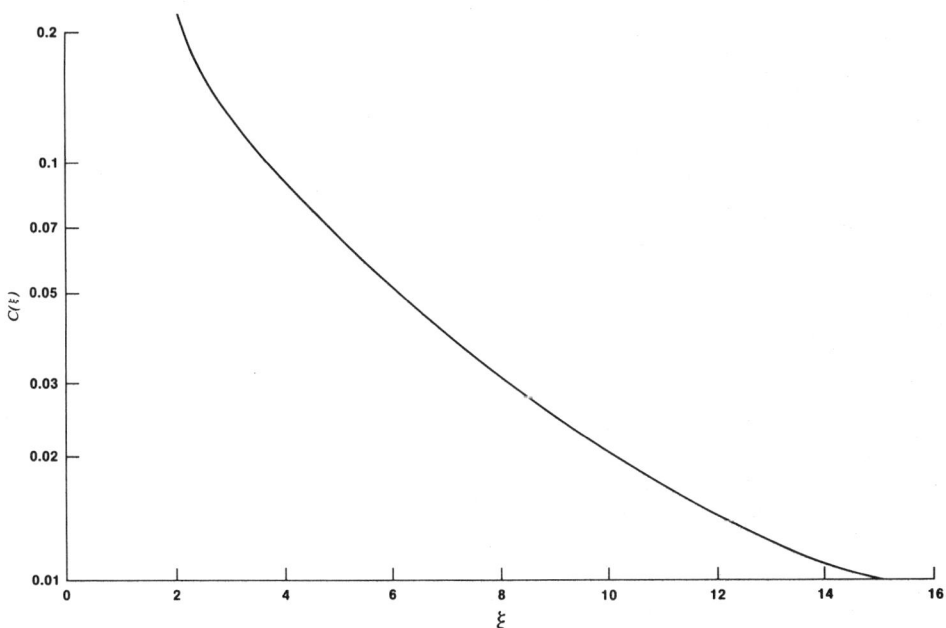

Fig. 5. The function $c(\xi)$ [taken from Rossi[4]].

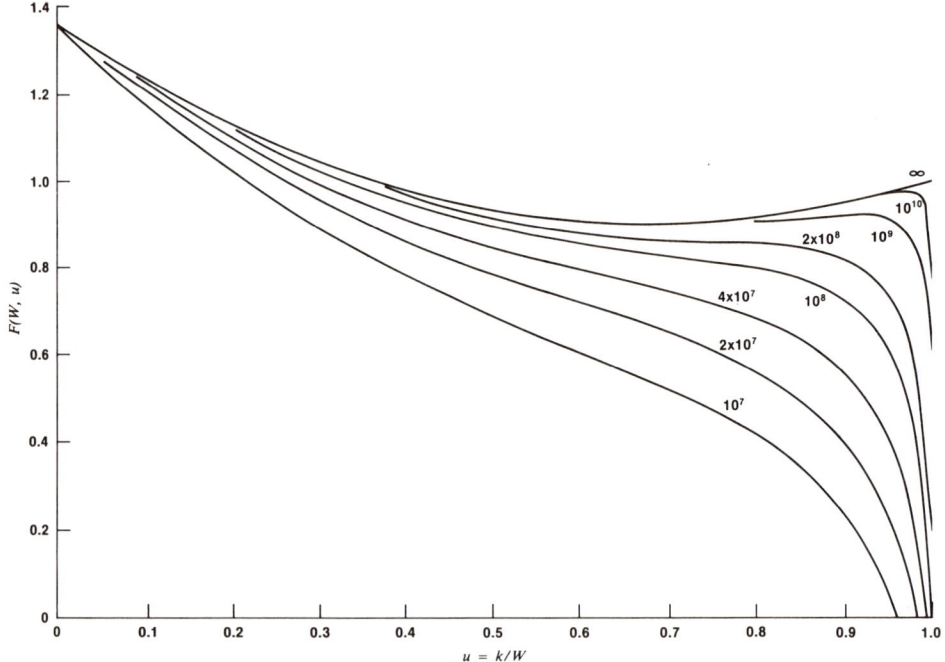

Fig. 6. Differential bremsstrahlung spectra per unit radiation length in lead as a function of total electron energies in eV (W = E + μ) [taken from Rossi[4]].

unaccompanied by radiation processes. The cross section for this process is

$$\sigma(\theta)\, d\omega = \frac{1}{4} z^2 r_e^2 \left(\frac{\mu}{\beta p}\right)^2 \left(1 - \beta^2 \sin^2 \frac{\theta}{2}\right) \frac{d\omega}{\sin^4(\theta/2)} ,$$

where

θ is the angle through which the electron is scattered,
$d\omega$ is the element of solid angle into which the electron is scattered, and
β is the electron velocity relative to the speed of light.[9]

The mean square angle of scatter after an electron has traveled x grams per cm^2 is

THE PHYSICS OF ELECTROMAGNETIC CASCADE

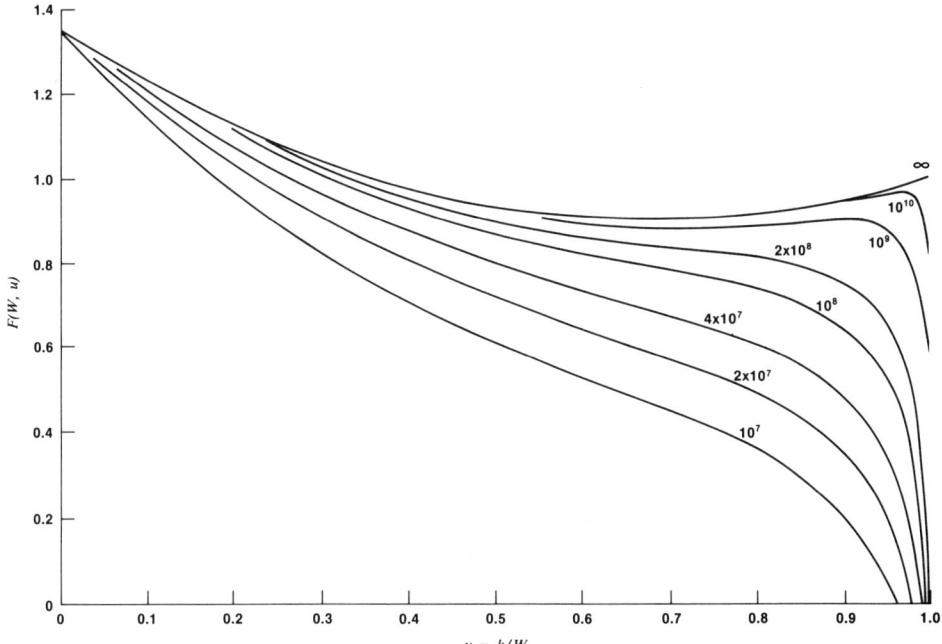

Fig. 7. Differential bremsstrahlung spectra per unit radiation length in air as a function of total electron energies in eV (W = E + μ) [taken from Rossi[4]].

$$<\theta^2> = 8\pi(Z^2/A) \; Lr_e^2(\mu/\beta p) \; \ln(\theta_2/\theta_1) \; x \quad ,$$

where

θ_2 is the maximum angle of scatter, and

θ_1 is the minimum angle of scatter,

if x is small enough so that energy loss during the traversal is negligible.

Each of these angles is related to its corresponding impact parameter; that is,

$$\theta_2 = (\hbar c/p)(b_{min})^{-1},$$

and

$$\theta_1 = (\hbar c/p)(b_{max})^{-1}.$$

The maximum scattering angle and the minimum impact parameter, and the minimum scattering angle and the maximum impact parameter, correspond. As in the radiation case noted above, the smaller the impact parameter, the greater the momentum transfer. In the case of elastic scattering, this necessitates a larger scattering angle.

If we define a constant in units of energy:

$$E_s = \mu(4\pi/\alpha)^{1/2} = 21 \text{ MeV},$$

Then we can rewrite the equation for $<\theta^2>$,

$$<\theta^2> = (E_s/p\beta)^2 \times [4Z^2(L/A)(r_e^2/\alpha) \ln(\theta_2/\theta_1)^{1/2}].$$

Next, we observe that b_{min} is the nuclear radius, and

$$b_{min} = r_0 A^{1/3} = 1.28\, A^{1/3} F.$$

The maximum impact parameter is the <u>atomic</u> radius, so that

$$b_{max} = a_0 Z^{-1/3},$$

where a_0 is the Bohr radius.

Substituting the appropriate impact parameters into the expression for the angles

$$(\theta_2/\theta_1)^{1/2} = (a_0/r_0)^{1/2}(Z/A)^{-1/3} ,$$

letting $Z/A = \sim 1/2$ (a good approximation for most elements).

$$(\theta_2/\theta_1)^{1/2} = 181 \, Z^{-1/3} ,$$

the expression for $<\theta^2>$ becomes

$$<\theta^2> = (E_s/p\beta)^2 \times [4 \, z^2 \, (L/A)(r_e^2/\alpha)\ln 181 \, Z^{-1/3}] ,$$

or, looking back to the earlier definition of radiation length, X_R,

$$<\theta^2> = (E_s/p\beta)^2 (x/X_R) .$$

Finally, the probability of finding an electron between θ and $\theta + d\theta$, due to Coulomb scattering, is

$$P(\theta) \, d\theta = (2/<\theta^2>) \, e^{-\theta^2/<\theta^2>} \theta d\theta$$

when θ is small.

For more accurate accounts of screening and the nuclear size effect, see Molière[10,11] and Cooper.[12]

Ionization Energy Loss

The energy lost per g/cm^2 of path by a particle through inelastic collisions with the electrons of the medium is [13,14]

$$-\left(\frac{dE}{dx}\right)_{coll} = \frac{L}{A} z^2 \frac{Z}{\beta^2} 2\pi \, r_e^2 \, \mu B ,$$

$$B = \ln \frac{2\mu \beta^2 E_m}{I^2(1-\beta^2)} - 2\beta^2 - \delta + B_{el},$$

$$E_m = 2\mu \frac{W^2 - M^2}{2W\mu + M^2 + \mu^2},$$

$$B_{el} = -\left[\beta^2 + 2(1 - \beta^2)^{\frac{1}{2}}\right]\ln 2 + 1 + \beta^2,$$

$$\delta = \begin{cases} \ln \alpha n(p/\mu)^2/I^2 - 1 & [\alpha n(p/\mu)^2/I^2 \leq e], \\ 0 & [\alpha n(p/\mu)^2/I^2 < e], \end{cases}$$

where

- I is the ionization potential of the medium ($= \sim 13\,Z$),
- z is the charge of the particle ($= 1$ for negatrons and positrons),
- δ is the "density effect", a term introduced to account for the reduction in ionization caused by the polarization of the medium due to the passage of a fast charged particle,
- M is the mass of the charged particle in MeV/c^2 ($= .511$ MeV for electrons and positrons),
- E_m is the maximum kinetic energy that can be transferred by the charged particle to the electrons of the medium,
- n is the electron density of the medium per cm^3, and
- B_{el} includes spin and exchange effects and applies only to electron stopping.^{2}

When the colliding particle is a positron, $M = \mu$. When the colliding particle is a negatron, the faster of the two particles coming out of the collision is assumed to be the primary (since the two electrons cannot be distinguished after the collision). Then

$$E_m = \frac{1}{2}(W - \mu) = \frac{1}{2}E$$

Positron Annihilation

At high energies, the behavior of the positron is essentially the same as the behavior of the electron. With decreasing energy however, the probability of the positron's combining with a free electron at rest becomes increasingly important. This combination involves the annihilation of the electron and positron resulting in the emission of two photons. The emission at high energies is polarized along the direction of the positron flight. Since the energy in the center of mass is shared equally, one photon will carry nearly all the energy in the forward lab. direction, and the other will appear in the lab. system with an energy of about 0.511 MeV. At low energies, the two photons are emitted at $180°$ from one another, but the angular distribution of the vector joining them is isotropic.

The cross section for the process is[15]

$$\sigma_{an} = \pi r_e^2 \frac{1}{\gamma + 1} \left\{ \frac{\gamma^2 + 4\gamma + 1}{\gamma^2 - 1} \ln\left[\gamma + (\gamma^2 - 1)^{1/2}\right] - \frac{\gamma + 3}{(\gamma^2 - 1)^{1/2}} \right\},$$

$$\gamma = (1 - \beta^2)^{-1/2}.$$

At low energies, this reduces to

$$\sigma_{an} = \pi r_e^2 \beta^{-1},$$

so it can be seen that annihilation is the ultimate fate of all positrons traveling in normal matter (i.e., not antimatter).

The Chudakov Effect

When an electron and positron are emitted from the pair-production process with a very small angle, the energy loss is not the sum of the energy losses calculated by the stopping-power formula discussed previously. It is less than this sum because the electromagnetic fields of the two particles interfere.[16] Hence, the energy loss is reduced by an amount which depends on the distance between the negatron-positron pair. In the special case where the created electron-positron pair have the same energy, the energy imparted to the medium is

$$-(dE/dx)_{pair} = (0.3072/\beta^2)[B + .5\beta^2 + 3\ln(W/\mu) - .13 - K_0(d/D)],$$

$$D = (4\pi r_e n)^{-1/2},$$

where

D is the shielding radius, and

K_0 is the modified Bessel function of the second kind.

Calculations and measurements by Yekutieli[17] carried out for photographic emulsion indicate that the ratio of the pair energy loss to the energy loss gotten by adding the individual stopping powers, ranges from o.46 at 15 μm to 0.86 at 200 μm, when the pair energy was 180 GeV.

The Landau-Pomeranchuk Effect

The bremsstrahlung process takes place during a finite length of time and over a finite track length. If, during this period, the electron is scattered by the electrical field of a nearby atom, the process is terminated. The effect increases with the density of the medium and with electron energy.

In the rest system of an electron with incident energy W, the time it takes to radiate a photon with frequency ν is $1/\nu$. In the lab system, the time is dilated by the Lorentz factor W/μ. The energy of the emitted photon is given by

$$k = \hbar\nu(W - k)/\mu ,$$

and the radiation time is

$$t_{lab} = \hbar W(W - k)/(\mu^2 k) .$$

Hence, the time it takes to shake the radiation photon free increases with electron energy and decreases as the emitted photon energy increases. It also follows that as the density of the medium increases, the number of "interruptions" of the process will increase.

Thus, the effect will be seen first on the reduction of the low-energy part of the bremsstrahlung spectrum, and most prominently in dense media. This is shown very clearly in Fig. 8 where calculations of this effect, carried out by Varfolomeev et al.[18], are shown, using the Bethe-Heitler formula discussed earlier and comparing it with calculations in carbon and (tungsten taking into account the Landau-Pomeranchuk effect).

The graph shows the mean number of bremsstrahlung photons produced above 0.1 MeV per radiation length in carbon and tungsten, both accounting for the Landau-Pomeranchuk effect, and using only the Beth-Heitler formulation. In addition, the relative radiation loss per radiation length was calculated in carbon and tungsten. The number of photons falls off sharply in both media, and most quickly and most radically in tungsten, which is the denser medium.

However, as we have seen before, prior to discussion of the Landau-Pomeranchuk effect, the radiation energy loss being given by

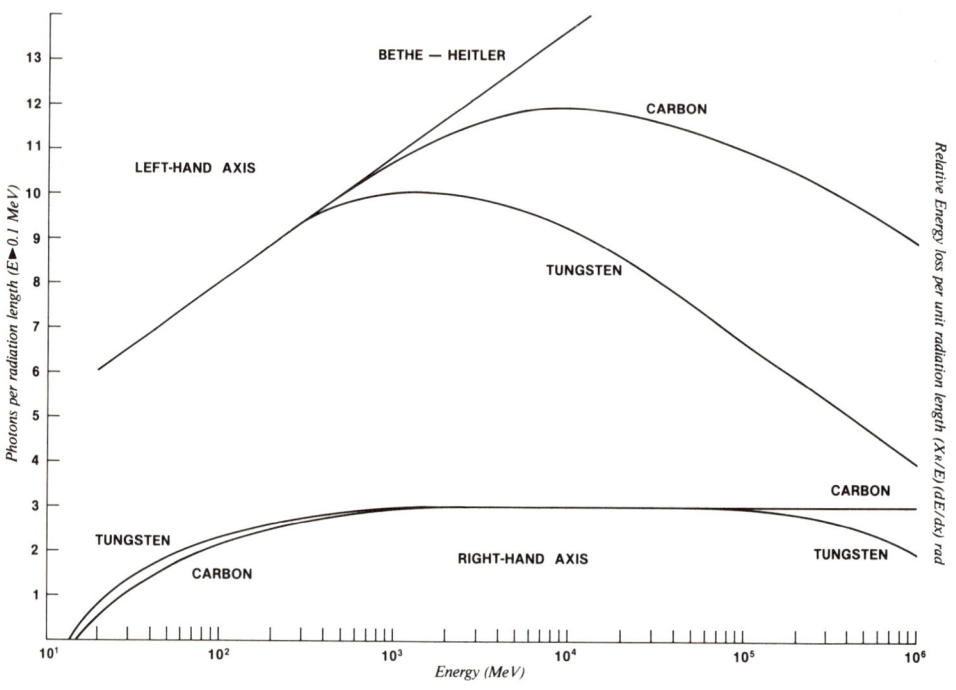

Fig. 8. Average number of photons above 0.1 MeV produced per radiation length in carbon and tungsten calculated accounting for the Landau-Pomeranchuk effect and comparing it with the Beth-Heitler formulation – on top. Below, the relative energy loss in carbon and tungsten [from Varfolomeev et al.[18]].

$$(dE/dX) = (E/X_R)$$

at sufficiently high energies, the relative energy loss per radiation length is

$$(X_R/E)(dE/dx) = 1 .$$

This holds quite well for carbon above a TeV and only fails for tungsten above about 100 GeV. A large reduction in the number of

THE PHYSICS OF ELECTROMAGNETIC CASCADE

low-energy photons while retaining the same mean energy loss is at the expense of low-energy photon production.

Čerenkov and Transition Radiations

When a charged particle traverses a medium with a refractive index N, if

$$N\beta > 1,$$

then Čerenkov radiation is emitted. The angle of emission of the photon is

$$\theta = \arccos(\beta N)^{-1}.$$

When produced in dielectric media, Čerenkov radiation generally lies in the optical range.

Transition radiation occurs when a charged particle passes from one medium to another with a different dielectric constant. The intensity of the transition radiation may considerably exceed that of the bremsstrahlung and at very low energies. Varfolomeev et al.[18] have calculated both the bremsstrahlung and transition radiation produced in carbon and tungsten plates 0.05 radiation lengths thick irradiated by 200 GeV electrons. The emitted spectra calculated by Varfolomeev et al. are shown in Fig. 9.

The photon yield per unit length is approximately

$$\frac{d^2 n(\nu)}{dL d\nu} = \frac{2}{\pi} \frac{z^2}{hc a \nu} \left[2 \ln\left(\frac{\nu_p}{\nu}\right) \gamma - 1 \right],$$

$$L = .0765 \; 2 \, Z/A = \sim .0765,$$

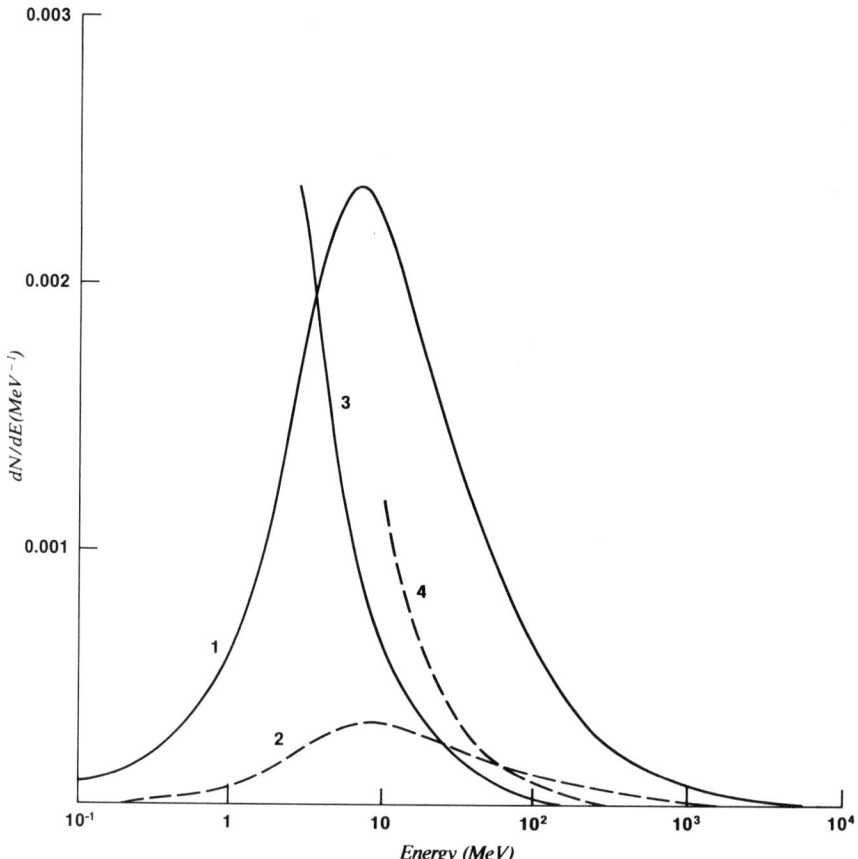

Fig. 9. Bremsstrahlung (1, 2) and transition - radiation (3, 4) spectra from 200 GeV electrons in carbon (1, 3) and tungsten (2, 4) plates 0.05 radiation lengths thick [from Varfolomeev et al.[18]].

$$\nu_p = (4\pi n_e e^2/m_e)^{1/2} \quad ,$$

$$\gamma = (E + \mu)/\mu \quad ,$$

where

THE PHYSICS OF ELECTROMAGNETIC CASCADE

a   is the plate thickness,

ν is the photon frequency,

ν_p is the plasma frequency,

n_e is the electron density, and

m_e is the electron mass.

PHOTON PROCESSES

Pair Production

As noted in the earlier description of cascade showers, bremsstrahlung and pair production are the two major processes which make possible shower propagation. When a photon passes near a nucleus, it may disappear and be replaced by two electrons, a negatron and a positron. The origin of these two newly created particles can be described by the Dirac theory. In Dirac's theory, negatrons have positive and negative energy states given by $W^{\pm} = \pm (\mu^2 + p^2)^{1/2}$. The available energy states are shown in Fig. 10. A photon with a sufficiently high energy may raise a negatron from a negative energy state to a positive energy state. The energy threshold for this process is obviously 2μ since, as we can see both from Fig. 10 and the equation for $W^{\pm}$, there are no states between $+\mu$ and $-\mu$. A difference of 2μ or 1.022 MeV is required.

This process is closely related to the bremsstrahlung process, and the equations describing them are very similar. In the radiation process, an electron makes a transition between two states in the positive region shown in Fig. 10, emitting a photon. In the pair production process, a photon is absorbed while raising a negative energy electron to a positive energy state.

The cross sections are related by

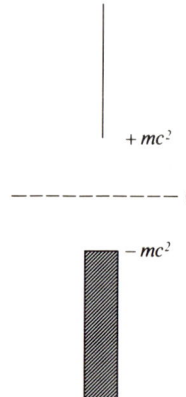

Fig. 10. Diagram of positive and negative electron energy states in pair production [from Fermi et al.[1]].

$$\sigma_{pp}(k, W) = \sigma_r(W, k)(W/k)^2$$

Setting $v = W/k$ where W is the total energy of one of the created pairs, one obtains

$$\sigma_{pp}(k, W) = 4\alpha Z^2 r_e^2 \, (dE/k) \, G(k, v), \quad \alpha = 1/137 ,$$

where

$$G(k, v) = [v^2 + (1 - v)^2 + \tfrac{2}{3} v(1 - v)]$$
$$\left[\ln\left(\frac{2k}{\mu}\right) v(1 - v) - \tfrac{1}{2}\right]$$

for $\xi \gg 1$ (no screening),

$$G(k, v) = [v^2 + (1 - v^2) + \tfrac{2}{3} v(1 - v)]$$
$$\left[\ln\left(\frac{2k}{\mu}\right) (1 - v) - \tfrac{1}{2} - c(\xi)\right]$$

for $2 < \xi < 15$,

$$G(k, \nu) = [\nu^2 + (1-\nu)^2]\left[\frac{f_1(\xi)}{4} - \frac{1}{3}\ln Z\right] + \frac{2}{3}\nu(1-\nu)\left[\frac{f_2(\xi)}{4} - \frac{1}{3}\ln Z\right]$$

for $\xi < 2$, and

$$G(k, \nu) = [\nu^2 + (1-\nu)^2 + \frac{2}{3}\nu(1-\nu)]\ln(183Z^{-1/3}) - \frac{1}{9}\nu(1-\nu)$$

for $\xi \simeq 0$ (complete screening),

and the functions ξ, f_1, and f_2 have the same meanings as in the bremsstrahlung equations.

The functions $G(k, \nu)$ are shown for various energies in eV for air and lead in Figs. 11 and 12, respectively.[4] The functions are rather flat, and essentially material independent, especially at high energies.

The Compton Effect

Photon-electron scattering is called Compton scattering. As indicated in Fig. 13, an incident photon collides with an atomic electron and recoils with a longer wavelength and a lower energy. The cross section for an incident photon of energy

$$k_0 = h\nu_0 = hc/\lambda$$

per steradian to be scattered into an angle θ is

$$d\sigma_c = \frac{1}{2}r_e^2 \frac{k^2}{k_0^2}\left(\frac{k_0}{k} + \frac{k}{k_0} - \sin^2\theta\right)d\Omega \quad,$$

where k is the recoiling photon energy. The energy and angle of the

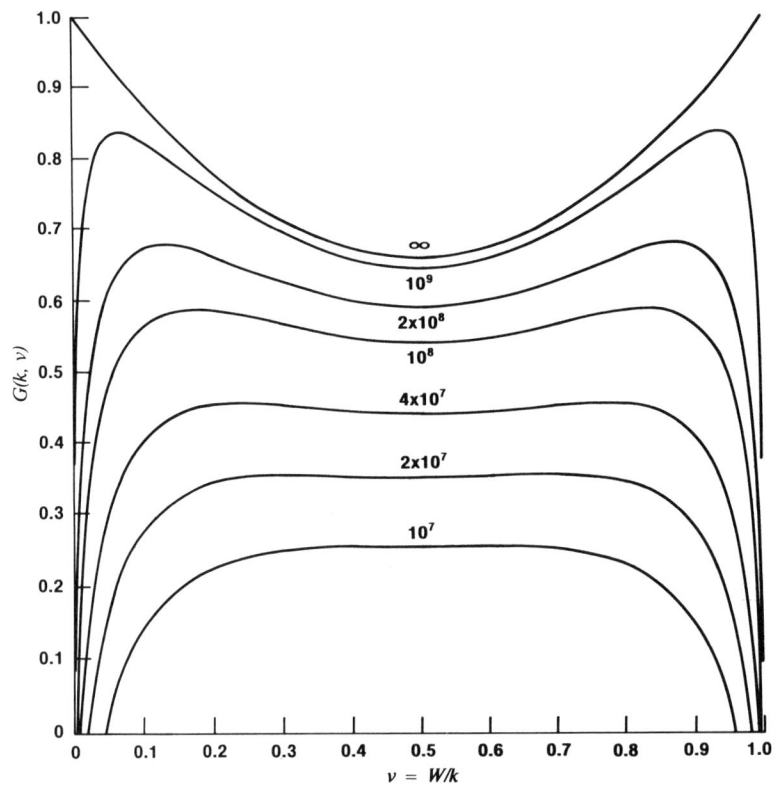

Fig. 11. Differential position or negatron spectra per unit radiation length in air as a function of photon energies in eV [from Rossi[4]].

scattered photon are related by

$$k = \frac{k_0}{1 + q(1 - \cos\theta)},$$

$$q = \frac{k_0}{\mu},$$

and, of course,

$$\lambda_{scattered} = hc/k .$$

THE PHYSICS OF ELECTROMAGNETIC CASCADE   165

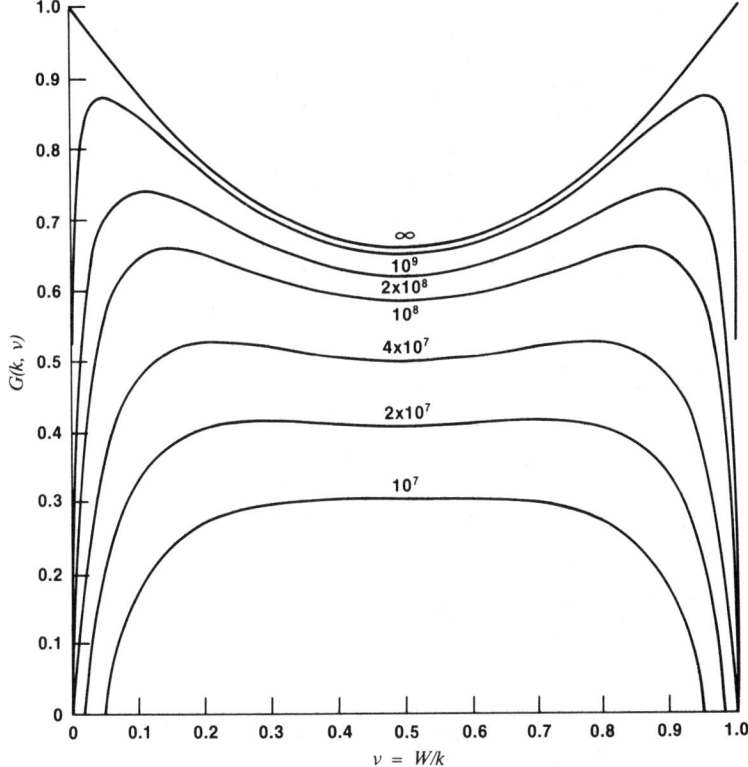

Fig. 12. Differential positron or negatron per unit radiation length in lead as a function of photon energies in eV [from Rossi[4]].

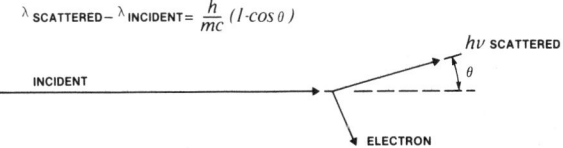

Fig. 13. Diagram of Compton scattering [from Fermi et al.[1]].

The energy and angular distributions of the scattered electrons and photons can be obtained from these formulae. The total cross section is obtained from integration over energy and angle,

$$\sigma_c = nr_e^2 \frac{1}{q} \left\{ (1 - \frac{2(q+1)}{q^2}) \ln(2q+1) + \frac{1}{2} + \frac{4}{q} - \frac{1}{2(2q+1)^2} \right\},$$

where

n is the electron density per gram, and

σ_c is the Compton cross section in cm^2/g.

The minimum energy a photon may have after colliding with an electron is

$$k_{min} = \frac{\mu k_0}{\mu + 2k_0},$$

and is often referred to as the single scattering discontinuity. Because a photon with energy k_0 cannot recoil with an energy less than k_{min} after a single collision with an electron, a discontinuity exists in scattered photon spectra produced by monochromatic primaries after passing through a small number of mean free-paths of matter.

The Photoelectric Effect

At energies below those where Compton scattering is important, the most significant photon process is the photoelectric effect. In this process, a photon is absorbed by an atom and an electron is ejected. The electron carries away all the energy of the absorbed photon less the energy which bound the electron to the atom. The interaction takes place primarily with the tightly bound K electrons of the innermost shell. When the energy of a photon falls below the binding energy of a given shell, an electron from that shell cannot be ejected. Hence, the cross section for the photoelectric effect at low energies exhibits a series of absorption edges which are character-

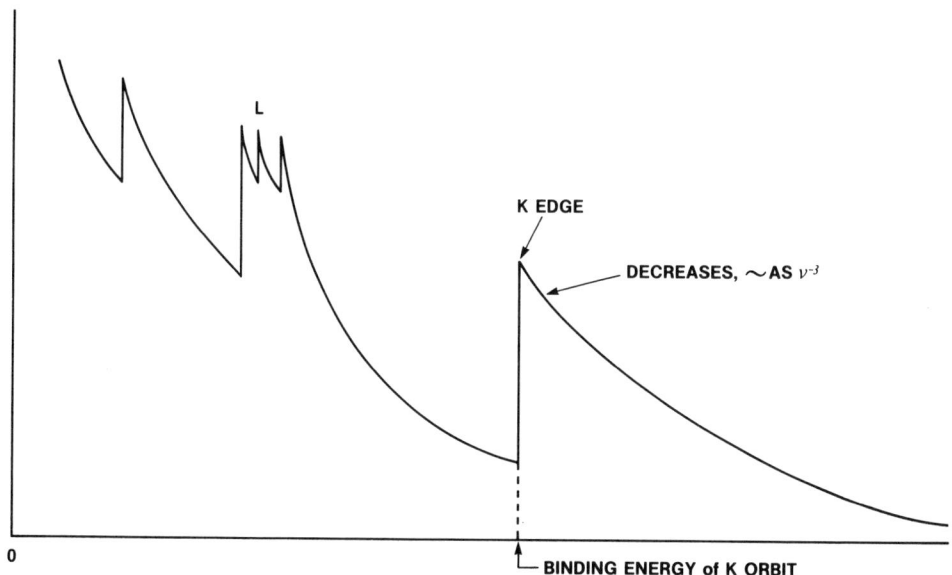

Fig. 14. Diagram of photoelectric cross section [from Fermi et al.[1]].

istic of the material.

In Fig. 14 is shown a drawing of the photoelectric cross section including the K and L edges. The L edges are threefold since the L shell has three subshells.

SUMMARY

In Figs. 15 and 16 are shown the major processes affecting electrons and photons expressed per radiation length of lead. In Fig. 15, the processes are given as a function of

$$\frac{X_R}{E}\left(\frac{dE}{dx}\right)_i,$$

the fractional energy loss per unit radiation length for each process i. In Fig. 16, the cross sections are given as

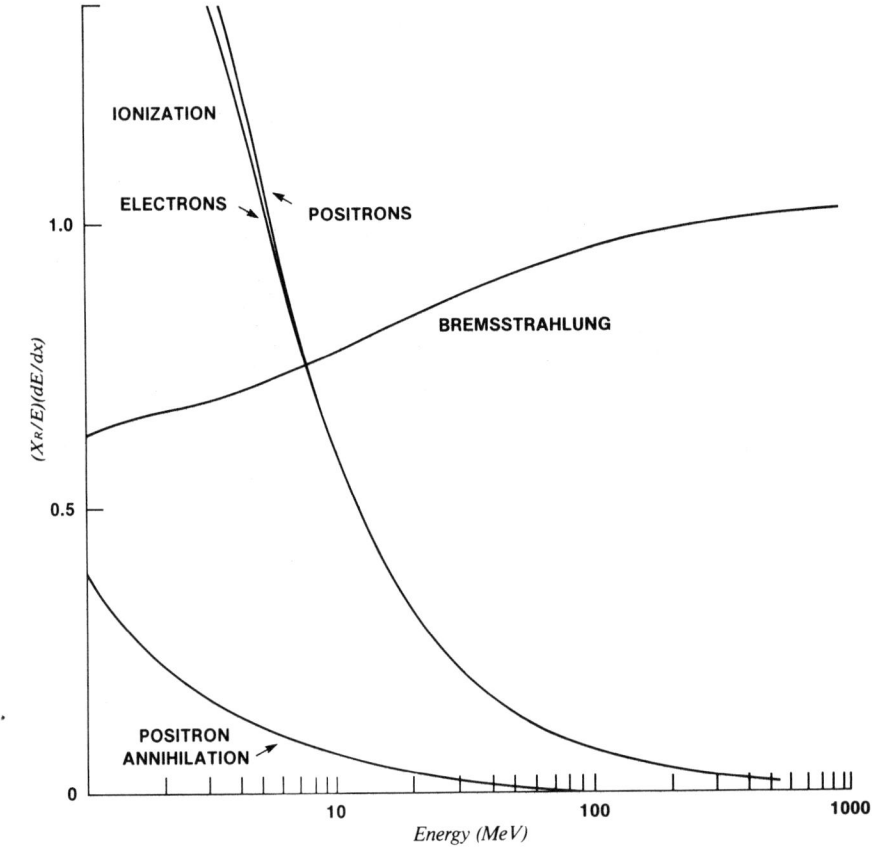

Fig. 15. Comparison of various processes to relative energy loss of electrons in lead per unit radiation length as a function of energy [from N. Brash-Schmidt et al.[21]].

$$(\sigma_i L/A) X_R = \Sigma_i X_R ,$$

where σ_i is the cross section in cm^2 per atom for each process i, and Σ_i is the macroscopic cross section in cm^2/g for each process i.

The critical energy, which has been shown to be equal to 800/Z, is apparent where the bremsstrahlung curve crosses the ionization

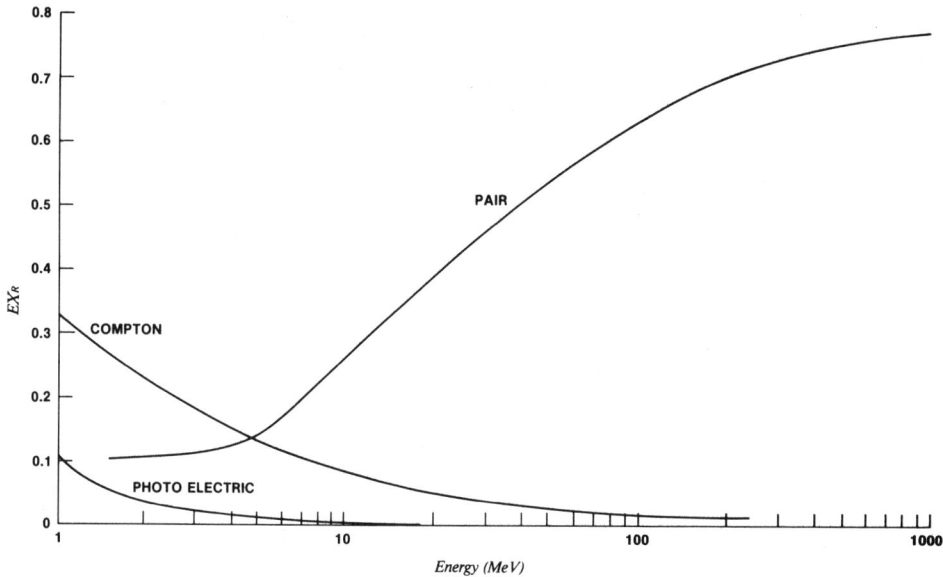

Fig. 16. Comparison of the cross sections per unit radiation length of photons in lead per unit radiation length as a function of energy [from N. Barash-Schmidt et al.[21]].

curve near 10 MeV in Fig. 15. The overwhelming dominance of bremsstrahlung and pair production above the critical energy is quite evident. Finally, the near equality of the pair production and radiation processes at high energies discussed in the introduction to this paper can be clearly seen by comparing the two figures.

Most of the processes discussed above have been written down by Beck[22] and by Ford and Nelson[23] in a form suitable for computer use. The equations in that report contain a number of refinements not discussed in this paper due to limit of space and time.

REFERENCES

1. E. Fermi, J. Orear, A. W. Rosenfeld, and R. A. Schluter. <u>Nuclear Physics</u> (University of Chicago Press, Chicago, 1951).

2. W. Heitler. The Quantum Theory of Radiation (The Clarendon Press, Oxford, 1954.

3. B. Rossi and K. Greisen. Revs. Mod. Phys. 13, 240 (1941).

4. B. Rossi. High-Energy Particles (Prentice-Hall, Inc., New York, 1952).

5. S. Hayakawa. Cosmic Ray Physics (Wiley-Interscience, New York, 1969).

6. J. C. Butcher and H. Messel. Nucl. Phys. 20, 15 (1960).

7. R. R. Wilson. Phys. Rev. 86, 261 (1952).

8. H. A. Bethe and W. Heitler. Proc. Roy. Soc. (London), A146, 83 (1934).

9. N. F. Mott. Proc. Roy. Soc. (London), A124, 425 (1929).

10. G. Molière. Z. Naturforsch 2a, 133 (1947).

11. G. Molière. Z. Naturforsch 3a, 78 (1948).

12. L. N. Cooper and J. Rainwater. Phys. Rev. 97, 492 (1955).

13. R. M. Sternheimer. Phys Rev. 88, 851 (1952).

14. T. W. Armstrong and R. G. Alsmiller, Jr. Nucl. Instrum. & Methods 82, 289 (1970).

15. P. A. M. Dirac. Proc. Camb. Phil. Soc. 26, 361 (1930).

16. A. E. Chudakov. Izv. Akad. Nauk. USSR 19, 650 (1955).

17. G. Yekutieli. Nuovo Cimento 5, 1381 (1957).

18. A. A. Varfolomeev, V. I. Glebov, E. I. Denisov, and A. S. Khlebnikov, Soc. J. Nucl. Phys. 23, 317 (1976) [translated from Yad. Fiz. 23, 604 (1976)].

19. M. L. Ter-Mikaelyan. Nucl. Phys. 24, 43 (1961).

20. O. Klein and Y. Nishina. Z. Physik 52, 853 (1929).

21. N. Barash-Schmidt, A. Barbaro-Galtieri, C. Bricman, R. L. Crawford, C. Dionisi, R. J. Hemingway, C. P. Horne, R. L. Kelly, M. J. Losty, M. Mazzucato, L. Montanet, A. Rittenberg, M. Roos, T. G. Trippe, G. P. Yost, and F. E. Armstrong. Phys. Lett. 75B, 1 (1978).

22. H. L. Beck. "Energy Spectra from Coupled Electron-Photon Slowing Down," U.S. Energy Research and Development Report HASL-309 (1976).

23. R. L. Ford and W. R. Nelson. "The EGS Code System, Computer Programs for the Monte Carlo Simulation of Electromagnetic Cascade Showers (Version 3)," Stanford Linear Accelerator Center Report SLAC-210 (1978).

LECTURE 12: SOLUTION OF THE ELECTROMAGNETIC CASCADE SHOWER PROBLEM BY ANALOG MONTE CARLO METHODS - EGS

Walter R. Nelson [*]

Health and Safety Division
CERN
Geneva 23, Switzerland

INTRODUCTION

Over a decade has passed since H. H. Nagel visited SLAC and "planted the seed" to what is now called EGS (Electron Gamma Shower) ---or more appropriately, the EGS Code System. Nagel's original program was one of three during the early 1960's aimed at solving the electromagnetic cascade shower problem by Monte Carlo simulation (Nagel, 1963, 1964, 1965). The other two codes that were developed during this period were by Zerby and Moran (1962a,b, 1963) and by Messel and his colleagues (1962, 1970).

Even though it was readily available to the scientific community, it became quite clear at the very beginning that the Nagel code had rather limited use as it stood. In fact, most of the problems that could be solved by any one of the three codes mentioned above, had been solved and published. Many interesting problems still remained to be solved, but Nagel's code, because of the way it was structured, was too limited to be of real use without considerable effort on the part of the user. So a program was initiated to develop the Nagel

---

[*] On leave of absence from SLAC (Stanford University), 1978/79.

code into a general, multi-media, modularized, and versatile form. Although considerable progress had been made in this direction, the most significant achievements in the EGS code were made just prior to and after the conversion to the MORTRAN language*.

Very slowly and deliberately then, the EGS Code System has evolved into the Version 3 form that we have today, and with it we feel that we now have a means of simulating almost any electron-photon transport problem. That is to say, the structure of the code, with its global features, modular form, and structured programming, is readily adaptable to virtually any interfacing scheme that is desired on the part of the user.

One must be careful, however, to choose the proper code for any given problem. Because EGS transports electrons, positrons, and photons in a very intrinsic manner, the code is, by its very nature, much slower than programs that simulate the average behavior of cascades. Furthermore, it is often more efficient to design a program for the problem at hand, and we shall have the opportunity in subsequent lectures to learn about shower codes that are so designed. We present the EGS code at this time in the course in order to demonstrate the so-called "analog" Monte Carlo approach to solving electromagnetic cascade showers.

SIMULATING THE PHYSICAL PROCESSES---AN OVERVIEW

The shower generation process has been adequately described in the previous lecture, so we will not need to present it here. In

---

* Starting with Version 2, the EGS Code System has been written in an extended FORTRAN language known as MORTRAN (Cook and Shustek, 1975). The reasons for converting to MORTRAN are listed in the introduction to Chapter 4 of SLAC-210, which is the definitive document on the EGS Code System (Ford and Nelson, 1978).

some approaches to solving radiation transport problems, the Boltzmann equation is written down for a system, and from it a Monte Carlo simulation of the system is derived. An excellant example of this is the neutron-photon transport program called MORSE. Such a method will give correct average quantities, such as fluences, but may not correctly represent fluctuations in the real situation due to variance reduction techniques which have been introduced. The reader is referred to Chapter 3 in the book by Carter and Cashwell (1975) for details of this particular approach.

In EGS we have taken a different and more simple-minded approach in that we attempt to simulate the actual physical processes as closely as possible. We have not introduced any variance reduction techniques, so that fluctuations in the Monte Carlo results should truly be representative of real-life fluctuations. For the design of high energy particle detectors, this is an important consideration. On the other hand, fluctuations are not usually of interest in radiation shielding-type problems, and the addition of optional variance reduction techniques would make some calculations more efficient. In anticipation of this, we have provided a weighting parameter as part of the state function description of each particle that is transported in the shower, and we can, in principle, introduce importance sampling techniques into an EGS problem, if needed.

The correct simulation of an electromagnetic cascade shower can be decomposed into a simulation of the transport and interactions of a single particle, along with some necessary bookkeeping. In EGS, a last-in-first-out stack is used to store the properties of particles which have yet to be transported. Initially, only the incident particle is on the stack---or to put it more correctly, the properties of the incident particle are stored in the first position of the corresponding arrays. The basic strategy is to transport the top particle until an interaction takes place, or until its energy drops below a predetermined cutoff energy, or until it enters a particular region of space. In the latter two cases, the particle is taken off the stack and simulation resumes with the new top particle. If an interaction takes place, and if there is more than one product particle, the particle with the lowest energy is put on top of the stack. When a particle is removed from the stack and no more remain, the simulation of the shower (history) is complete. Many of the physical processes that are represented in the EGS code have been described in the previous lecture, and we will not discuss them here. Suffice it to say that, given the total cross sections, branching ratios, final state joint density functions, and an endless supply of uniformly distributed random numbers, it is possible to simulate an electromagnetic shower.

Due to the statistical nature of the Monte Carlo method, the accuracy of the results will depend on the number of shower histories run. Generally, the relative errors will be proportional to the inverse square root of the number of histories (Shreider, 1966). Thus, to cut errors in half, it is necessary to run four times as many histories. Also, for given cutoff energies, the computer time for a shower history is slightly more than linear in the energy of the incident particle. The point to make here is that Monte Carlo calculations can be very time consuming. It is for this reason that the computational task of the EGS Code System is divided into two parts. First, a Preprocessor code for EGS (PEGS) uses theoretical (and sometimes empirical) formulas to compute the various physical quantities needed and prepares them in a form for fast numerical evaluation. Then EGS uses this data, along with user supplied data and routines, to perform the actual simulation. The reader is advised to consult SLAC-210 (in particular, Chapter 5) for details about PEGS.

SOLUTION OF ELECTROMAGNETIC CASCADE SHOWER PROBLEM

GENERAL IMPLEMENTATION

As shown in Fig. 1, the EGS code itself consists of two "user-callable" subroutines, HATCH and SHOWER, which in turn call the other subroutines in the EGS code---some of which call two "user-written" subroutines, HOWFAR and AUSGAB. The latter determine the geometry and output (scoring), respectively. The user communicates with EGS by means of various COMMON variables. To use EGS, the user must write a MAIN program and the subroutines HOWFAR and AUSGAB. Usually, MAIN will perform any initialization needed for HOWFAR and/or AUSGAB, and will set the values of certain EGS COMMON variables which specify such things as names of the media to be used, the desired cutoff energies, and the unit of distance (e.g., centimeters, radiation lengths, etc.). MAIN then calls the HATCH subroutine which "hatches" EGS by doing necessary once-only initialization and by reading the

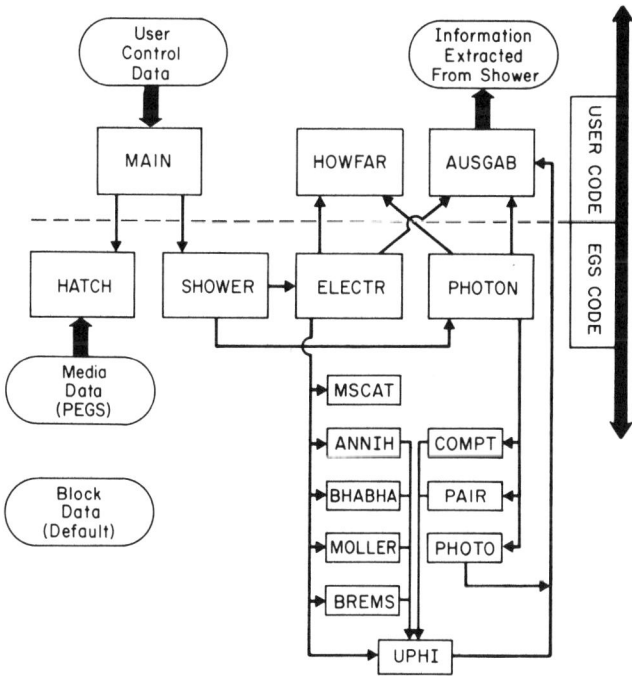

Fig. 1. Flow control with user using EGS by means of User Code.

material data prepared (earlier) by PEGS for the media requested. With this initialization completed, MAIN then calls SHOWER as many times as desired. Each call to SHOWER results in the generation of one Electron-Gamma Shower history. The arguments to SHOWER specify the parameters of the incident particle. Thus, the user has the freedom to use any source distribution he desires. Figure 1 illustrates the flow of control and data when a user-written program (i.e., "User Code") interfaces with the EGS code. Details and rules for writing such user programs are given in Chapter 4 of SLAC-210. We will not repeat them in this lecture, but instead, we will provide a simple example in the next section to illustrate how things are done.

UCSAMPLE---AN EXAMPLE OF A "COMPLETE" USER CODE

In studying the code listing for UCSAMPLE (see Table 1), the reader should not concern himself too much with items that look strange or confusing, such as the MORTRAN macros, but should concentrate on the flow of things. The example that we give simulates a cascade initiated by a 1 GeV electron incident (normally) on a

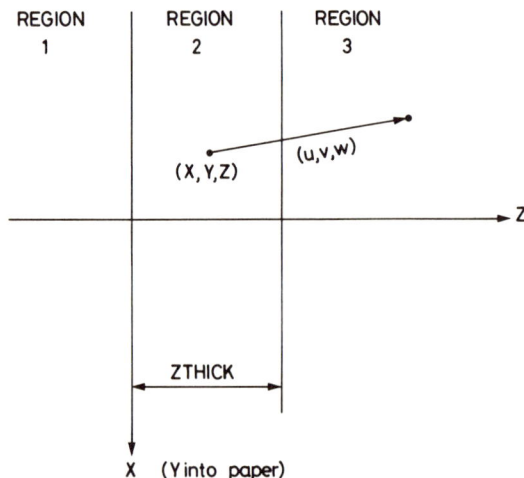

Fig. 2. Geometry layout used in subroutine HOWFAR of UCSAMPLE.

Table 1.

Listing of the User Code UCSAMPLE

```
%E    "EJECT (START NEW PAGE) IN MORTRAN LISTING"
"*****************************************************"
"************************* UCSAMPLE *******************"
"*****************************************************"
"*** MAIN ***"
"************"

"STEP 1.  USER-OVER-RIDE-OF-EGS-MACROS"

%*$MXREG'='20'  "OVER-RIDING THE MAXIMUM NUMBER OF REGIONS"

"DECLARATIONS:"

COMIN/BOUNDS,MEDIA,MISC,RANDOM,USEFUL/;   "COMMONS NEEDED"
COMMON/PASSIT/ZTHICK; "SLAB THICKNESS....NEEDED IN HOWFAR"
COMMON/LINES/NLINES,NWRITE; "TO KEEP TRACK OF LINES-PRINTED"
COMMON/TOTALS/ESUM($MXREG);  "FOR ENERGY CONSERVATION CHECK"
$ENERGY PRECISION EI,ESUM,EKIN,TOTKE,ETOT; "DOUBLE PRECISION"

"CREATE A TEMPORARY ARRAY AND DEFINE THE MEDIA, NEXT"
INTEGER TEMP(24,2)/$S'FE',22*' ',$S'AIR AT NTP',14*' '/;

"STEP 2.  PRE-HATCH-CALL-INITIALIZATION"

NREG=3;  "THE NUMBER OF REGIONS---A LOCAL VARIABLE ONLY"
NMED=2;  "TWO MEDIA WILL BE USED"
DO J=1,NMED <DO I=1,24 <MEDIA(I,J)=TEMP(I,J);>>

MED(1)=0;  "REGION 1 IS VACUUM"
MED(2)=1;  "REGION 2 IS IRON"
MED(3)=2;  "REGION 3 IS AIR AT NTP"

"SET ENERGY CUTOFFS FOR EACH REGION NEXT"
DO I=1,NREG <ECUT(I)=100.0; PCUT(I)=100.0;>

DUNIT=-1; "DISTANCES WILL BE IN RADIATION LENGTHS (OF IRON)"

"STEP 3.  HATCH-CALL"

CALL HATCH;

"STEP 4.  INITIALIZATION-FOR-HOWFAR"

ZTHICK=3.0;  "SLAB THICKNESS IN CENTIMETERS"
ZTHICK=ZTHICK/RLC(1);  "CONVERT TO RADIATION LENGTHS"

"STEP 5.  INITIALIZATION-FOR-AUSGAB"

DO I=1,$MXREG <ESUM(I)=0.D0;> "ZERO THE ENERGY BALANCE ARRAY"
NLINES=0;  "INITIALIZE THE NLINES-COUNTER"
NWRITE=15; "THE NUMBER OF LINES TO PRINT OUT"
```

Table 1.

(continued)

```
"STEP 6.  DETERMINATION-OF-INCIDENT-PARTICLE-PROPERTIES"

IQI=-1;  "INCIDENT PARTICLE IS AN ELECTRON"
EI=1000.D0;  "INCIDENT ENERGY (TOTAL) IN MEV"
EKIN=EI-PRM;  "K.E. OF ELECTRON---PRM IS THE REST MASS"
XI=0.0; YI=0.0; ZI=0.0;  "COORDINATES OF INCIDENT PARTICLE"
UI=0.0; VI=0.0; WI=1.0;  "DIRECTION COSINES---ALONG Z=AXIS"
IRI=2;  "INCIDENT PARTICLE STARTS OUT IN REGION 2---IRON"
WTI=1.0;  "WEIGHT FACTOR---NOT USED IN CALCULATION, BUT"
 "         IS A PARAMETER IN SUBROUTINE SHOWER; HENCE DEFINE"
 "         AS UNITY"
IXX=987654321;  "RANDOM NUMBER GENERATOR SEED"
NCASES=10;  "NUMBER OF HISTORIES (CASES) TO RUN"
IARG=-1;  "IARG (HERE ONLY), INVENTED TO MARK THE"
 "         INCIDENT PARTICLES"

"STEP 7.  SHOWER-CALL"

OUTPUT;  (/,' SHOWER RESULTS:',///,7X,'E',14X,
  'Z',14X,'W',10X,'IQ',3X,'IR',2X,'IARG',/);

DO I=1,NCASES <
  IF(NLINES.LT.NWRITE) <
    OUTPUT EI,ZI,WI,IQI,IRI,IARG;
    (3G15.7,3I5);
    NLINES=NLINES+1;>

  CALL SHOWER(IQI,EI,XI,YI,ZI,UI,VI,WI,IRI,WTI);

"END OF SHOWER-CALL LOOP">

"STEP 8.  OUTPUT-OF-RESULTS"

TOTKE=NCASES*EKIN;  "TOTAL K.E. INVOLVED IN RUN"
OUTPUT EI,ZTHICK,NCASES,IXX;
(//,' INCIDENT TOTAL ENERGY OF ELECTRON=',F12.1,' MEV',/,
  ' IRON SLAB THICKNESS=',F6.3,' RADIATION LENGTHS',/,
  ' NUMBER OF CASES IN RUN=',I3,/,' LAST RANDOM NUMBER=',
  I12,//,' ENERGY DEPOSITION SUMMARY:',//);

"CALCULATE AND PRINT OUT THE FRACTION OF ENERGY"
"DEPOSITED IN EACH REGION"
ETOT=0.D0;
DO I=1,NREG <
  ETOT=ETOT+ESUM(I);
  ESUM(I)=ESUM(I)/TOTKE;  "FRACTION IN EACH REGION"

  OUTPUT I, ESUM(I);  (' FRACTION IN REGION',I3,'=',F10.7);
>

ETOT=ETOT/TOTKE;  "THE TOTAL FRACTION OF ENERGY IN RUN"

OUTPUT ETOT;  (//,' TOTAL ENERGY FRACTION IN RUN=',G15.7,/,
  ' WHICH SHOULD BE CLOSE TO UNITY');

STOP;
END;  "LAST STATEMENT OF MAIN"
```

Table 1.

(continued)

```
%E   "EJECT (START NEW PAGE) IN MORTRAN LISTING"

SUBROUTINE AUSGAB(IARG);
COMIN/EPCONT,STACK/;  "COMMONS NEEDED IN AUSGAB"
COMMON/LINES/NLINES,NWRITE;  "TO KEEP TRACK OF LINES-PRINTED"
COMMON/TOTALS/ESUM($MXREG);  "FOR ENERGY CONSERVATION CHECK"
$ENERGY PRECISION ESUM;  "DOUBLE PRECISION"

"KEEP A RUNNING SUM OF THE ENERGY DEPOSITED IN EACH REGION"
ESUM(IR(NP))=ESUM(IR(NP)) + EDEP;

"PRINT OUT THE FIRST NLINES OF STACK INFORMATION, ETC."
"BUT, ONLY FOR PHOTONS THAT ARE DISCARDED IN REGION 3"

IF(NLINES.LT.NWRITE) <
  IF(IARG.EQ.3.AND.IQ(NP).EQ.0.AND.IR(NP).EQ.3) <
    OUTPUT E(NP),Z(NP),W(NP),
    IQ(NP),IR(NP),IARG;  (3G15.7,3I5);

    NLINES=NLINES+1; >>

RETURN;
END;  "LAST STATEMENT OF SUBROUTINE AUSGAB"


%E   "EJECT (START NEW PAGE) IN MORTRAN LISTING"

SUBROUTINE HOWFAR;
COMIN/EPCONT,STACK/;  "COMMON NEEDED IN HOWFAR"
COMMON/PASSIT/ZTHICK;  "SLAB THICKNESS DEFINED IN MAIN"

IF(IR(NP).NE.2) <IDISC=1; RETURN;>

"MIGHT AS WELL SET DNEAR NEXT"
DNEAR(NP)=AMIN1(Z(NP),ZTHICK-Z(NP));

IF(W(NP).EQ.0.0) <RETURN; "PARTICLE GOING PARALLEL TO PLANES">

"CHECK FORWARD PLANE FIRST SINCE SHOWER HEADING THAT WAY"
"MOST OF THE TIME"
IF(W(NP).GT.0.0) <DELTAZ=(ZTHICK-Z(NP))/W(NP); IRNEXT=3;>
"OTHERWISE, PARTICLE MUST BE HEADING IN BACKWARDS DIRECTION"
ELSE <DELTAZ=-Z(NP)/W(NP); IRNEXT=1;>
"NOW CHECK WITH USTEP AND RESET THINGS IF NECESSARY"
IF(DELTAZ.LE.USTEP) <USTEP=DELTAZ; IRNEW=IRNEXT;>

RETURN;
END;  "LAST STATEMENT OF SUBROUTINE HOWFAR"

"*********************************************************"
"********************* END OF UCSAMPLE *******************"
"*********************************************************"
```

3 cm, semi-infinite slab of iron (see Fig. 2). The upstream region
of the slab is vacuum and the downstream region is air at NTP. A
particle is discarded whenever its total energy falls below a pre-
set cutoff energy of 100 MeV. The stack variable information* E(NP),
Z(NP), W(NP), IQ(NP), IR(NP), plus the IARG parameter of AUSGAB
(discussed later), is outputted on the line printer (first 15 lines
only) for photons reaching Region 3. The maximum number of regions
is changed from 2000 (default) to 20 by means of an over-ride macro
at Step 1. Furthermore, the distance unit (DUNIT) is changed (at
Step 2) so that distances are in radiation lengths of iron rather
than in centimeters (the default choice).

A total of 10 cases of incident electrons is run and the total
energy fraction for each region is summed and printed out at the
end of the run for an energy balance check. Finally, the last random
number (integer) that was used is printed out for future (statistically different) runs of the same problem. The code listing for
UCSAMPLE is also the one given in Chapter 4 of SLAC-210, and the
interested reader may refer to that document for a more step-by-step
description of the various features that are employed.

In the MAIN program, we have divided things into steps that
correspond to those in the EGS User Manual. Subroutines HATCH and
SHOWER are called in Steps 3 and 7 respectively, which are also
depicted in Fig. 1. Subroutine HOWFAR has no arguments, but utilizes
information provided by COMMON/EPCONT/ and COMMON/STACK/, both of
which will be correctly inserted as ANSI-Standard FORTRAN coding---
via the MORTRAN macro expansion of the pattern "COMIN/EPCONT,STACK/;"
---prior to the FORTRAN compilation step. The stack variable infor-

---

* By "stack variable information", we mean those quantities that
specify the "state" of the particle being transported, such as its
energy, position, direction, type, etc.

mation is brought in by means of COMMON/STACK/, and other necessary quantities are provided by COMMON/EPCONT/.

In addition, subroutine AUSGAB has the argument, IARG, which permits the user to "score" information about the shower history depending on the reason AUSGAB is currently being called. IARG normally takes on one of the five values shown in Table 2, and these are the ones invoked (by default) in UCSAMPLE. However, the user has the capability of "switching-on" other calls to AUSGAB, and these are indicated in Table 3. We shall have the opportunity to look at a kerma-absorbed dose problem that makes use of some of these additional AUSGAB-call situations in Lecture 17.

Table 2.
Reasons for Calling AUSGAB (default)

IARG	Situation
0	Particle is going to be transported by distance TVSTEP.
1	Particle is going to be discarded because its energy is below the cutoff ECUT (for charged particles) or PCUT (for photons)---but its energy is larger than the corresponding PEGS cutoff AE or AP, respectively.
2	Particle is going to be discarded because its energy is below both ECUT and AE (or PCUT and AP).
3	Particle is going to be discarded because the user requested it (in HOWFAR usually).
4	A (fluorescent) photon is going to be discarded with the binding energy of a photoelectron.

Table 3.

Additional Reasons for Calling AUSGAB

IARG	Situation
5	Particle has been transported by distance TVSTEP.
6	A bremsstrahlung interaction has occured and a call to BREMS is about to be made in ELECTR.
7	Returned to ELECTR after a call to BREMS was made.
8	A Møller interaction has occured and a call to MOLLER is about to be made in ELECTR.
9	Returned to ELECTR after a call to MOLLER was made.
10	A Bhabha interaction has occured and a call to BHABHA is about to be made in ELECTR.
11	Returned to ELECTR after a call to BHABHA was made.
12	An in-flight annihilation of the positron has occured and a call to ANNIH is about to be made in ELECTR.
13	Returned to ELECTR after a call to ANNIH was made.
14	A positron has annihilated at rest.
15	A pair production interaction has occured and a call to PAIR is about to be made in PHOTON.
16	Returned to PHOTON after a call to PAIR was made.
17	A Compton interaction has occured and a call to COMPT is about to be made in PHOTON.
18	Returned to PHOTON after a call to COMPT was made.
19	A photoelectric interaction has occured and a call to PHOTO is about to be made in PHOTON.
20	Returned to PHOTON after a call to PHOTO was made (assuming NP is non-zero).
21	Subroutine UPHI was just entered.
22	Subroutine UPHI was just exited.

SOLUTION OF ELECTROMAGNETIC CASCADE SHOWER PROBLEM

Obviously, UCSAMPLE is a fairly simple example of a User Code. It was designed specifically to demonstrate how the user interfaces with EGS. In the next section, we will make several comparisons of EGS with experiments and other Monte Carlo calculations in order to see how well the code does.

COMPARISON BETWEEN EGS CALCULATIONS AND VARIOUS EXPERIMENTS AND OTHER MONTE CARLO RESULTS

Comparison of EGS with Longitudinal and Radial Experiment at 1 GeV in Water and Aluminum

An experiment was performed by Crannell et al (1969) to measure the three-dimensional distribution of energy deposition for 1 GeV electron showers in water and aluminum. The water target consisted of a steel tank containing 8000 liters of distilled water. The incident beam entered the water through a 0.13 mm aluminum window centered on the square end of the tank (122 cm x 122 cm x 460 cm). The aluminum target, on the other hand, consisted of plates varying in thickness from 0.64 to 2.5 cm. The plates were pressed together to simulate a solid target (61 cm x 61 cm x 180 cm). Details concerning the detectors used and other related items are in the publication and will not be discussed here. Differential, as well as integral, energy block diagram data obtained from this experiment afford a good opportunity to test the EGS Code System---this is particularly of interest to us for two reasons:

1.) A reasonably good comparison was made by Alsmiller and Moran (1969), who used the Zerby and Moran (1962a,b, 1963) code mentioned at the beginning of this lecture.

2.) Crannell indicates in the paper that the Nagel code does not give radial distributions in agreement with the experiment---and EGS3 descends from the Nagel code.

A comparison of the Crannell data with EGS is given in Figs. 3a and 3b for water and aluminum, respectively. The agreement is

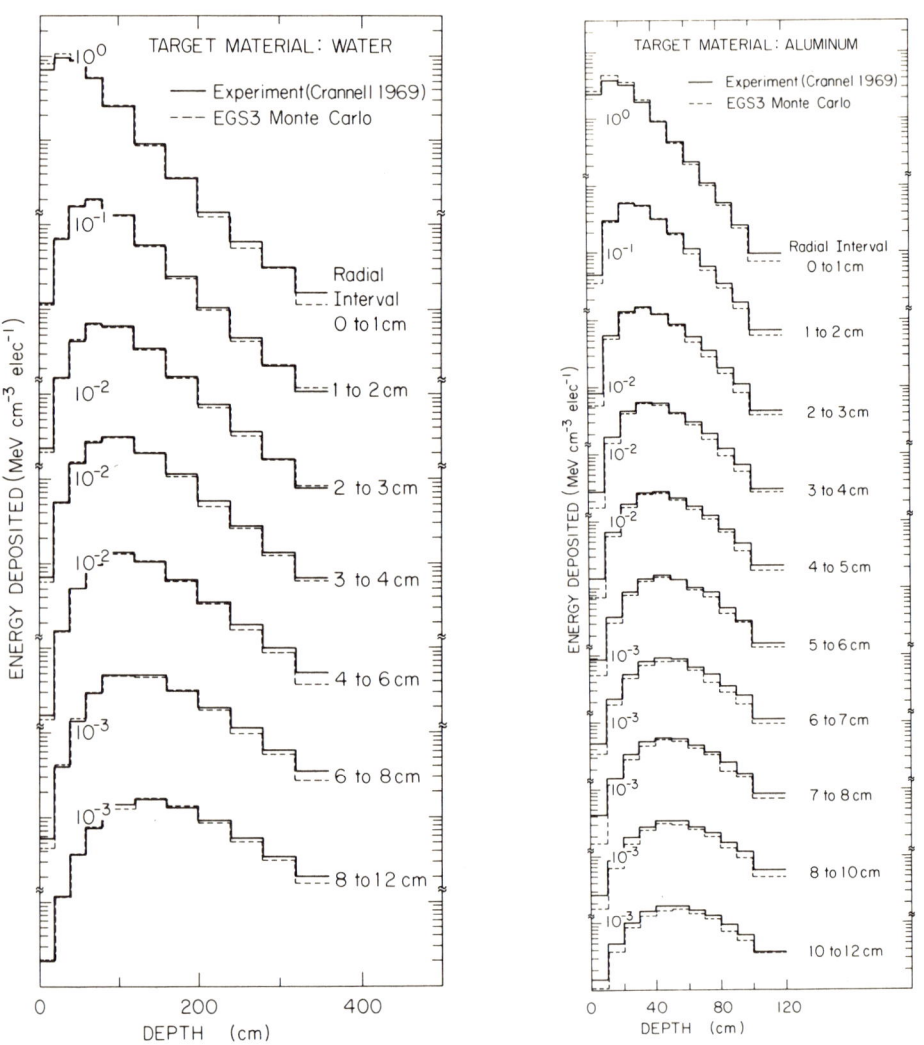

Figs. 3a and 3b. Comparison of EGS with Crannell et al (1969) shower experiment in water and aluminum, respectively.

extremely good everywhere for the water case, and reasonably good for the aluminum experiment. The slight discrepancy at large radii in the aluminum comparison might be accounted for by the fact that the detector used, CaF(Eu), was not as "well-matched" with the absorber as in the water experiment where an anthracene detector was used.

SOLUTION OF ELECTROMAGNETIC CASCADE SHOWER PROBLEM

Track Length Calculations

Track length calculations are most easily done with EGS by summing the step length (TVSTEP) in subroutine AUSGAB each time a simple transport (IARG=0) takes place. In the case of photon track lengths, the calculation is simplified because the photon does not lose energy during transport. Charged particles, on the other hand, lose energy along the entire step, and the method of scoring is correspondingly more complicated. We have provided the details of the track length scoring method in SLAC-210 for those who are interested. In Figs. 4 and 5 we compare the EGS results with those by Zerby and Moran, where we see that the two codes agree quite well with one another, and with a formula by Clement (1963).

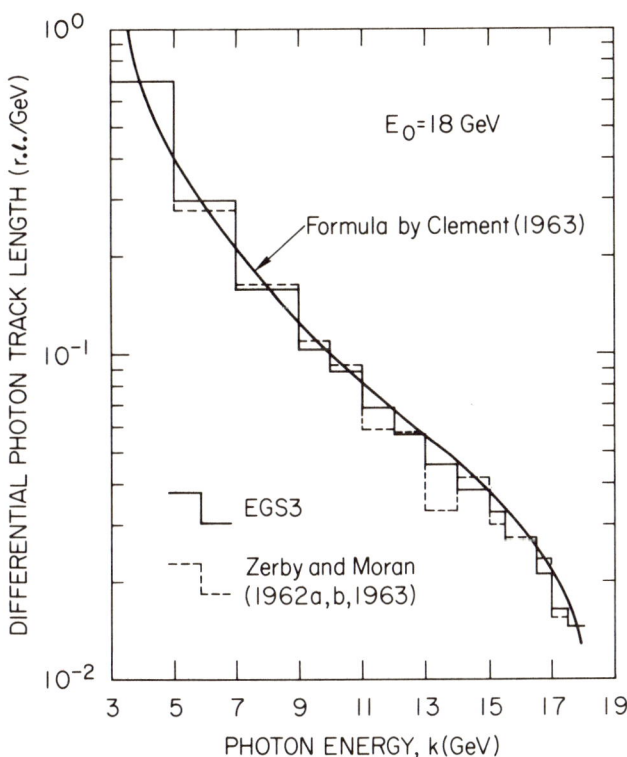

Fig. 4. Differential photon track length. Comparison of EGS with Zerby and Moran (1962a,b, 1963) and with Clement (1963).

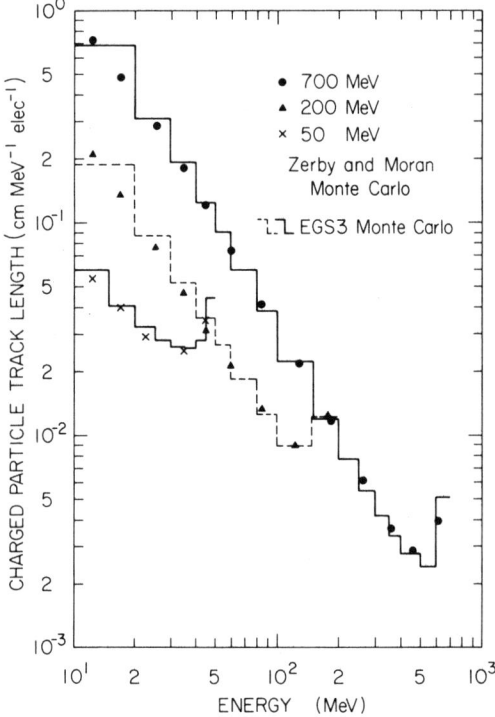

Fig. 5. Differential charged particle track length. Comparison of EGS with Zerby and Moran (1962a,b, 1963) results for three incident electron energies.

Multiple Scattering of 15.7 MeV Electrons in Thin Foils

An experiment that is considered by many to be a benchmark for testing electron multiple scattering at low energies is that by Hanson et al (1951), who found excellent agreement with the theory of Molière (1947, 1948). Details of the experiment are left to the reader. In Fig.6, we present calculational results obtained with a gold foil thickness of 18.66 mg/sq.cm. In addition to the experimental results (solid line), we have plotted both EGS2 and EGS3 histograms*.

---

* EGS3 is more versatile than EGS2. The physics is the same, however.

SOLUTION OF ELECTROMAGNETIC CASCADE SHOWER PROBLEM

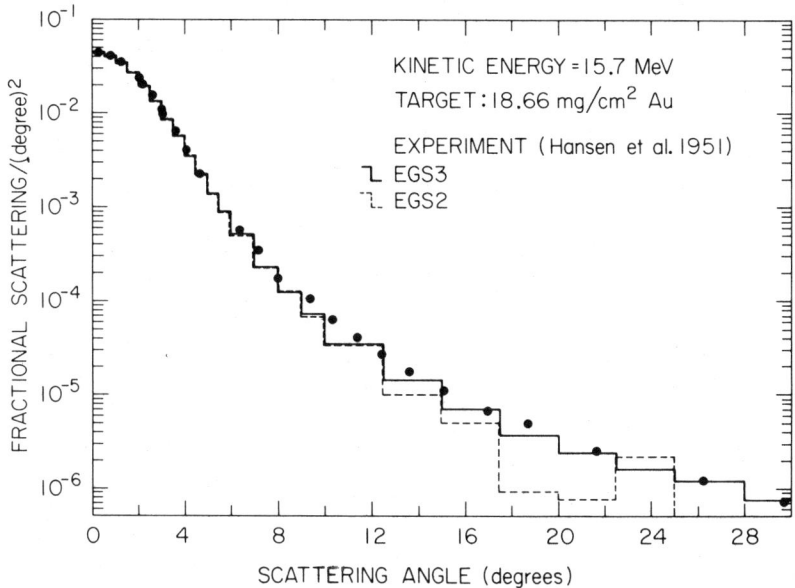

Fig. 6. Multiple scattering of 15.7 MeV electrons in a gold foil. Comparison of EGS Monte Carlo (Versions 2 and 3) with the experiment by Hanson et al (1951).

In Fig. 6, EGS3 is observed to agree with experiment in a smooth manner over the entire angular range. EGS2, on the other hand, calculates less scattering after 12 degrees and is somewhat erratic at the tail. This is <u>not</u> caused by statistical variations, but is due to the fact that the multiple scattering reduced angle is sampled from a discrete distribution in EGS2. The sampling of the reduced angle is done in a continuous fashion in EGS3, and the result is quite apparent in the figure.

The above examples illustrate some of the things that EGS is capable of doing. We shall have a chance to see other such examples in Lectures 13, 16, and 17.

COMMUNICATION WITH EGS BY THE USER

To summarize up to this point, the user communicates with EGS by means of:
1.) Subroutines HATCH --- to establish media data,
    SHOWER --- to initiate the cascade,
    HOWFAR --- to specify the geometry,
    AUSGAB --- to score and output results;
2.) COMMON blocks --- by changing values of variables.

In addition, although we have not really discussed it as yet, the user has the ability to re-define certain features of EGS by means of "macro definitions". A simple example would be the best way to demonstrate how one does this. Consider the translation of a particle from position A to B

XB = XA + U*USTEP;   YB = YA + V*USTEP;   ZB = ZA + W*USTEP;,

where XB and XA are the final and initial x-coordinates, etc., U,V,W are the direction cosines at position A, and USTEP is the transport step for translation. It would be quite easy to include such statements within the subroutines ELECTR and PHOTON---the charged particle and photon transport routines, respectively. However, if the user needed to transport particles in magnetic fields, he would then have to get into subroutine ELECTR itself to make the necessary changes. To avoid having to do this and, hence, to avoid having many different user-versions of EGS floating around, we have made use of the macro-facility of MORTRAN as illustrated by this example. Within subroutine ELECTR at the point where charged particles are translated, we have included the statement $CHARGED-TRANSPORT;. This "macro-pattern" is recognized in the MORTRAN step and a "macro-replacement" is inserted in its place. The default macro-definition that we provide is

%'$CHARGED-TRANSPORT;'='XB=XA+U*USTEP; ...etc. for YB, ZB...', where the %-symbol is used as a MORTRAN macro control character (see Cook and Shustek (1975)). The user, however, can over-ride this default macro by inserting (e.g., magnetic field transport)

%'$CHARGED-TRANSPORT;'='CALL MAGTRN;'

in Step 1 of his User Code. He would, of course, also have to supply subroutine MAGTRN, but he would be able to use the EGS code without making permanent changes to it. Duplication of data sets, and the problems associated with such practice, is thereby avoided.

Taking things one step further, let us assume that MAGTRN is a well established FORTRAN program that the user wishes to use with EGS. Can he do so without re-writing MAGTRN in MORTRAN? The answer is yes---one simply places a %F before the MAGTRN routine and a %M after it, and inserts the deck with the other statements of the User Code.

A LIST OF THE FEATURES PROVIDED BY THE EGS CODE SYSTEM

The EGS Code System (Version 3) can be distinguished from Nagel's original code in a number of ways, the most noteworthy being:

1.) Showers can be simulated in any element, compound, or mixture. That is, PEGS creates data to be used by EGS by using a cross section table for elements 1 through 100.
2.) Photons and charged particles are transported in random rather than in discrete steps resulting in a faster code.
3.) Positrons may annihilate either in flight or at rest, and their annihilation quanta are followed to completion.
4.) Electrons and positrons are treated separately and the appropriate formulas are used (e.g., Møller and Bhabha, respectively). Exact, rather than asymptotic formulas, are used.
5.) Sampling schemes have been made more efficient.

6.) The dynamic range of charged particles has been extended so that showers can be initiated and followed in the energy range of 1.5 MeV to 100 GeV (total energy).

7.) The dynamic range of photons has been extended and lies between 1 keV and 100 GeV.

8.) The output data from the preprocessing code (PEGS) is in convenient form for direct use by EGS.

9.) PEGS control input uses the NAMELIST read facility. In addition to the options needed to produce data for EGS, PEGS contains options to plot any of the physical quantities used in EGS, as well as to compare sampled distributions produced by the TESTSR code with theoretical spectra (see Chapter 6 of SLAC-210).

10.) PEGS constructs piecewise-linear fits over a large number of energy intervals of the cross section and branching ratio data, as opposed to high-order polynomial fits over a small number of intervals. This makes EGS run faster.

11.) EGS is a subroutine package with <u>user interface.</u> This allows the user great flexibility in the way he uses EGS without requiring him to be familiar with the internal details of the code. This also reduces the likelihood that user edits will introduce bugs into the code.

12.) A main program and user routines, which simulate the options used in Versions 1 or 2, are available.

13.) The geometry is specified by a "user-written" subprogram. Routines are available for some commonly used geometries (e.g., planes, cylinders, cones, boxes).

In particular, for Version 3 versus Version 2:

14.) The control logic in the charged particle transport subroutine, ELECTR, has been greatly simplified.

15.) The control logic in both the charged particle and photon transport subroutines (ELECTR and PHOTON, respectively) has been modified in order to make interaction at a boundary impossible.

16.) As a result of 14.) and 15.) above, it should now be possible to implement importance sampling into EGS without further <u>internal</u> changes to the system itself. Examples that come to mind include the production of secondary electron beams at large angles, photon energy deposition in relatively small (low Z) absorbers, and deep penetration (radial and longitudinal) calculations associated with shower counter devices.

17.) Provision has been made for allowing the density to vary continuously in a region.

18.) The multiple scattering reduced angle is now sampled from a continuous rather than discrete distribution. This is done for arbitrary step sizes, provided that they are not too large to invalidate the theory. An immediate application of this is the simplification mentioned in 14.) above.

19.) A new subroutine (PHOTO) has been added in order to deal with the photoelectric effect in a manner comparable to the other interaction processes. This should facilitate in developing more general photoelectric routines. For example, rather than simply depositing the energy of fluorescent photons, as is the case in the versions of EGS up to now, these photons can be followed as well. A practical use for this is envisioned in the design of shields for detectors that are placed in the synchrotron radiation associated with electron-positron storage rings, as well as various engineering applications and solutions to problems arising because of this synchrotron radiation.

20.) Additional calls to AUSGAB, bringing the total from 5 to 23, have been made possible in order to allow for the extraction of additional information without requiring the user to edit the EGS code proper. For example, the user can now determine the number of various collision types (Compton, Møller, etc.) by means of the user-written codes.

CONCLUSIONS

The EGS Code System has been discussed as an example of the analog Monte Carlo approach to solving electromagnetic cascade shower problems. The code is shown to be much more generally useful than the original Nagel program or the codes published by either Messel et al or by Zerby and Moran. Another code, ETRAN, is quite popular for solving shower problems, particularly in the low energy realm (i.e., less than 100 MeV). This code was written by Berger and Seltzer (1970), and a later version called SANDYL (Colbert, 1973) contains a fairly general geometry. ETRAN appears to be accurate and is available from the Radiation Shielding Information Center. It has been used in a number of low energy bremsstrahlung studies, and we will refer back to it in Lecture 17.

REFERENCES

Alsmiller, R. G., Jr., and Moran, H. S., 1969, Calculation of the energy deposited in thick targets by high-energy (1 GeV) electron-photon cascades and comparison with experiment, Nucl. Sci. Eng., 38:131.

Berger, M. J., and Seltzer, S. M., 1970, Bremsstrahlung and photo-neutrons from thick tungsten and tantalum targets, Phys. Rev., C2:621.

Carter, L. L., and Cashwell, E. D., 1975, Particle-transport simulation with the Monte Carlo method, TID-26607, National Technical Information Service, U. S. Department of Commerce, Springfield, Virginia.

Clement, G., 1963, Differential path length of the photons produced by an electron of very high energy in a thick target, Comptes Rendus, 257:2971; translated for the Stanford Linear Accelerator Center, SLAC-TRANS-141 (1972).

Colbert, H. M., 1973, SANDYL: A computer program for calculating combined photon-electron transport in complex systems, Sandia Laboratories (Livermore) Report Number SCL-DR-72019.

Cook, A. J., and Shustek, L. J., 1975, A user's guide to MORTRAN2, Stanford Linear Accelerator Center Computation Research Group Report Number CGTM-165.

Crannell, C. J., Crannell, H., Whitney, R. R., and Zeman, H. D., 1969, Electron-induced cascade showers in water and aluminum, Phys. Rev., 184:426.

Ford, R. L., and Nelson, W. R., 1978, The EGS code system: Computer programs for the Monte Carlo simulation of electromagnetic cascade showers (Version 3), Stanford Linear Accelerator Center Report Number SLAC-210.

Hanson, A. O., Lanzl, L. H., Lyman, E. M., and Scott, M. B., 1951, Measurement of multiple scattering of 15.7 MeV electrons, Phys. Rev., 84:634.

Messel, H., Smirnov, A. D., Varfolomeev, A. A., Crawford, D. F., and Butcher, J. C., 1962, Radial and angular distributions of electrons in electron-photon showers in lead and in emulsion absorbers, Nucl. Phys., 39:1.

Messel, H., and Crawford, D. F., 1970, Electron-photon shower distribution function, Pergamon Press, Oxford.

Molière, G. Z., 1947, Theorie der Streuung schneller geladener Teilchen I. Einzelstreuung am abgeschirmten Coulomb-Feld, Z. Naturforsch, 2a:133.

Molière, G. Z., 1948, Theorie der Streuung schneller geladener Teilchen II. Mehrfach- und Vielfachstreuung, Z. Naturforsch, 3a:78.

Nagel, H. H., and Schlier, C., 1963, Berechnung von Elektron-Photon-Kaskaden in Blei für eine Primärenergie von 200 MeV, Z. Physik, 174:464.

Nagel, H. H., 1964, Die Berechnung von Elektron-Photon-Kaskaden in Blei mit Hilfe der Monte-Carlo Methode, Inaugural-Dissertation zur Erlangung des Doktorgrades der Hohen Mathematich-Naturwissenschaftlichen Fakultät der Rheinischen Friedrich-Wilhelms-Universität zu Bonn.

Nagel, H. H., 1965, Elektron-Photon-Kaskaden in Blei: Monte Carlo Bechnungen für Primärelektronenenergien zwischen 100 und 1000 MeV, Z. Physik, 186:319; translated for the Stanford Linear Accelerator Center, SLAC-TRANS-28 (1965).

Shreider, Y. A. (Editor), 1966, The Monte Carlo method, Pergamon Press, New York.

Zerby, C. D., and Moran, H. S., 1962a, Studies of the longitudinal development of high-energy electron-photon cascade showers in copper, Oak Ridge National Laboratory Report Number ORNL-3329.

Zerby, C. D., and Moran, H. S., 1962b, A Monte Carlo calculation of the three-dimensional development of high-energy electron-photon cascade showers, Oak Ridge National Laboratory Report Number ORNL-TM-422.

Zerby, C. D., and Moran, H. S., 1963, Studies of the longitudinal development of electron-photon cascade showers, J. Appl. Phys., 34:2445.

LECTURE 13: SOME EXAMPLES FOR THE APPLICATION
OF THE MONTE CARLO CODE EGS

Herbert Dinter

Deutsches Elektronen-Synchrotron DESY
Hamburg
Federal Republic of Germany

INTRODUCTION

From a long list of applications of the code EGS[1] at DESY, I will select three topics because they are directly connected with the protection of men and material against radiation. Furthermore, two of them imply the possibility to test EGS by comparing its data with experimental results.

The first example deals with problems concerning synchrotron radiation. Scattering coefficients of low-energy photons are derived for design and construction of experimental set-ups at high-energy electron accelerators. The results may be compared with those of another shower code. Another possibility for the application of EGS is a shielding experiment performed at DESY. The doses measured thereby can be interpreted if one takes into consideration the energy spectra of the particles obtained by EGS. The last chapter is dedicated to dose calculations in and behind absorbers. Some results may be compared with experimental values from other DESY experiments.

PHOTON SCATTERING AT LOW ENERGIES

The high-energy photons of the synchrotron radiation spectrum can seriously disturb the experiments performed at high-energy electron accelerators. In order to protect experimental arrangements from this background radiation, it is necessary for the design of beam transport systems near experiments to have a detailed information of the interaction of low-energy photons with matter. In particular, it is important to know the fraction of scattered ("reflected") or backscattered radiation as well as its energy spectrum and angular distribution when hitting a target like the wall of a beam pipe or the edge of a collimator.

The Compton effect and the photoeffect are the two mechanisms which contribute most to the scattering process in the energy region below 1 MeV. So EGS may be used to produce scattering coefficients depending on several parameters as material, thickness or energy. For simplicity and for flexibility, the coefficients are calculated for monoenergetic incident beams rather than for the continuous spectrum of the synchrotron radiation. These EGS-data will underestimate the real yield of photons as the code (in the version 1 which was used here) does not transport fluorescence quanta following a photoelectric absorption process. One has to consider these quanta separately[2].

In order to get values of practical importance, the calculations were performed for two simple geometries. So they can be can be adapted to real arrangements. The energy cut for photons was 20 keV and for electrons 1.5 MeV. The first geometry is the "reflection" of photons at a thin plate. A beam of monoenergetic photons impinges on that plate under a small angle and all photons scattered back into the half-room in which the beam hits the target are investigated. The number of scattered photons delivers a "reflection coefficient" when normalized to one incident photon. Such coefficients are plotted in Fig.1.

EXAMPLES FOR APPLICATION OF THE MONTE CARLO CODE EGS 199

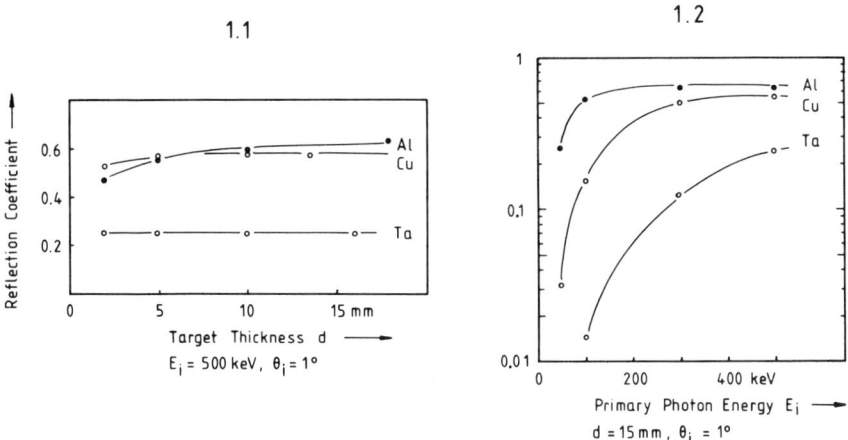

Fig.1   Reflection coefficients as a function of the target thickness and of the incident photon energy.

The second geometry is a two dimensional infinite plate of varying thickness positioned perpendicularly to the primary photon beam. From these calculations, one can derive "backscatter coefficients". These coefficients are defined as: number of photons leaving the target under angles between $90°$ and $180°$, relative to the primary beam, normalized to one incident photon.   The dependence of these backscatter coefficients on target thickness and primary energy is shown in Fig.2. In addition to the values calculated by EGS there are some points indicated in Fig.2.2 showing the results of Berger and Raso[3]. The agreement is excellent.

DOSE AND SHIELDING PARAMETERS OF ELECTRON-PHOTON STRAY RADIATION FROM A HIGH-ENERGY ELECTRON BEAM

Some years ago an experiment was performed at DESY for the investigation of parameters for the shielding of PETRA being under construction at that time[4].

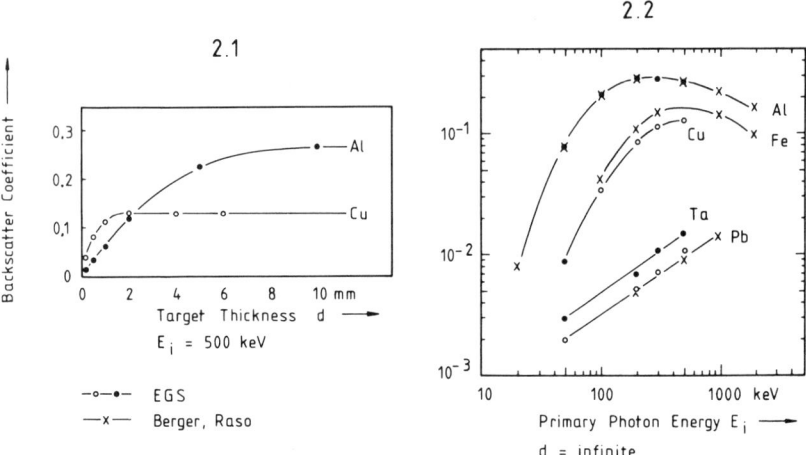

Fig. 2 Backscatter coefficients as a function of the target thickness and of the incident photon energy.

The experiment consisted of two parts. From the first one, we obtained the angular distribution of the absorbed dose of the stray radiation coming from various target geometries. These doses may be used as source terms in practical shielding calculations, as the target arrangements correspond to realistic situations. The aim of the second part was to receive attenuation coefficients of the absorbed dose for this stray radiation for several shielding materials. In both cases absorbed doses were measured by thermoluminescence dosimetry. There was no experimental identification of particles and no detection of energy spectra.

The experimental arrangement was quite simple. An electron beam of 5 GeV and 1 cm in diameter hits a target consisting of an iron plate which was 5 cm high and 70 cm long. The thickness was variable between 0.2 cm and 10 cm, and the angle ϕ between target and beam could be varied between $0°$ and $\pm 90°$. The angle θ is the angle of observation. Both angles, ϕ and θ, are defined as negative at the left side of the beam, and as positive at the right one (see Fig. 3).

EXAMPLES FOR APPLICATION OF THE MONTE CARLO CODE EGS

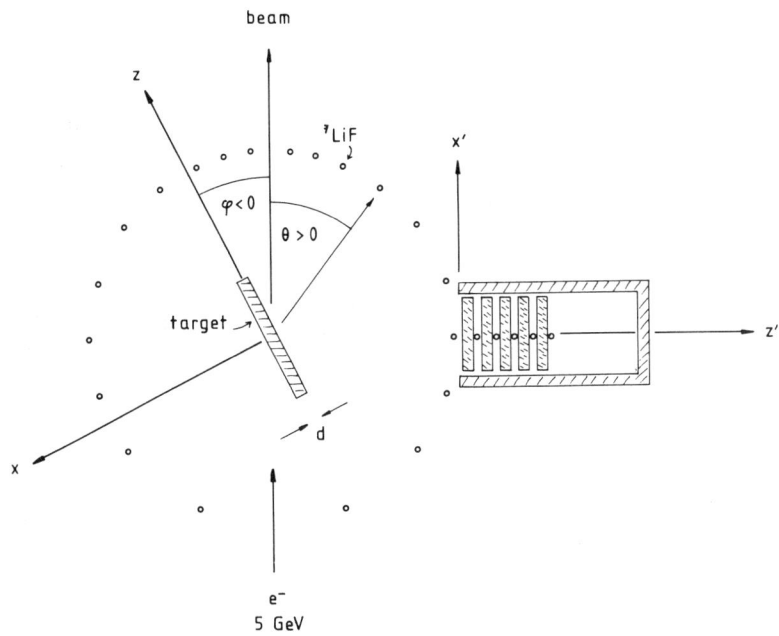

Fig. 3 Geometry of the source term and shielding experiment.

In order to be able to compare dose values with EGS results we used a code in addition to EGS which converted the fluence of marked particles to dose. This code demands a "detection-plane" at a certain distance from the target, and selects those particles, previously marked and discarded by EGS, which hit that plane. Each hitting particle gives, in connection with an energy and particle dependent conversion coefficient, a contribution to the dose "measured" by the detection plane. After summing up all contributions, the result is a dose averaged over that plane. (In the next chapter the problems of dose calculations are discussed in more detail). Interactions of the particles of the stray radiation with air between target and detector are neglected for the calculation.

In Fig. 4, two examples of the angular distribution of measured doses are shown for two typical target geometries. In Fig. 4.1, the distri-

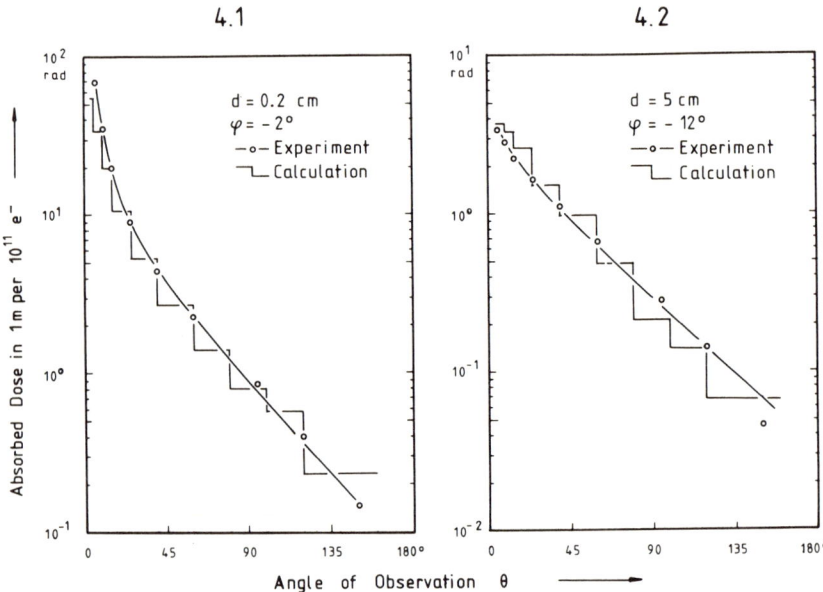

Fig. 4 Angular distribution of absorbed dose of stray radiation for a thin-target and a thick-target arrangement.

bution is plotted as obtained with a target of thickness 0.2 cm and under $-2°$ to the beam. This may stand for the case when the beam is grazing a beam pipe. In the second example (Fig.4.2) the target thickness is 5 cm (it may represent a collimator). The step curve gives the result of EGS and the dose calculations, and it agrees very well with the measured points.

The conversion coefficients for the calculation are taken from ref.8, and give the dose at the maximum of the depth dose curve in water. The agreement with the measured surface dose means that the particles of the stray radiation produce the maximum of the absorbed dose near the surface. This is an indication that the particles are of relatively low energy.

EXAMPLES FOR APPLICATION OF THE MONTE CARLO CODE EGS

Having now good arguments that both codes give reasonable results, one also may calculate energy spectra for electrons and photons, and may assume that they will agree with experimental spectra. In Fig. 5, the results are shown for two angles of observation, $30°$ and $90°$, and for the target configuration d = 0.2 cm, $\phi = 2°$. The detector plane is 40×40 cm^2 at a distance of 1 m from the target, and the intensities correspond to $1 \cdot 10^{11}$ electrons at 5 GeV. The energy cut of EGS was 20 keV for photons, and 1.5 MeV for charged particles. The predominance of particles of very low energy is surprisingly strong. Even at $30°$ most of the electrons and positrons (90 %) have energies below 20 MeV and 65 % of the photons have energies below 0.55 MeV. The annihilation photons are isotropic, as expected.

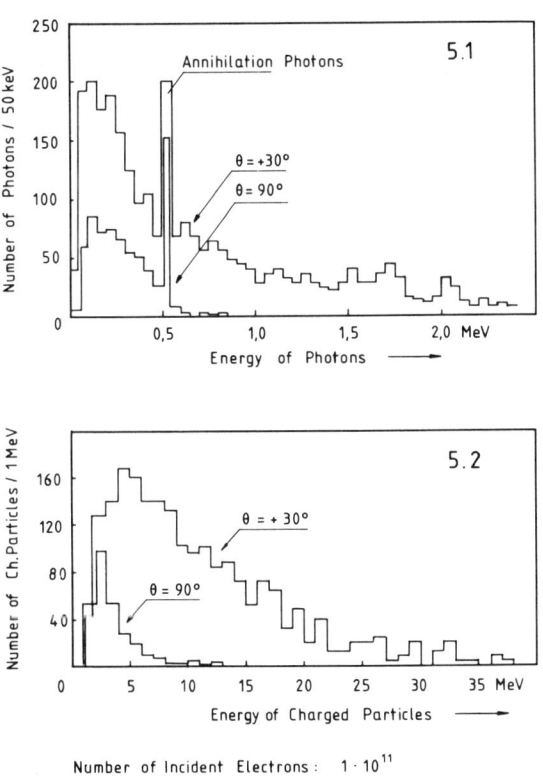

Fig. 5. Energy spectra of stray radiation for photons and for charged particles.

The second part of the experiment was dedicated to attenuation coefficients of some shielding materials. At a distance of 1 m and at 90° (or 2 m at 30°), a stack of absorber plates, 40 x 40 cm² and with a total thickness of about 20 radiation lengths, was arranged, and in small gaps between these plates the doses were measured.

In order to calculate these doses EGS was used twice. At first, it produced, for the actual geometry, marked particles which leave the target. A second code selected those particles which hit the first absorber plate, and transformed their entrance coordinates into the coordinate system of the absorber. Now EGS was used for a second time, taking these selected particles as incoming particles. The stack of absorber plates was treated as one compact target. The gaps for the dose measurements were simulated by planes, predefined within the target, where the particles are marked when crossing them.

The agreement of the results is good for forward angles (30°) as may be seen in Fig.6 for the example of an iron absorber. The first point at x = 0 is identical with the source term whose agreement is already known by fig.4. The strong decrease of intensity within the first few centimeters is also confirmed by the calculation, and, as one can see now, it is resulting from charged particles (of low energy). The slope above 5 cm agrees well with the minimum absorption coefficient, especially the slope of the photon contribution to the dose. The charged particles above 5 cm are presumed to be Compton electrons. Although there are only a few of them, they contribute most to the dose. If the angle of observation increases (90°), the agreement becomes worse, as one can see for a lead absorber, as a typical example, in Fig.6.2. Source term and spike are reproduced well by the calculation, but at greater depth, the calculated doses are too low, and the decrease of the attenuation curve is too strong. One reason for this disagreement may be the cut-off energy for charged particles. Electrons from photoeffect or Compton effect with energies just below 1.5 MeV can leave the back of an absorber plate (e.g. the range of an electron of 1 MeV kinetic energy in iron is 0.8 mm), and contribute to the measured doses whereas they are not transported by EGS and therefore do not contribute to the calculated doses. This effect becomes more distinct when the energy of the photons becomes lower, i.e. at greater angles.

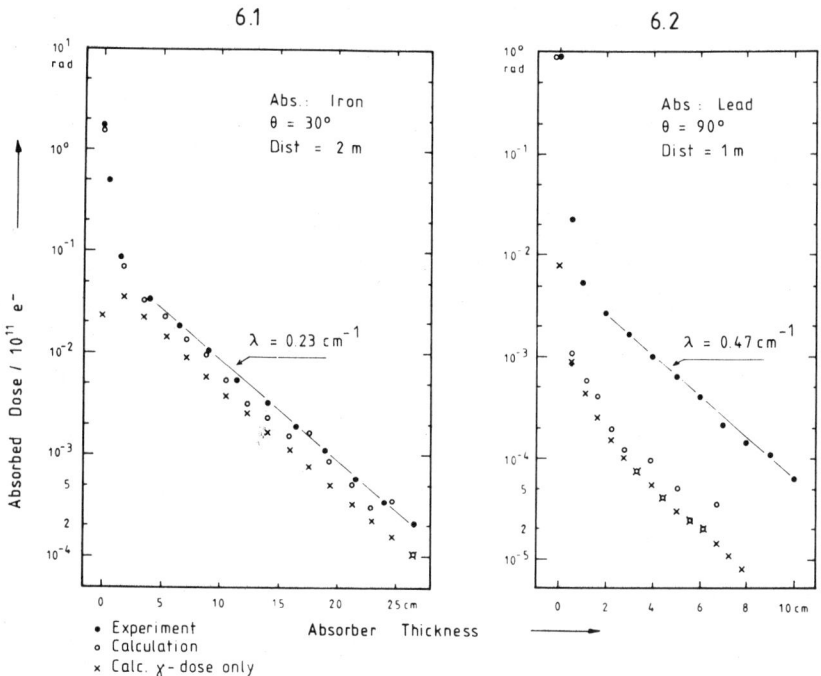

Fig.6 · Absorption curves of stray radiation for two absorbers and two angles of observation.

ELECTRON-PHOTON CASCADES WITHIN AND BEHIND ABSORBERS

A subject of great importance for radiation protection is the design of beam dumps and beam shutters. Sometimes they have to be dimensioned in such a way that the safety of persons is guaranteed behind them while they are bombarded by an electron beam.

For this reason, some years ago a cascade experiment was performed at DESY[5,6] and we tried to simulate this experiment by EGS[7]. In order to measure the absorbed energy within the absorber, many small phosphate glass detectors were distributed in the target volume. A second aim of the experiment was to measure the maximum dose behind absorbers which is present in the axis of the cascade as a function of the absorber thickness. For this measurement, an ionization chamber of very small dimensions (3 x 3 x 3 mm^3) filled with air under normal conditions was used.

The experimental values of the absorbed doses in the interior of targets are displayed as isodose curves (indicated by symbols) in Fig.7. These doses do not represent tissue equivalent doses, but correspond to the target material, copper or lead, respectively. The calculated doses are displayed by solid lines for the same isodose values without any adaption. The agreement is good, especially for small radii. The deviation far outside the shower axis may result from particles of very low energy accompanying the electron beam (halo).

Here is the point where some remarks should be made concerning our dose calculations: The code, EGS, delivers energy block-diagrams for the absorbed energy within a target as well as a series of marked particles which (for example) leave the target. For each particle, its energy, position, direction and charge are noted. The best way to calculate a dose resulting from EGS-particles at a certain point would be to select those particles which hit a detector volume positioned at that point and filled with tissue equivalent material, and to produce an energy block-diagram within this volume. The deposited amounts of energy (in units of MeV) easily can be transformed to units of absorbed dose (e.g. rad), as shown in the previous example (Fig.7). Since there is only a limited computing time available, the detector must not be too small.

EXAMPLES FOR APPLICATION OF THE MONTE CARLO CODE EGS

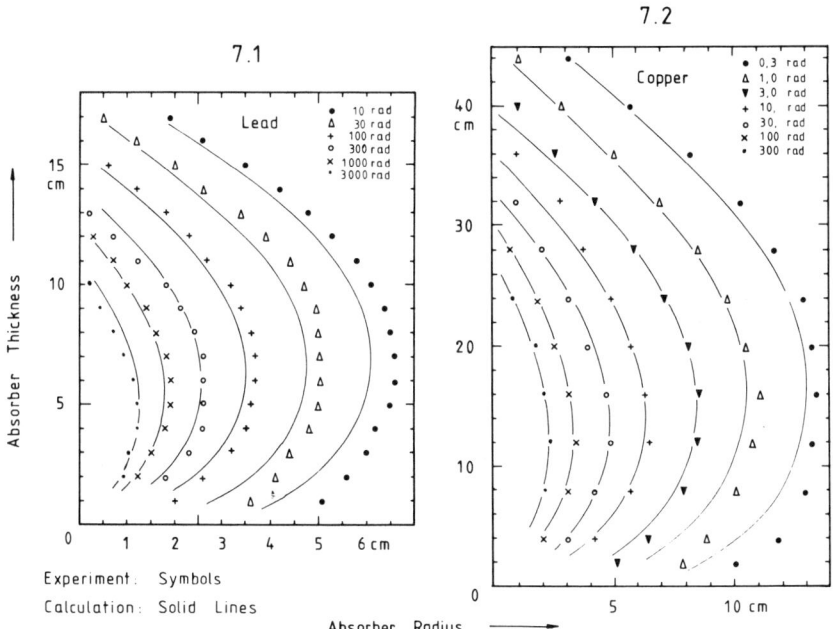

Fig.7 Curves of equal absorbed doses within two different absorber materials.

In order to calculate doses measured by very small detectors as e.g. used in the second part of the cascade experiment, one may assume that the energy of the particles traversing the detector is constant (although depositing energy there). With this approximation, it is easy to calculate absorbed doses using mass energy absorption coefficients for the gamma dose and stopping power values for the dose caused by charged particles. We applied this (somewhat crude) method to calculate the dose measured by the tiny ionization chamber of the experiment and the results are shown in Fig.8. In Fig. 8.1, the calculation is based on the stopping power of electrons in air. The dose contribution of the photons was neglected because of the small size of the detector. The indicated doses result from restricted stopping power-calculations, taking into account that a certain amount

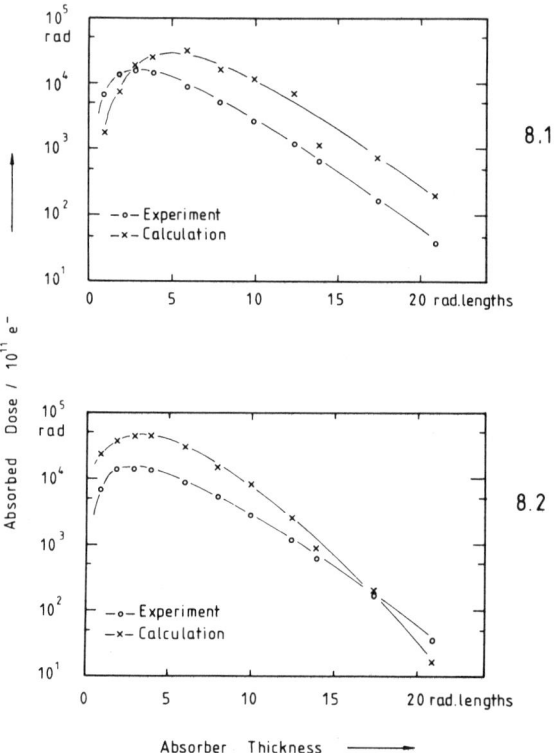

Fig.8 Absorbed doses behind a copper target.
    8.1 Doses calculated by means of Restricted Stopping Power compared with experimental values.
    8.2 Doses calculated by means of conversion coefficients compared with experimental values.

of energy escapes from the detector in form of δ-particles. The two curves seem to be shifted against each other by approximately 2 radiation lengths, and the doses of the calculated curve are higher, especially in the decreasing part, up to a factor of 4. Fig.8.2 shows the same experimental curve being compared with a curve resulting from calculations with the conversion coefficients for depth doses[8].

As a conclusion, one could state that the calculation of absorbed dose is connected with some difficulties. The best results are obtained by calculating an energy block-diagram within a detector volume. If this is not applicable because of the intensity of the particle flux, the two other possibilities give fairly good results:
For the calculation of surface doses produced by high-energy particles, as behind a thin absorber, one should prefer the method using the stopping power of electrons. If one is interested in maximal doses of the depth dose curve the conversion coefficients of ref.8 may be applied. This method also gives good results for surface doses when the doses are produced by particles of low energy (Fig.5) as shown in the chapter before.

SUMMARY

The application of EGS at DESY has shown that the code is a very valuable tool for all questions concerning electron-photon cascades. In most examples presented here, calculated and measured values of absorbed dose have been compared. So, a disagreement need not necessarily result from EGS, especially as we have seen that dose calculations are sometimes very difficult.

There seem to be two weak points in EGS (at least in version 1):
The first is that fluorescence-gamma quanta following a photoelectric process are not transported, and the second one is that the minimum cut-off energy for charged particles for some special applications seems to be not small enough.

The code converting EGS particles to dose values is still somewhat crude and it has more than two weak points. Nevertheless, having in mind their limits, both codes are useful for the interpretation and prediction of data for radiation protection.

REFERENCES

1. R.L.Ford, W.R.Nelson, SLAC Report No. 210, June 1978
2. K.U.Scholz, DESY Internal Report PET-78/03
3. M.J.Berger, D.J.Raso, Rad. Research $\underline{12}$ (1960) 20
4. H.Dinter, K.Tesch, Nucl. Instr. Meth. $\underline{143}$ (1977) 349
5. G.Bathow, E.Freytag, K.Tesch, Nucl. Phys. $\underline{B2}$ (1967) 669
6. G.Bathow, E.Freytag, R.Kajikawa, M.Köbberling, K.Tesch, Nucl. Phys. $\underline{B20}$ (1970) 592
7. H.Dinter, DESY Internal Report D3/28, Juli 1978
8. Recommendations of the International Commission on Radiological Protection, ICRP Publication 21, 1973

LECTURE 14: CALCULATION OF THE AVERAGE PROPERTIES OF ELECTRO-

MAGNETIC CASCADES AT HIGH ENERGIES (AEGIS)

A. Van Ginneken
Fermi National Accelerator Laboratory
Batavia, Illinois 60510
USA

INTRODUCTION AND MOTIVATION

AEGIS[1] is a Monte Carlo program which simulates electromagnetic (EM) showers. It is intended mainly for applications to problems involving multi-GeV incident particles. In the high energy domain, analogue calculations tend to be very slow. Their execution time per incident particle grows linearly with energy. For the most energetic electrons and photons produced at present day accelerators (> 100 GeV), one can barely manage to simulate showers in this fashion because of computer time limitations. As to the future of analogue calculations, it is amusing to speculate whether improvements in computer speed and organization can keep pace with increases in particle accelerator energy.

The main emphasis of AEGIS is on calculating energy deposition by the EM, cascade although it can be readily extended to other applications. The calculation of energy deposition is directly related to that of (for example) radiobiological dose, heating effects and radiation damage. The program traces the full three dimensional development of the EM shower. Heterogeneous materials, arbitrary geometry and magnetic fields are easily included by stepwise tracking the particles through the target. These facets of the program, along with others which are common to many similar programs, will not be further discussed here.

The main feature of AEGIS, vis-a-vis analogue calculations is the introduction of biases (importance sampling). This not only decreases execution times but simplifies coding and computer storage requirements considerably. The particular bias applied in AEGIS, which will be more fully explained below, aims at broad applicability of the program; that is, to study the effects of the cascade (primarily energy deposition) without any <u>a priori</u> information. From the moment more becomes known about the answer to the problem, the introduction of special biases which favor those chains of events having important consequences will lead to an improved program. It is an added advantage of AEGIS over analogue calculations that the introduction of such special biases is easy: they can be merely substituted for the "general" bias present in the program. In actual practice, the time and effort required to devise and code such tailor-made biases is seldom justified, especially when a longer run with the more general code will serve almost as well. Nonetheless, for some problems severe special biasing may offer the only way to get meaningful results, though great care is needed in devising such biasing and in interpreting the results. However, even in these circumstances the general program may be useful since the approach to a good set of biases is often gradual: starting out with a general calculation such as described here and then, based on its results, progressively introducing more and more refined biases (indeed, this line of reasoning leads to the well known conclusion that to devise a perfect set of biases one must know the complete solution to the problem).

A disadvantage of these biased calculations versus analogue calculations is that they are generally not well suited to studying "fluctuation problems". This includes important applications such as

detector response and particle correlations in experimental high
energy and nuclear physics. Problems encountered in radiation
damage and radiation protection are seldom of this type. Note
that this is a disadvantage but not a fundamental limitation. For
example, to generate the probability distribution of the light
emitted by a set of scintillators within a calorimeter device is
conceptually a rather simple matter; that is, essentially repeating
the experiment by means of an analogue calculation. The same
problem is much less transparent in a biased calculation such as
AEGIS, though not unthinkable. Perhaps the higher energies at
future accelerators or the large detectors planned for cosmic ray
experiments will prompt new ways to deal with these problems.

MODEL AND METHOD OF CALCULATION

The only particles considered in the calculation are electrons
and photons. Their dominant interactions which govern the EM
shower behavior are well established.[2] For electrons and positrons,
energy loss by ionization, multiple Coulomb scattering and
bremsstrahlung are included. For photons, Compton scattering and
pair production are considered. All other interactions are
neglected. In bremsstrahlung, as well as in pair production, effects
of recoil of nuclei or atomic electrons are neglected while the
outgoing particles assume the direction of the incident particle.

Fig. 1 is a symbolic representation of an EM shower. The
electrons are represented by straight lines and the photons by
wavy lines. Vertex B denotes a bremsstrahlung event, P is pair
production and C is Compton scattering. Fig. 1 could be a representation of a simulated shower in an analogue Monte Carlo calculation of (for example) the energy depositon in a steel block.

An alternate way to do this calculation would be to trace
only one particle emanating from each vertex. Since there are
only two outgoing particles per interaction, one could flip a coin

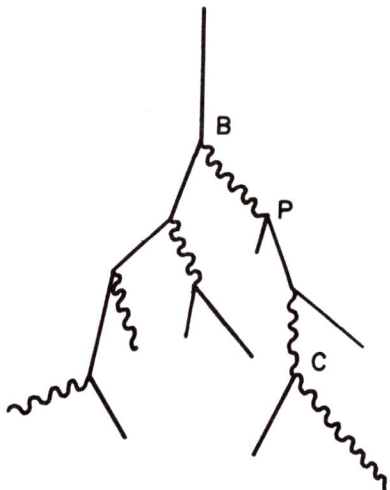

Fig. 1   Schematic representation of an electromagnetic cascade. Straight lines represent electrons. Wavy lines represent photons. B indicates bremsstrahlung, P is pair production and C is Compton scattering.

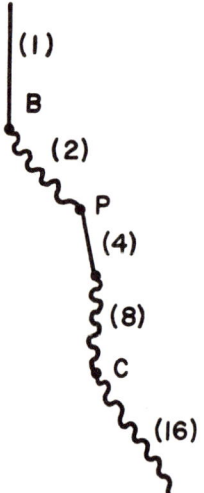

Fig. 2   A sub-cascade taken from Fig. 1. The numbers in parenthesis next to each trajectory indicate the weights.

to decide which of the two to follow. Fig. 2 represents a possible outcome of such a series of trials on the shower of Fig. 1. To compensate for the particles that are neglected, a weight is assigned to each particle that is followed. In this case, the weight is particularly easy to calculate: it increases by a factor of two at each vertex (as indicated besides each trajectory in Fig. 2).

It is clear that if we run each calculation long enough, the results will be the same. One advantage of the scheme of Fig. 2 is apparent: only one particle is to be remembered at any time. Likewise one disadvantage is apparent: one can no longer extract information about the correlations.

More could be said for and against either scheme, but in general, the simple scheme of Fig. 2 is not very satisfactory. This is true particularly in calculations of energy deposition. The main trouble is that too much time is spent calculating the effects of the first few generations. For example, the incident electron will generally emit many soft photons because that is where the bremsstrahlung cross section is largest. These soft photons quickly dissipate their energy via Compton scattering and so a lot of time is devoted to calculating energy deposition in a shallow region (of, say, a few mm^3 in steel) where the beam strikes. Occasionally there would be a deeply penetrating string of particles, and since this will generally involve higher generations, their representative particles will carry a large weight which compensates for their lesser occurrence. Without going into details, it can be seen that unless one is prepared for a long run on the computer, such random occurrences of large weights will lead to large fluctuations and hence large errors in the calculation wherever these higher generations are of importance.

A straight analogue calculation (Fig. 1) has somewhat the opposite problem. An energetic particle (typically produced in the first few generations) is more important than a low energy member of the cascade in almost all imaginable applications, this for the simple reason that an energetic particle can produce many low energy particles in subsequent generations. Yet in the analogue calculation, they are all treated equally.

In describing the solution employed in AEGIS, it is to be noted that the model assumes exactly two outgoing particles per event. Under the assumptions of the calculation, selecting the energy of one of the particles completely determines the event (except for Compton scattering where an azimuthal angle also needs to be specified). Based on the above arguments, a reasonable prescription for a general calculation is to apply a bias linear in E, the energy of the outgoing particle. This will accomplish the desired over-sampling of energetic particles. Another way of seeing this is by noting that each particle participating in a shower creates its own "subshower" in which the particle's energy is dissipated. The above bias will therefore tend to follow the flow of energy in the shower (and hence energy deposited). By comparison, the scheme of Fig. 2 would tend to follow the "particle flow" of the shower.

This leads to two selection schemes: (1) choose the particle's energy with a probability (roughly) proportional to the differential cross section, $d\sigma/dE$, and then decide whether to follow either this particle or its outgoing partner with a probability proportional to each particle's energy, (2) first choose the outgoing particle type and use a selection function (roughly) proportional to $Ed\sigma/dE$. One could generalize this to include any combination of the two-step selection functions with product (roughly) equal to $Ed\sigma/dE$. This might actually be preferred over (1) or (2), for example, if it simplified the coding. In AEGIS, such a mixed approach has not been explored and all selection functions are of type (1) or (2). In particular, for bremsstrahlung and Compton scattering, type (1) is used, while pair production is of type (2). In the latter case, the outgoing particle types (e^+, e^-) have equal selection probability.

The term "roughly" is used above to indicate that the correspondence need not be exact, since the weights will compensate for this.

Under the assumptions of the model, one only has to select two variables at each vertex (plus at most an azimuthal angle). These variables are type j (e^+, e^-, γ) and energy E_j. The weight of the particle selected is then readily calculated as

$$w = P(E_j)/S(j,E_j), \qquad (1)$$

where P denotes the actual physical probability(i.e., the cross section normalized to unity). S denotes the selection probability of the particle.

It is perhaps instructive to apply eq. (1) to the selection schemes mentioned above. Note that in an analogue calculation, S and P are identical, and hence the weight is always unity. In the method where one particle is selected at random with equal probability (Fig. 2), one has

$$S(j,E_j) = (1/2)P(E_j), \qquad (2)$$

and therefore the weight is always equal to two.

For the first type of AEGIS selection scheme, one may write

$$S(j,E_j) = P'(E_j) \, (E_j/E_o), \qquad (3)$$

where E_o is the incident energy and $P'(E_j)$ [$\simeq P(E_j)$] represents the first step of the procedure, which is equivalent to selecting the event in "roughly" analogue fashion. The second step introduces the factor, (E_j/E_o). It is clear that in this case

$$w \simeq E_o/E_j. \qquad (4)$$

When this is rewritten as

$$wE_j \simeq E_o, \qquad (5)$$

it expresses rough conservation of energy at each vertex. For the second type of selection scheme, used in pair production:

$$S(j, E_j) = (1/2) \, Q(E_j), \qquad (6)$$

where the factor of (1/2) is incurred in selecting particle type, and

$$Q(E_j) \simeq kE_j \cdot P(E_j). \qquad (7)$$

This leads once more to eq. (5) via the normalization condition

$$\int kE_j P(E_j) dE_j = 1, \qquad (8)$$

by noting that

$$\int E_j P(E_j) dE_j = (1/2) E_o. \qquad (9)$$

Equation (9) expresses the fact that in pair production, the average energy of each member equals half the photon energy (neglecting recoil).

The correctness of eq. (5) depends directly on the correctness of eq. (7). In formulating eq. (7), or any similar expression, one seeks a compromise between optimization of the bias (i.e. correctness of eq. (7)), simplicity of code, considerations of computer time and storage requirements, as well as desired accuracy of the result. It is therefore difficult to write down general rules about the "tolerance" allowed in formulating selection functions, and almost always the process will involve a certain amount of trial and error. As a first step, one seeks agreement between actual and desired selection functions to within a factor of two over most of the domain, and to within a factor of ten everywhere. Being in general agreement to better than ten percent is not especially desirable. It must be stressed that the above statements are meant only to convey to the reader a qualitative impression of the problem and not as strict guidelines.

In the program AEGIS, the approximation of the selection function to $E_j \cdot P(E_j)$ (i.e., the degree to which energy is conserved at each vertex) is typically 25%. This must not be confused with the global energy balance of the calculation which is nearly perfect for a reasonable number of incidents.

The selection procedures employed for each process are now described in turn as practical illustrations of the above discussion. They are not to be regarded as the "right answers" to textbook problems, but instead, as one choice from among many possible ones. It is certainly not claimed that they are the "best" choice but only that they seem to work.

BREMSSTRAHLUNG

The incomplete screening cross sections of Bethe and Heitler are assumed:[2]

$$d\sigma/d\nu = Z(Z+1)\alpha r_e^2 \nu^{-1} \quad (10a)$$

$$\times \{(1+E^2/E_0^2)[\phi_1(\gamma)-4\log Z/3]-(2E/3E_0)[\phi_2(\gamma)-4\log Z/3]\} \quad \text{for} \quad (\gamma<2)$$

and

$$d\sigma/d\nu = Z(Z+1)\alpha r_e^2 \nu^{-1} \quad (10b)$$

$$\times 4(1+E^2/E_0^2-2E/3E_0)[\log(2E_0 E/mc^2 \nu)-1/2-c(\gamma)] \quad \text{for} \quad (\gamma>2),$$

where Z is the atomic number, α the fine structure constant, r_e the classical electron radius, ν the radiated photon energy, mc^2 the electron rest energy, E_0 and E are the incident and outgoing electron energy, and $\gamma \equiv (100\ mc^2 \nu/EE_0 Z^{1/3})$. The functions $\phi_1(\gamma)$, $\phi_2(\gamma)$ and $c(\gamma)$ are taken from Ref. 2 and are stored in tabular form in the program. The functions $\phi_1(\gamma)$, $\phi_2(\gamma)$ are such that $\phi_1(\gamma) \approx \phi_2(\gamma)$ to within a few percent. Above $\gamma = 0.8$, they are virtually identical. The function $c(\gamma)$ is small everywhere. Both forms (10a) and (10b) of $d\sigma/d\nu$ are equal at $\gamma \sim 2$ as required to make $d\sigma/d\nu$ continuous.

The selection function employed is

$$S(\nu) = N(E_0)\nu^{-1}[1-\nu/E_0+3\nu^2/4E_0^2]f(\gamma), \quad (11)$$

where $N(E_0)$ is a normalizing factor such that the integral of (11) over all permissible values of ν is unity, and where $f(\gamma)$ is defined by

$$f(\gamma) = B - \gamma[B - \log(200/0.8Z^{1/3}) + 1/2]/0.8 \quad \text{for} \quad (\gamma \leq 0.8) \quad (12a)$$

and $f(\gamma) = \log(2E_o E/mc^2 \nu) - 1/2 \quad \text{for} \quad (\gamma > 0.8),$ (12b)

with $B = 5.75 - (\log Z)/3$. It is apparent by inspection that $S(\nu)$ is an approximate form of $d\sigma/d\nu$ (up to normalization). So the first method is used, and after ν is selected according to (11), a uniform random number r is generated and compared with ν/E_o to decide whether to follow the photon ($r < \nu/E_o$) or the electron. The particle is then assigned a weight

$$W_{\gamma,e} = S^{-1}(\nu)(d\sigma/d\nu)/P_{(\gamma \text{ or } e)} \quad (13)$$

where $P_\gamma = \nu/E_o$ and $P_e = 1 - P_\gamma$ are the probabilities of choosing a photon or electron, respectively.

COMPTON SCATTERING

The Klein-Nishina differential cross section with respect to the outgoing photon is taken as

$$d\sigma/d\nu = \pi r_e^2 (mc^2) \nu_o^2$$
$$\times [\nu/\nu_o + (m^2 + 2m\nu_o)/\nu_o + (\nu_o^2 - 2m\nu_o - 2m^2)/\nu_o\nu + m^2/\nu^2], \quad (14)$$

where ν is between the limits $\nu_o mc^2/(mc^2 + 2\nu_o)$ and ν_o, the incident photon energy. The selection of ν is made in an unbiased way: during initialization of the program each term in the square brackets of (14) is integrated between the ν limits for a number of different ν_o and these results are stored in tabular form. Selection of ν then proceeds by first choosing a term from among the four and then using a probability distribution directly proportional to that term. The correctness of this is easily seen by thinking of how one would proceed if each term were to represent a physically different process. The decision to follow the outgoing photon or electron is made as in the bremsstrahlung case. The weighting factor in each case is then

$w_\gamma = \nu_o/\nu$ or $w_e = \nu_o/(\nu_o - \nu)$.

PAIR PRODUCTION

Similar to bremsstrahlung the incomplete screening formulae are assumed:

$$d\sigma/dE_+ = Z(Z+1)\alpha r_e^2 \nu_o^{-3} \quad (15a)$$

$$\times \{(E_+^2 + E_-^2)[\phi_1(\gamma') - (4/3)\log Z] + (2/3)E_+E_-[\phi_2(\gamma') - (4/3)\log Z]\}$$
for $(\gamma' \leq 2)$

and

$$d\sigma/dE_+ = Z(Z+1)\alpha r_e^2 \nu_o^{-3} \quad (15b)$$

$$\times \{[E_+^2 + E_-^2 + (2/3)E_+E_-][\log(2E_+E_-/mc^2\nu) - 1/2 - c(\gamma')]\} \quad \text{for} \quad (\gamma' > 2),$$

where $\gamma' \equiv 100\nu_o mc^2/(E_+E_- Z^{1/3})$ and E_+ and E_- are the energies of the positron and electron. Note that they appear symmetrically in (15). The selection function used is

$$S(E) = M(\nu_o)[(4/3)E^3/\nu_o^3 + E/\nu_o - (4/3)E^2/\nu_o^2]g(\gamma'), \quad (16)$$

where E represents either E_+ or E_-, with $M(\nu_o)$ a normalizing factor, and $g(\gamma')$ defined as

$$g(\gamma') = B - \gamma'[B - \log(200/0.8Z^{1/3}) + 1/2]/0.8 \quad \text{for} \quad (\gamma' \leq 0.8),$$

or

$$g(\gamma') = \log[2E(\nu_o - E)/\nu_o mc^2] - 1/2 \quad \text{for} \quad (\gamma' > 0.8).$$

Note that S(E) is roughly proportional to $Ed\sigma/dE$. This is an example of the second method. Following selection of E, the particle is assigned to represent an electron or positron with equal probability. In either case, the weight is

$$w = 2S^{-1}(E)(d\sigma/dE). \quad (17)$$

ACCURACY

If the selection functions were exactly proportional to $Ed\sigma/dE$, then energy would be conserved exactly at each vertex. If, in addition, the energy deposition part of the calculation is

performed exactly, then all of the energy of the incident particle
will be deposited (at least in an infinitely large target) and no
more. But, if the proportionality of the selection function to
$Ed\sigma/dE$ is only approximate, then the total energy deposited will
only approximately equal the incident energy. However, it must be
emphasized that, on the average (i.e., averaged over many events),
energy is conserved. Using the above procedures, the total energy
deposited for each incident particle calculated by AEGIS follows a
normal distribution with a mean equal to the incident energy and a
standard deviation of about 7% of the incident energy. This is
quite a bit better than need be and indicates that perhaps cruder
approximations to $d\sigma/dE$ or $Ed\sigma/dE$ could be substituted for the
selection functions if there is anything to be gained by it.

Comparisons with experiment and other calculations are quite
close (see Ref. 1 for details). Computer time required per
incident particle varies logarithmically with incident energy and
ranges from about 15 msec at 1 GeV to about 40 msec at 1000 GeV on
an IBM 370/195 computer.

REFERENCES

1. A. Van Ginneken, "AEGIS. A Program to Calculate the Average
    Behavior of Electromagnetic Showers", Fermilab FN-309 (1978).
2. See e.g. H. A. Bethe and J. Ashkin, "Passage of Radiations
    through Matter" in Experimental Nuclear Physics, Vol. I,
    E. Segre, Ed., Wiley, N.Y. (1953); also see Lecture 11 by
    K. O'Brien in this volume.

LECTURE 15: ELECTRON DOSIMETRY USING MONTE CARLO TECHNIQUES

Keran O'Brien

Environmental Measurements Laboratory
U.S. Department of Energy
New York, New York 10014
USA

INTRODUCTION

The energy response of thermoluminescence dosimeters (TLDs) to electrons is of great practical importance since the response of a dosimeter to any form of ionizing radiation depends, essentially, on electron energy deposition. However, measurements of TLD electron energy response are in generally poor agreement with one another. In Fig. 1 are shown two recent sets of measurements by Holt et al.[1] and by Paliwal and Almond[2], for LiF. The results are expressed as f_γ^e, the response of the TLD relative to the cobalt-60 γ-ray dose absorbed in water. Only those TLDs exposed in a polystyrene "phantom" at a depth of 1.5 cm are exhibited in the graph. The energy ordinate is the electron energy incident on the outer surface of the polystyrene. The large discrepancy in the shape and magnitude of the two response curves was rather disconcerting and has provoked much discussion in the literature. It has even been suggested that the discrepancy is the result of differences in the chemical composition of the dosimeters.[3]

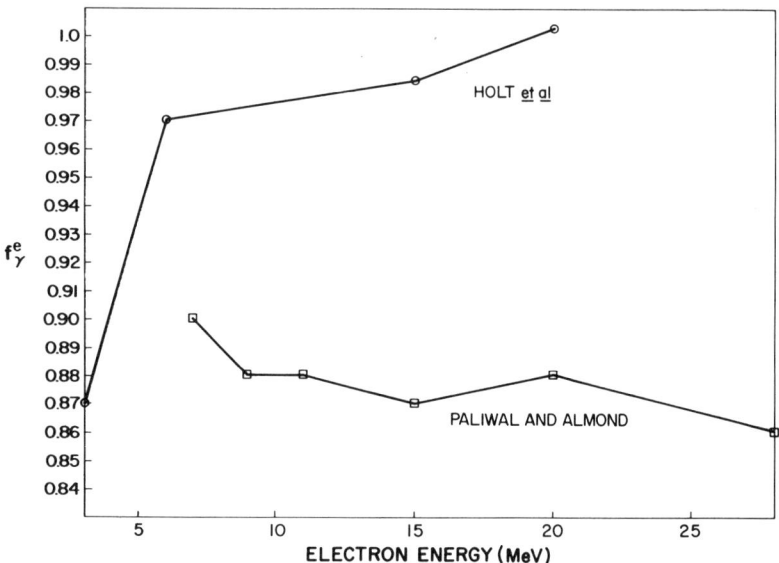

Fig. 1. A comparison of two recent sets of measurements of the energy response of LiF to high-energy electrons. The response is relative to cobalt-60 gamma rays absorbed in water. The dosimeters were exposed to the electron beam at a depth of 1.5 cm in polystyrene.

Detailed Monte Carlo calculations of electron-photon transport were carried out which accounted for the discrepancy and which are discussed below. The cause of the discrepancy is shown to lie in the conversion of the dose registered in air in an ionization chamber used by one of the groups[1] to dose in water. Both sets of measurements can be accounted for, and the Paliwal-Almond measurements give the correct dependence of TL response to electron energy.

DESCRIPTION OF THE MONTE CARLO CODE

The Monte Carlo calculations performed for this study were carried out using a modified version of the one-dimensional slab-geometry code CASCADE of Beck.[4] As CASCADE was originally designed to be applicable to electrons and photons down to 1 MeV and as it was desired to apply it to detectors smaller than a millimeter in size, the code was modified to reduce the lower energy limit down to 10 keV.

Beck has given a detailed description of his code[4], so only a brief description of the modified version will be given here.

Electrons (negatrons and positrons) are assumed to slow down continuously between bremsstrahlung reactions. Coulomb scattering is taken into account. The kinetic energy of any negatron, positron, or photon which falls below 10 keV is assumed to be absorbed directly. In addition, a positron with energy below 10 keV is assumed to emit two 511 keV photons isotropically. Photons may interact by the photoelectric, Compton or pair-production processes. All secondary photons, negatrons (except delta rays) and positrons generated either by the primary radiation or by other secondaries are followed until their kinetic energy falls below 10 keV, or until they leave the medium. Delta rays are not generated in these calculations.

Differential and total cross sections were taken from Beck,[4,5] except for the photoelectric cross section, which is a slightly modified version of that given by Diffey,[6] changed, where necessary, to improve agreement with Hubbell.[7]

The collision stopping powers were calculated using Heitler,[8] and the density effect formula is that of Armstrong and Alsmiller.[9] Ionization potentials were taken from Dalton and Turner[10] and Turner et al.[11] Finally, the transport-corrected scattering cross section and the moments of the multiple scattering distribution were taken from Lewis,[12] utilizing the Thomas-Fermi atom.[13]

THE MONTE CARLO CALCULATIONS

Paliwal and Almond exposed 0.9 to 1.1 cm LiF crystals, 0.17 to 0.19 cm thick, to high-energy electron beams, while the crystals were embedded at a depth of 1.5 cm in a polystyrene phantom. The dosimeters were calibrated by exposing a Fricke dosimeter to cobalt-60 γ rays at the same place in the phantom. As the Monte Carlo code is one-dimensional, the dosimeters were represented by their thickness only. Since the dose was relative to water and the calibration procedure was not specified, the Fricke dosimeter was simulated by pure water, giving the dose to water directly.

Table 1. Monte Carlo simulation of the experiment of Paliwal and Almond

Particle	Energy (MeV)	Calculated dose per unit incident particle per cm^2 (MeV g^{-1})	
		LiF*	H_2O*
^{60}Co gamma rays	1.17 and 1.33	0.0309	0.0366
Electrons	28	1.70	2.23
	18	1.72	2.25
	15	1.73	2.25
	11	1.73	2.26
	9	1.69	2.22
	7	1.71	2.18

*All standard deviations are less than 1%. All data satisfy the Shapiro-Wilk normality criterion at better than the 1% level.

The function of the calibration is to give the so-called TL response per rad. The light output of the LiF crystal is compared with a standard dosimeter of known response exposed to cobalt-60 gamma rays in the same location. It is assumed here that the TL response of the LiF dosimeter is proportional to absorbed energy.

The calibration was simulated by calculating the dose from 1.17 and 1.33 MeV photons normally incident on the idealized polystyrene phantom. The results are shown in Table 1. The ratio of the water to lithium fluoride dose is 1.18 ± 0.011. The photon energy spectrum in the center of the LiF crystal is shown in Fig. 2 when irradiated by 1.33 MeV gamma rays from cobalt-60. There is substantial photon build-up, 1.18 photons for each photon incident, and considerable dispersion of the spectrum from the initially monochromatic primary photons.

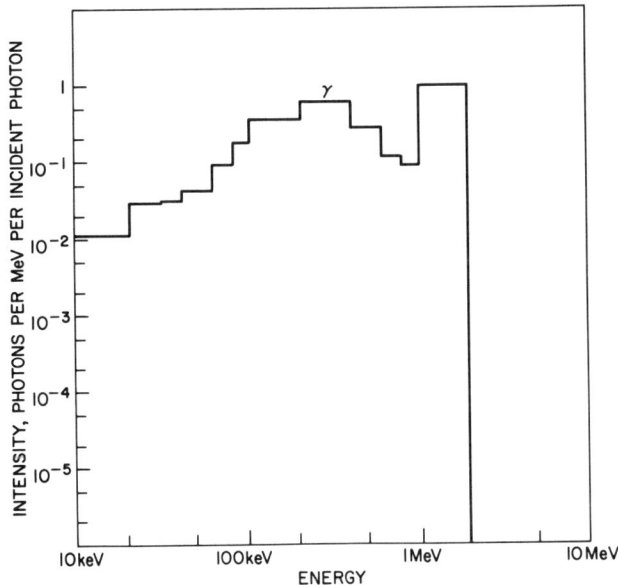

Fig. 2. Photon spectra in the LiF crystal at a depth of 1.5 cm in polystyrene when irradiated by the 1.33 MeV photon from cobalt-60.

Table 2. Comparison of Monte Carlo simulation with the measurements of Paliwal and Almond

Incident electron energy (MeV)	Monte Carlo calculations		Paliwal and Almond measurements		% Dev.*
	f_γ^e	$\pm \sigma(\%)$	f_γ^e	$\pm \sigma(\%)$	
28	0.893	1.4	0.86	2.0	-3.8
18	0.903	1.8	0.88	0.8	-2.8
15	0.912	1.9	0.87	3.4	-4.8
11	0.903	2.1	0.88	1.7	-2.6
9	0.901	2.0	0.88	1.6	-2.4
7	0.925	2.0	0.90	2.2	-2.8
Mean	0.906	2.2	0.878	1.5	-3.2

*% Dev. = $\left[\dfrac{\text{exp.} - \text{calc.}}{\text{exp.}}\right] \times 100$.

The calculation of electron energy deposition by the electrons was carried out in the same fashion. These results are also shown in Table 1. The combined results are shown in Table 2 and compared with the experimental data.

The Monte Carlo calculations agree well with the measurements, lying a nearly constant 3% above them, which is well within the combined stated errors.

In Figs. 3 - 5 are exhibited the electron and photon spectra at the center of the LiF dosimeter when the polystyrene phantom is irradiated by 7 and 28 MeV electrons, respectively. Paliwal and Almond calculated the effective energy at the dosimeter using a formula due to Harder.[14] The location of this point is indicated in both figures by an H. In both cases, the electron spectrum has a peak

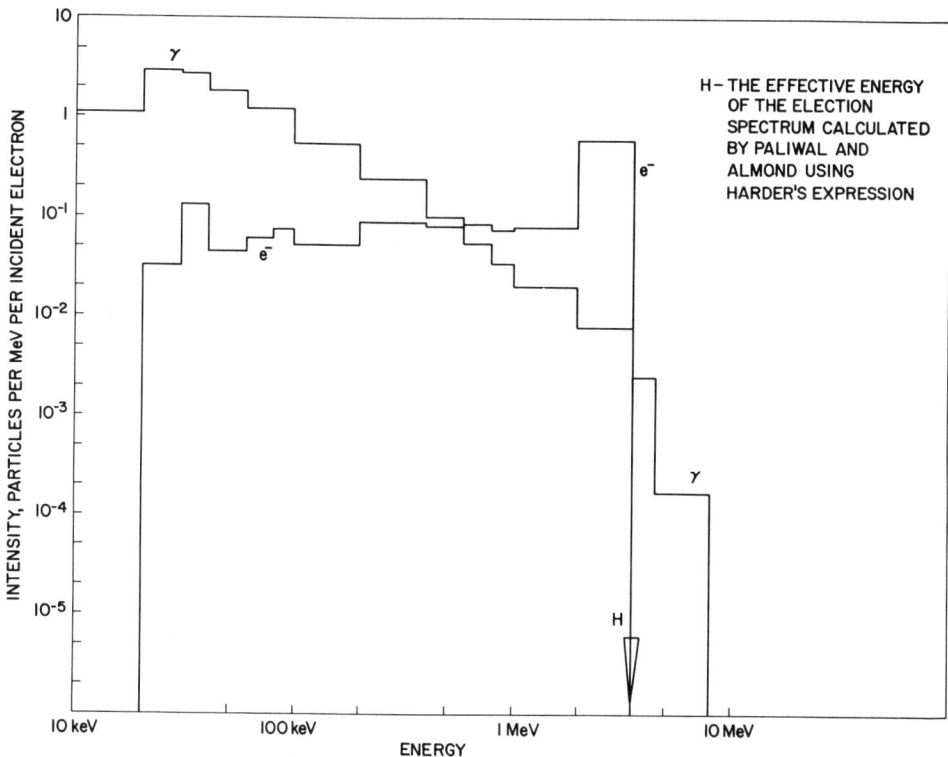

Fig. 3. Photon and electron spectra in the LiF crystal at a depth of 1.5 cm in polystyrene when irradiated by 7 MeV electrons.

in the neighborhood of the calculated effective energy, but both spectra have been significantly broadened. A large secondary photon component exists in both cases. Figures 3 - 5 indicate that, although Harder's expression may be useful in locating the peak of the transmitted electron spectrum, it should be used with care as the spectra at depth are nowhere near monochromatic.

Holt et al.[1] carried out two classes of experiment. The first (to be called the "A" series) is comparable to the measurements of Paliwal and Almond discussed and simulated above. LiF rods were exposed at depths 0.5 and 1.5 cm in polystyrene to electron energies

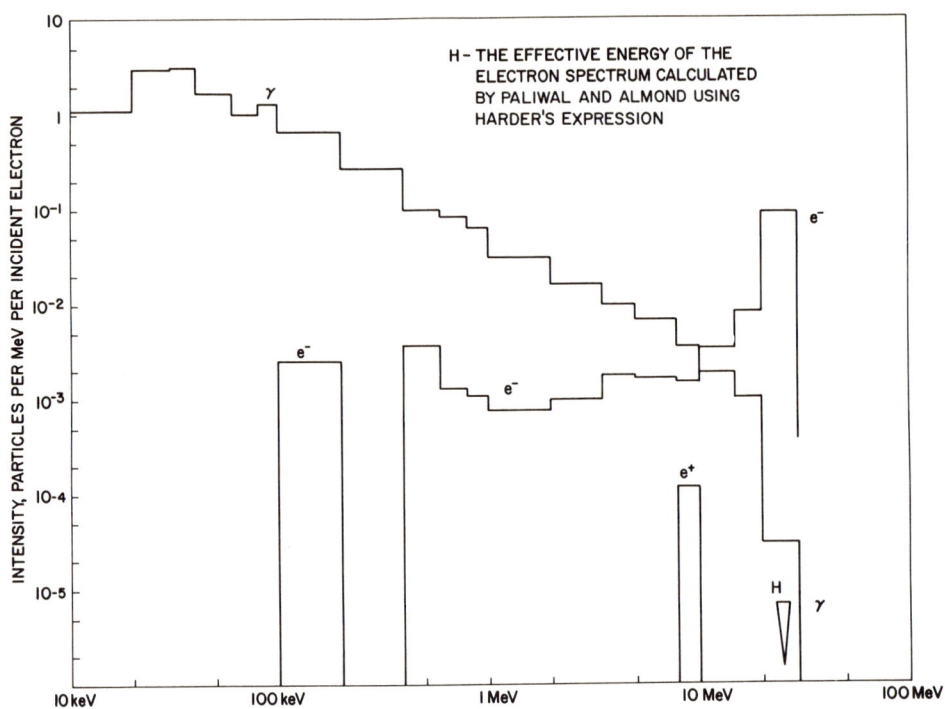

Fig. 4. Photon and electron spectra in the LiF crystal at a depth of 1.5 cm in polystyrene when irradiated by 28 MeV electrons.

from 3 to 20 MeV (it is the 1.5 cm depth measurements of series A which are plotted in Fig. 1). They were calibrated by exposure to a pancake ionization chamber the response of which was made equivalent to the dose in water by an explicit procedure which was duplicated in the Monte Carlo simulations. The second (the "B" series) class of experiment was performed with a 20 MeV electron beam. The electron energy incident on the dosimeter was varied by moving the exposure position in the polystyrene phantom, and using Harder's expression to calculate the electron energy corresponding to the depth.

The calibration was simulated as above, save that instead of a Fricke dosimeter, the pancake ionization chamber was simulated by a

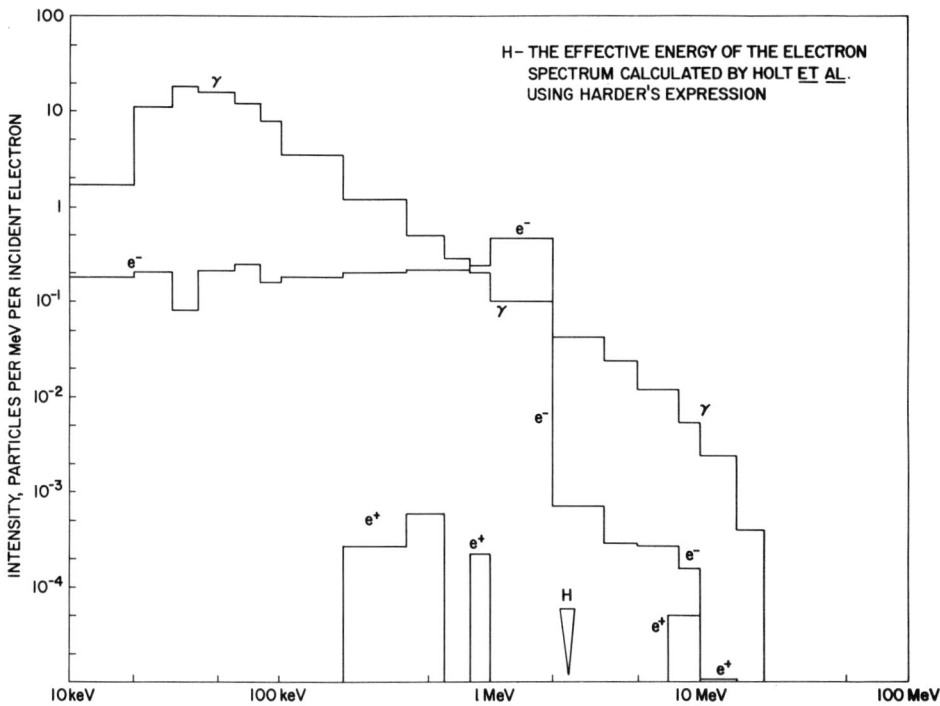

Fig. 5. Photon and electron spectra in the LiF crystal at a depth of 1.5 cm in polystyrene when irradiated by 28 MeV electrons.

thin (2.4×10^{-4} g/cm^2) sheet of air. The composition of air was taken from CIRA.[15] In this case the calibration was performed at 0.5 cm depth in polystyrene, as was the Monte Carlo simulation. The LiF rods in these calculations were represented by a LiF sheet with a thickness equal to the rod diameter. The ionization calculations were converted to the equivalent water values by duplicating the procedures of Holt et al., as exemplified by the relationship

$$D_{H_2O} = \frac{(\mu/\rho)_{H_2O}}{(\mu/\rho)_{(CH)}} {}_m S^{(CH)}_{air} D_{air},$$

where

$D_\alpha[\alpha = H_2O, (CH)]$ are the doses in water and polystyrene,

$(\mu/\rho)_\alpha$ are the mass energy absorption coefficients in the two media, and

$_mS_{air}^{(CH)}$ is the ratio of the mass stopping powers of polystyrene to air.

The values of μ/ρ were obtained from Hubbell.[7] The value of $_mS_{air}^{(CH)}$ was calculated in the same way as the stopping powers described earlier.[8,9]

The ratio of the LiF to water dose so obtained is 1.16 ± 0.011, which is in agreement with the previous value of 1.18 ± 0.011 obtained directly under 1.5 cm of polystyrene.

The doses from electron radiation were obtained analogously. The calculations simulating the ionization chamber were, again duplicating Holt et al.'s procedure, transferred to water using the formula

$$D_{H_2O} = {_mS_{air}^{H_2O}} \, D_{air},$$

where $_mS_{air}^{H_2O}$ was evaluated at the energy obtained by Holt et al. from Harder's expression. These results are given in Table 3.

The combined results and the comparison with experiment are shown in Table 4. The agreement between calculation and measurement is reasonably good, lying within the combined stated errors, except for the last point of the "B" series. The large deviation of this

Table 3. Monte Carlo simulation of the experiment of Holt et al.

Particle	Dosimeter depth (cm)	Energy (MeV)	Calculated dose per unit incident particle per cm^2 (MeV g^{-1}) LiF*	AIR*
^{60}Co gamma rays	0.5	1.17 and 1.33	0.0319	0.0319
Electrons A	1.5	20	1.75	2.24
	1.5	15	1.77	2.24
	1.5	10	1.77	2.16
	1.5	8	1.74	2.07
	1.5	6	1.73	2.02
	0.5	3	1.89	2.20
Electrons B	1.5	20	1.75	2.24
	4.0	20	1.74	2.12
	6.5	20	1.57	1.84
	7.5	20	1.46	1.67
	8.5	20	1.22	1.11

*All standard deviations are less than 1%. All data satisfy the Shapiro-Wilk normality criterion at better than the 1% level.

point is due, in part, to the buildup of bremsstrahlung in the semi-infinite medium of the calculation. This is evident in Fig. 5 where the spectra in the center of the chip at its depth of 8.5 cm are shown. Since the mesurements were actually carried out in a 25 cm x 25 cm phantom, much of the photon radiation must have

Table 4. Comparison of Monte Carlo simulations with the measurements of Holt et al.

Incident electron energy (MeV)	Harder expression energy (MeV)	Monte Carlo calculations f_γ^e	$\pm \sigma(\%)$	Holt et al. measurements f_γ^e	$\pm \sigma(\%)$	% Dev.*
A 20	16.9	0.939	1.8	1.003	6	6.4
15	11.9	0.911	2.1	0.990	6	8.0
10	7.06	0.908	1.9	0.984	6	7.7
8	5.09	0.903	2.3	0.970	6	6.9
6	3.11	0.899	2.1	0.970	6	7.3
3	2.03	0.898	2.2	0.870	6	-3.2
B 20	16.9	0.939	1.8	1.008	6	6.8
20	11.7	0.911	2.1	1.010	6	6.9
20	6.50	0.940	2.0	0.975	6	3.6
20	4.43	0.936	2.3	0.983	6	4.7
20	2.35	1.1428	2.1	0.898	6	-27

*% Dev. = $\left[\dfrac{\text{exp.} - \text{calc.}}{\text{exp.}}\right] \times 100$.

leaked out before having penetrated to this depth.

The energy dependence of Holt et al.'s series of measurements is due to the use of the pancake chamber in combination with the assumption that all the electrons are concentrated at the energy given by Harder's expression. As Figs. 3-5 demonstrate, the electron spectrum extends to very low energies. The mass stopping power ratios $_m S_c^\alpha$ are graphed in Fig. 6. $_m S_{air}^{H_2O}$ is strongly energy dependent. The average absolute values of the results will be inflated by this procedure, since this choice of $_m S_{air}^{H_2O}$ will be too large

ELECTRON DOSIMETRY USING MONTE CARLO METHOD

whenever the "Harder energy" lies above 2 MeV.

It is possible that the accuracy of the low-energy Monte Carlo calculations is adversely affected by the representation of the array of lithium fluoride rods as a single sheet. This geometry may result in some additional energy dependence not reflected in this calculation.

Paliwal and Almond's results are largely unaffected by these considerations since, as Fig. 6 shows, $_mS^{LiF}_{H_2O}$ is essentially independent of energy.

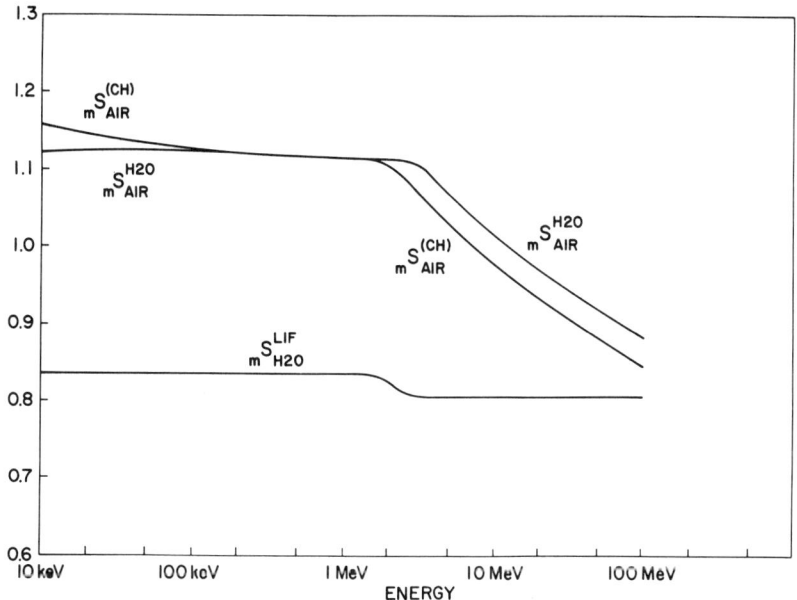

Fig. 6. Stopping power ratios for polystyrene (CH) to air, water to air, and LiF to water, calculated using Heitler[8] and Armstrong and Alsmiller.[9]

CONCLUSION

The apparent discrepancy between the two different sets of measurements made by Holt et al.[1] and by Paliwal and Almond[2] can be tracked down and accounted for by simulating the experimental procedures with the Monte Carlo technique, and evaluating the assumptions made in the course of the various steps in the analysis. It is apparent that, since the radiation transport processes associated with high-energy electrons in matter are complex, the use of cavity theory in the analysis or interpretation of results may lead to serious error.

REFERENCES

1. J. G. Holt, G. R. Edelstein, and T. E. Clark, Phys. Med. Biol. 20:559 (1975).
2. B. Paliwal and P. R. Almond, Phys. Med. Biol. 20:547 (1975).
3. M. G. Gantchew and K. Toushlekova, Phys. Med. Biol. 21:300 (1976).
4. H. L. Beck, "A Monte Carlo Simulation of the Transport of Electrons and Photons in Matter," USAEC Report HASL-213 (1969).
5. H. L. Beck, "Energy Spectra from Coupled Electron-Photon Slowing Down," USERDA Report HASL-309 (1976).
6. B. L. Diffey, Int. J. Appl. Radiat. Isotopes 26:492 (1974).
7. J. H. Hubbell, "Photon Cross Sections, Attenuation Coefficients, and Energy Absorption Coefficients from 10 keV to 100 GeV," National Bureau of Standards Report USNRDS-NBS 29 (1969).
8. W. Heitler, The Quantum Theory of Radiation, The Clarendon Press, Oxford, 1954.
9. T. W. Armstrong and R. G. Alsmiller, Jr., Nucl. Instr. and Meth. 82:289 (1970).
10. P. Dalton and J. E. Turner, Health Phys. 15:257 (1968).

11. J. E. Turner, P. D. Roecklein, and R. B. Vora, Health Phys. 18: 159 (1970).
12. H. W. Lewis, Phys. Rev. 78:526 (1950).
13. R. D. Birkhoff, Encyclopedia of Physics, S. Flugge, ed., 34:53, Springer, Berlin (1958).
14. "Radiation Dosimetry: Electrons with Initial Energies Between 1 and 50 MeV," Report 21, ICRU Publications, Washington, D.C. (1972).
15. CIRA, "COSPAR International Reference Atmosphere, Report of the Preparatory Group for an International Reference Atmosphere" (Accepted at the COSPAR Meeting in Florence, April 1961), North Holland, Amsterdam (1961).

LECTURE 16: APPLICATION OF EGS TO DETECTOR DESIGN

IN HIGH ENERGY PHYSICS

Walter R. Nelson [*]

Health and Safety Division
CERN
Geneva 23, Switzerland

INTRODUCTION

As we have seen in a previous lecture, the EGS Code System is typical of analog Monte Carlo programs that simulate electromagnetic cascades. A number of examples were presented in order to demonstrate how EGS3 could be used to predict certain experimental results, as well as the the results of other independent Monte Carlo calculations. We have also seen how EGS1, which only differs from EGS3 in that it is less versatile and user-oriented (i.e., the physics is the same), can be used rather effectively in problems dealing with: 1) photon scattering at low energies (i.e., synchrotron radiation); 2) cascades within and behind absorbers (i.e., beam dumps); 3) dose and shielding parameters of stray electron-photon radiation from a high energy electron accelerator.

In the final two lectures on electromagnetic cascades, we will return to the EGS Code System and look at some solutions to problems in high energy physics, medical physics, and dosimetry. We will keep in mind, however, that we might not be solving the problem in

---

[*] On leave of absence from SLAC (Stanford University), 1978/79.

the most efficient way, and that one might be wise to use AEGIS or some other code instead of EGS. In fact, for problems involving low energy photon scattering (e.g., ducting and streaming), it might be better to use MORSE or a similar type code that has importance sampling methods as an integral part of the code. The present lecture will look specifically at the application of EGS3 to the design of radiation detectors---especially those used in high energy physics.

CONVERSION EFFICIENCY OF LEAD FOR 30-200 MEV PHOTONS

We begin by describing an EGS simulation of a recent experiment (Darriulat, 1975) that measured the conversion efficiency for 44, 94, and 177 MeV photons incident upon lead, as shown in Fig. 1. By tagging the photons, the mean energy was determined to an accuracy of ± 4 MeV. The photon beam, with an area less than 15 cm x 15 cm, struck a lead plate of desired thickness (1 to 20 mm) and area (20 cm x 20 cm). Immediately following the lead was a large plastic scintillation detector, 28 cm x 40 cm in area and 5 mm thick. An event was counted as a conversion if more than 60 keV was deposited

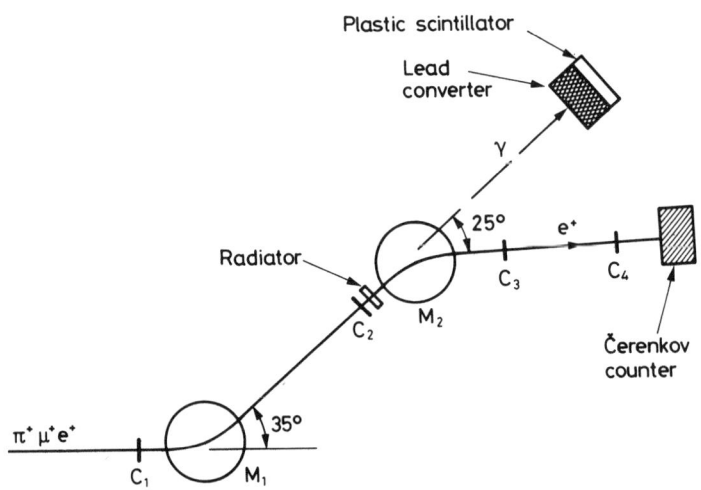

Fig. 1. Experimental arrangement of Darriulat et al (1975).

APPLICATION OF EGS TO DETECTOR DESIGN

in the scintillator for each incoming photon. To calculate the conversion efficiency with EGS, the geometry layout shown in Fig. 2 was used, consisting of four regions separated by three semi-infinite planes. Polystyrene, with a density of 1.032 g/cm^3 and consisting of hydrogen and carbon with an atomic ratio H/C=1.10, was used as the medium for plastic scintillator in region 3. PEGS was used in order to create the necessary material data with cutoff energies of 1.5 MeV and 0.1 MeV for electrons and photons, respectively. The actual User Code * that was used in this calculation is called UCCONEFF and is given in the EGS report (Ford and Nelson, 1978).

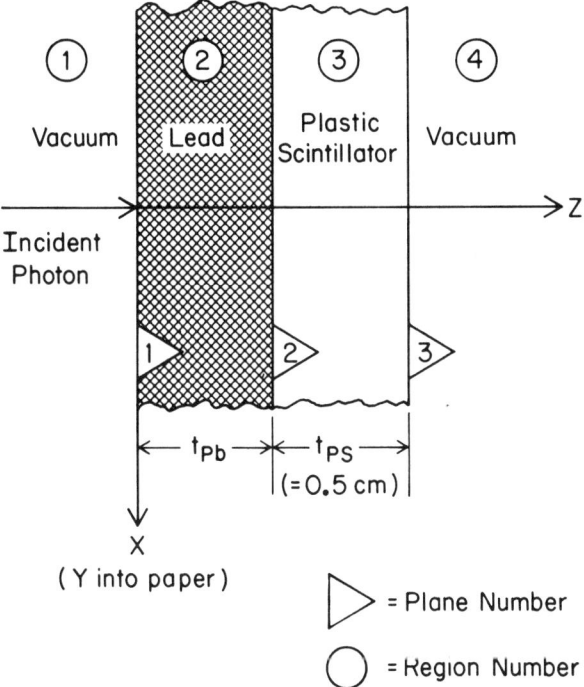

Fig. 2. Geometry layout used in HOWFAR for simulation of conversion efficiency experiment by Darriulat et al (1975).

---

* The reader is encouraged to review Lecture 12 of this course for an understanding of the role of the User Code, as well as for other EGS-related terminology.

The AUSGAB (scoring and/or outputing) subroutine was set up to sum the energy deposition in the plastic for each incident photon. Upon completion of a photon shower generation initiated by a CALL SHOWER statement in MAIN, a conversion event was scored provided that the energy sum in the plastic exceeded 0.060 MeV as dictated by the experiment. The conversion efficiency was calculated by dividing the event count by the total number of incident photons.

The results of the calculations are compared with the experimental data in Fig. 3. Needless to say, the agreement is extremely good over the entire lead thickness range for the two energies shown. In the text describing the experimental results, Darriulat et al point out that the energy spectrum in the scintillator showed

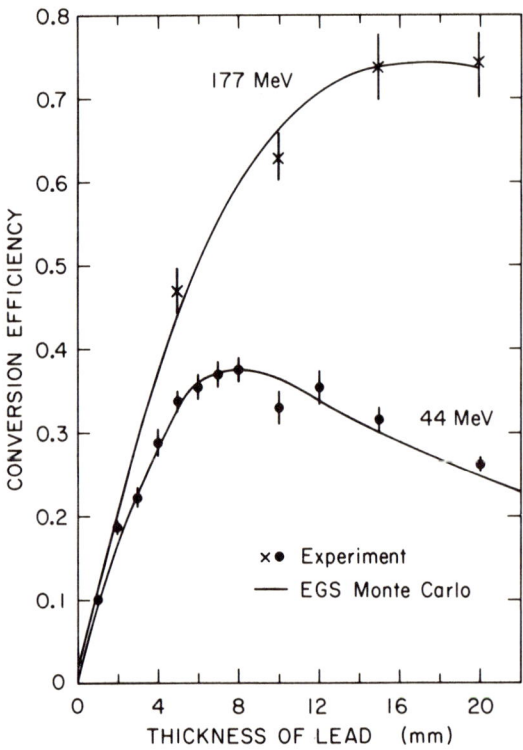

Fig. 3. Conversion efficiency data by Darriulat et al (1975) and comparison with EGS Monte Carlo simulation.

APPLICATION OF EGS TO DETECTOR DESIGN

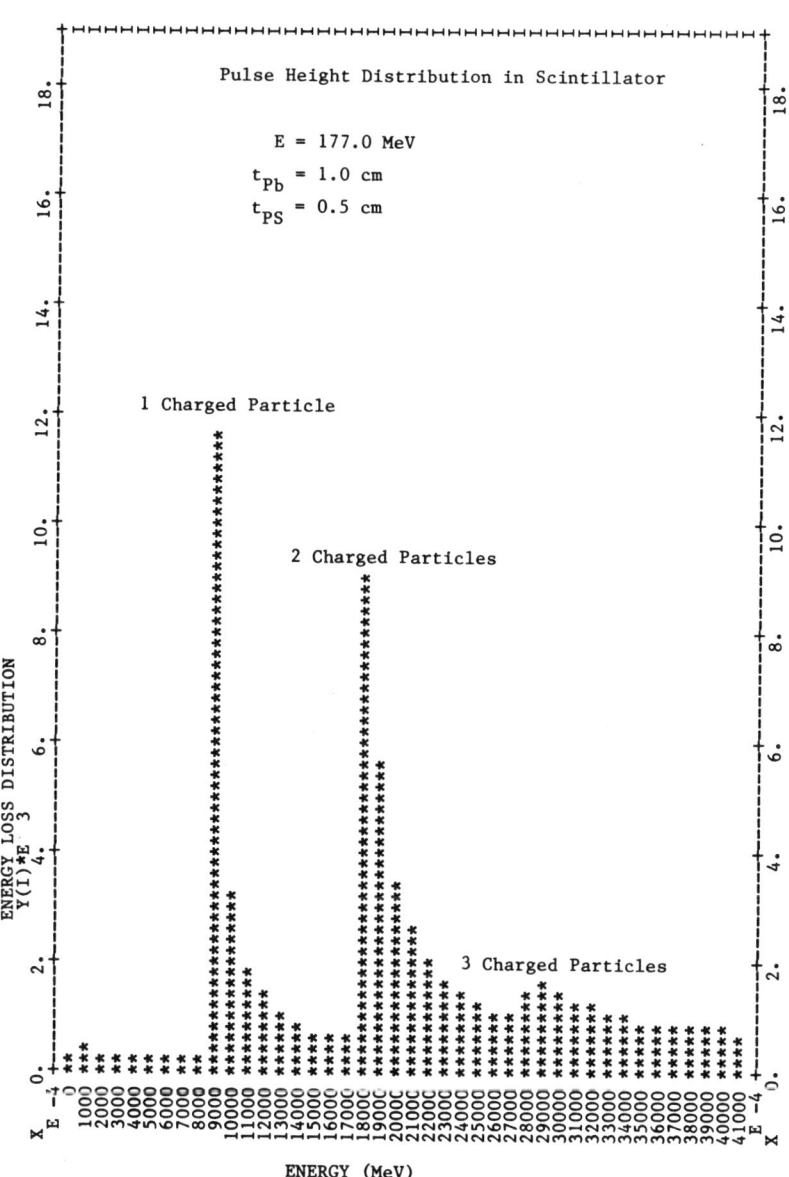

Fig. 4. EGS simulation of pulse height distribution in scintillator for the conversion efficiency experiment by Darriulat et al (1975).

characteristic peaks corresponding to the production of one, two, or three secondary electrons. To check out this observation, the total energy deposition in the scintillator per incident photon was histogrammed. Fig. 4 is representative of the results obtained and indeed shows three peaks. Assuming an energy loss of about 2 MeV-cm^2/g for a single electron traversing the plastic thickness (0.5 cm), the peaks are also seen to be in their expected energy locations of about 1, 2, and 3 MeV. In conculsion, EGS can predict photon conversion efficiencies rather well, at least in the energy range 30-200 MeV and for geometries similar to the one described here.

THE CHARGED COMPONENT OF 1 GEV ELECTRON SHOWERS

The longitudinal development of electromagnetic cascades has been analytically modeled and solved under various approximations (Rossi, 1952). Typically, the mean number of charged particles, when plotted as a function of the thickness of the absorber in radiation lengths, starts at unity, rises to a maximum, and exhibits a tail corresponding to attenuation of the photon component. The height of the curve at the maximum can be parameterized in terms of the ratio, E_0/E, where E_0 is the incident energy and E is the lower limit at which charged particles are considered in the estimate. The smaller the value of E the larger the height of the shower curve. Such curves can be obtained rather easily with EGS, and we shall illustrate this by comparing such a calculation with an experiment that was designed to sort the secondary charged particle component in terms of E.

The experiment (Drickey et al, 1968) was performed by observing the secondary charged particle tracks traversing a streamer chamber located around a stack of lead plates that were placed in a beam of 1 GeV positrons. The streamer chamber arrangement is given in Fig. 5 showing three sections with lengths 11, 13, and 15 cm. The upbeam section was used to ensure that one and only one electron was incident

APPLICATION OF EGS TO DETECTOR DESIGN

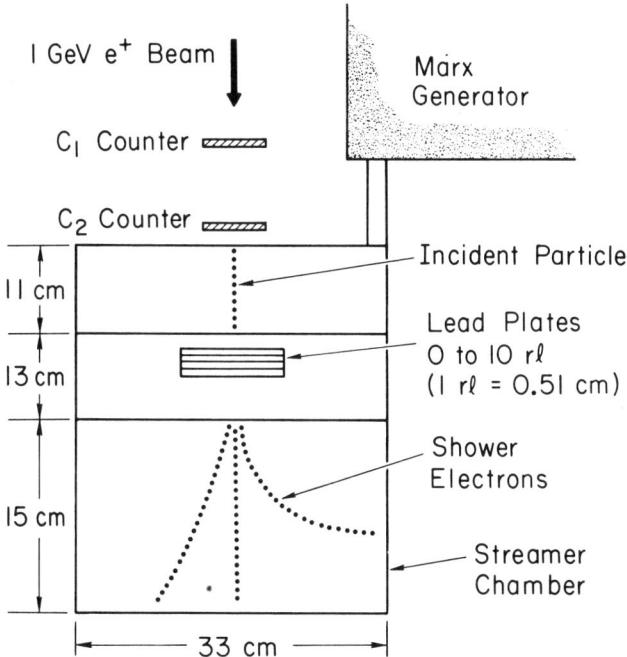

Fig. 5. Streamer chamber arrangement used in the experiment by Drickey et al (1968). Note: 12 cm chamber gap in the direction of normal to plane of figure; magnetic field in that direction also.

at a time, and also eliminated confusion between those cases when no charged particle emerged from the lead and when the chamber fired improperly. The downbeam surface of the lead converter was 4.5 cm upbeam from the entrance to the third section (15 cm long) in which the shower particles were observed and measured. Data were taken with lead thicknesses out to 10 radiation lengths. A magnetic field of 1665 G, perpendicular to the chamber plane shown in Fig. 5, was established (uniformly, we will assume) in the third section only. This provided a means of determining the momentum of the individual secondaries that were photographed in stereo. The 12 cm chamber gap (in the direction of the magnetic field vector), the chamber width (33 cm), and a scanning criteria---namely, tracks had to penetrate at least 2.54 cm into the third section in order to be scored---

essentially established the geometric acceptance for this experiment, although the helical trajectory of the particles was a contributing factor, as we shall soon see.

The geometry layout that we used in order to write the User Code is similar to the one used in the previous example (see Fig. 2) and the HOWFAR subroutines are identical to one another. Differences are accounted for in the MAIN section in the form of coordinate-axis identification, dimensions, and region-media definitions. Charged particles that reach the third chamber are discarded in subroutine HOWFAR and are further analyzed in AUSGAB---provided that they are within the lateral extents of the chamber. Having passed this test, trajectories are calculated (with and without magnetic field) and particles that leave the chamber prior to reaching the 2.54 cm Z-position are eliminated from further analysis. Those that remain are sorted by energy. Details of the calculation are given in the AUSGAB portion of the User Code called UCSPARK in the EGS report (Ford and Nelson, 1978).

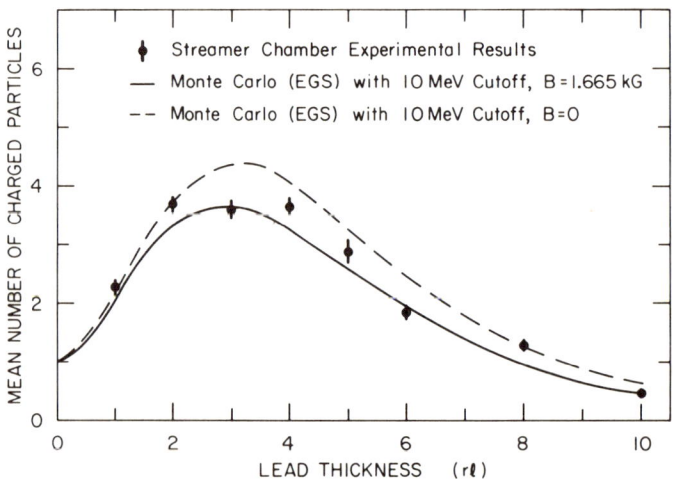

Fig. 6. Streamer chamber experimental and Monte Carlo results for secondary charged particles of 10 MeV or greater (with and without magnetic field).

Typical results are shown in Fig. 6. The experimental points essentially lie between the B=0 and B=1.665 kG curves obtained by EGS. The conclusion reached is that the EGS program does a reasonable job of simulating the Drickey experiment. Refinements could be made in the EGS analysis if additional information were provided by the experimentalists. In particular, details about the scanning criteria and a better understanding of the magnetic field (e.g., fringe-field effects) could possibly result in a closer agreement.

EGS MONTE CARLO SIMULATION OF A LARGE, MODULARIZED SODIUM IODIDE DETECTOR EXPERIMENT

The application of large, modularized NaI(Tl) detectors to physics experiments, particulary those involving the gamma-ray spectroscopy of the newly discovered psi particles at high energy electron-positron storage rings, has increased considerably during the last few years. A recent paper by Ford et al (1976) describes an experiment that was performed at SLAC to measure, among other things, the energy resolution of a typical detector array consisting of 19 NaI(Tl) hexagonal modules. Although each module itself cannot be expected to provide good energy resolution at high gamma-ray or electron energies, due to the transverse spread of energy in the the secondary electromagnetic shower (i.e., leakage), this problem is overcome in a detector made up of an array of modules. Each crystal is encapsulated in a hexagonal stainless steel container with a 0.020 inch wall thickness. The individual modules are optically coupled at one end to a 0.5 inch thick glass window, through which the crystal volume is viewed by a 3 inch diameter photomultiplier tube.

The properties of this composite detector were explored by Ford et al using electron beams in the range 0.1 to 4 GeV. The observed response of the modular array of 19 hexagons to electrons incident along the axis of the central module is summarized in Fig. 7. This figure shows not only the energy resolution obtained when the energies

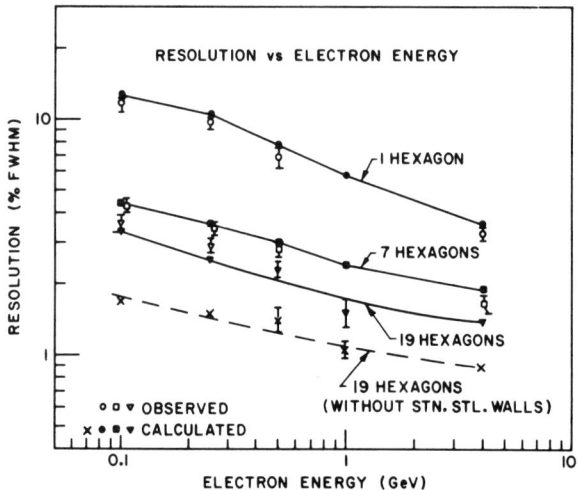

Fig. 7. Comparison between observed and EGS-calculated resolution versus energy (taken from Ford et al (1976)).

deposited in all 19 crystals are summed, but also those obtained when only the energies in the central 7 modules or in the central module alone are used. Also shown in this figure is the EGS Monte Carlo simulation of the electromagnetic shower in this detector. The agreement is observed to be very good. EGS takes into account both the energy leakage fluctuations from the detector volume and fluctuations due to energy absorption in the stainless steel walls surrounding each crystal module.

The 0.020 inch thick stainless steel walls cause undesirable effects when the trajectory approaches closely or intercepts these walls, as illustrated in Fig. 8. In this figure energy resolution is shown as a function of the displacement of the trajectory from the axis of an array of 7 modules. No significant loss in the resolution is experienced until the trajectory approaches within about 0.5 inch of the nearest wall. It can be seen that the agreement between measurement and calculation is very good and gives confidence in the ability of the EGS Code System to predict the

APPLICATION OF EGS TO DETECTOR DESIGN

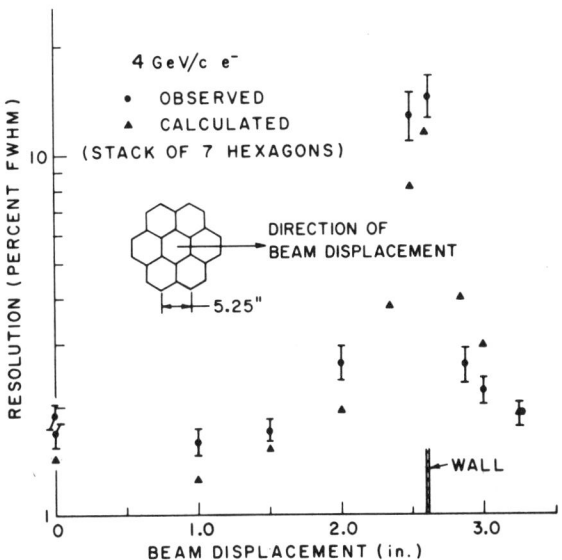

Fig. 8. Comparison between observed and EGS-calculated resolution as a function of the electron beam displacement (taken from Ford et al (1976)).

response of the detector for any incident trajectory. In conclusion, the resolution provided by assemblies of these hexagonal NaI(Tl) crystals is well accounted for by the EGS simulation of the electromagnetic cascade in the detector.

MULTIWIRE PROPORTIONAL COUNTERS AND LIQUID ARGON CHAMBERS

A recent study by Fischer (1978) has shown that the response of multiwire quantameters can be reliably predicted using the EGS code. A number of calculations were reported in this paper, but we will only discuss one of them. Energy resolutions were calculated by Fischer for the lead-MWPC sandwich of Nordberg (1971) and for the iron-MWPC sandwich of Ritson and Prepost (1976). Since the two quantameters were sufficiently different in material and size (i.e., thickness), it was felt that such a comparison would provide a

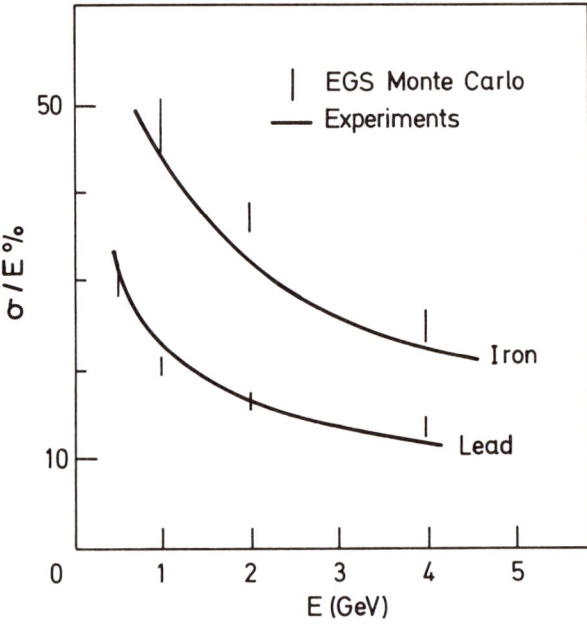

Fig. 9. Energy resolution data for two MWPC's and comparison with EGS simulation (taken from Fischer (1978)).

significant test of the Monte Carlo simulation. Fig. 9 has been taken from Fischer's paper and we see that the measured resolutions are well predicted by EGS.

The use of shower simulation programs has also proven to be rather useful in the design and optimization of lead/liquid argon shower counters. In a recent paper, Hitlin et al (1976) use EGS/PEGS as an aid in understanding the energy resolution, and for extrapolating the experimental results, of liquid argon detectors. In a similar fashion, EGS has been used by Bharadwaj et al (1978) in the design of a segmented liquid scintillation shower detector with a reasonable amount of success. The interested reader should refer to these papers for details of the experiments and the calculations that were done.

CONCLUDING REMARKS

In this lecture we have given several examples of the use of the EGS Code System in high energy physics. Three recent developments are worthy of mention at this time.

1.) EGS3 is capable, in principle, of simulating showers in magnetic fields and a group at SLAC has written a User Code to do just this.
2.) Background radiation to counters from synchrotron radiation in e^+e^- storage ring experiments can cause considerable problems. EGS has proven to be useful at SLAC in transporting low energy photons (including fluorescent quanta) and in aiding in counter shield design.
3.) EGS has recently been coupled with KASPRO at CERN in order to calculate e^+e^- yields from hadron beams striking targets.

REFERENCES

Bharadwaj, V. K., Cain, M. P., Caldwell, D. O., Denby, B. H., Eisner, A. M., Joshi, U. P., Lu, A., Morrison, R. J., Pfost, D. R., Summers, D. J., Yellin, S. J., Pellet, D. E., and Nash, T., 1978, An inexpensive large area shower detector with high spatial and energy resolution, Nucl. Instr. Meth., 155:411.

Darriulat, P., Gygi, E., Holder, M., McDonald, K. T., Pugh, H. G., Schneider, F., and Tittel, K., 1975, Conversion efficiency of Lead for 30-200 MeV photons, Nucl. Instr. Meth., 129:105.

Drickey, D. J., Kilner, J. R., and Benaksas, D., 1968, Charged component of 1-GeV electron showers in lead, Phys. Rev., 171:310.

Fischer, H. G., 1978, Multiwire proportional quantameters, CERN Internal Report Number CERN/EF/78-2; submitted to the Wire Chamber Conference, Vienna, 14-16 February, 1978.

Ford, R. L., Beron, B. L., Carrington, R. L., Hofstadter, R., Hughes, E. B., Kirkbridge, G. I., O'Neill, L. H., and Simpson, J. W., 1976, Performance of large, modularized NaI(Tl) detectors,

High Energy Physics Laboratory (Stanford University) Report Number HEPL-789; presented at the IEEE 1976 Nuclear Science Symposium and Scintillation and Semiconductor Symposium, New Orleans, LA., 20-22 October 1976.

Ford, R. L., and Nelson, W. R., 1978, The EGS code system: Computer programs for the Monte Carlo simulation of electromagnetic cascade showers (Version 3), Stanford Linear Accelerator Center Report Number SLAC-210.

Hitlin, D., Martin, J. F., Morehouse, C. C., Abrams, G. S., Briggs, D., Carithers, W., Cooper, S., Devoe, R., Friedberg, C., Marsh, D., Shannon, S., Vella, E., and Whitaker, J. S., 1976, Test of a lead/liquid argon electromagnetic shower detector, Nucl. Instr. Meth., 137:225.

Nordberg, M. E., 1971, A total energy shower detector using multi-wire proportional chambers, Cornell Laboratory Report Number CLNS-138.

Ritson, D. M., and Prepost, R., 1976, A proposal for a lepton total energy detector at PEP, Proposal PEP-6 (SLAC).

Rossi, B., 1952, High-energy particles, Prentice-Hall, Inc., Englewood Cliffs, New Jersey.

LECTURE 17:   APPLICATION OF EGS AND ETRAN TO PROBLEMS IN

MEDICAL PHYSICS AND DOSIMETRY

Walter R. Nelson *
Health and Safety Division
CERN
Geneva 23, Switzerland

INTRODUCTION

In this final lecture dealing with electromagnetic cascades we will look at a few applications of Monte Carlo programs to problems of interest in medical physics and dosimetry. In particular, we will consider two areas: 1) bremsstrahlung production from medical accelerators; 2) photon dosimetry at medium to low energies.

LOW ENERGY BREMSSTRAHLUNG FROM HIGH-Z TARGETS

In recent years there has been an increase in the production of electron linacs that operate in the energy range from 10 to 35 MeV. Of particular interest to the therapeutic radiologist are machines that produce bremsstrahlung beams that have a hard spectrum, are spatially flat, and have a low electron contamination. The design of the target and the associated field flattener can be a difficult job for the radiation physicist. To complicate matters, most clinical accelerators make use of collimating jaws and monitoring devices, such as ionization chambers, so it is rather difficult to design such machines to meet the specifications stated above.

---
* On leave of absence from SLAC (Stanford University), 1978/79.

Monte Carlo programs can be used in order to determine the energy-angle distribution of bremsstrahlung produced by electron beams incident on targets, and this information is of particular use to the accelerator designer, as well as the physicist working on dosimetry. Such calculations have been done using both EGS and ETRAN, and we will compare them with each other, as well as with experimental data, in the following sections.

Comparison of EGS with ETRAN

ETRAN is a Monte Carlo electron-photon transport code written by Berger and Seltzer (1970). A later version of this program, which contains a fairly general geometry package, is known as SANDYL (Colbert, 1973). ETRAN is complementary to the other shower codes that have been discussed in that it treats the low energy processes (down to 1 keV) in greater detail. Instead of using the Moliere (1947, 1948) multiple scattering formulation, ETRAN uses the Goudsmit-Saunderson (1940a,b) approach which avoids the small angle approximations. The code also treats fluorescence, the effect of atomic binding on the atomic electrons, and energy-loss straggling. Because of the greater detail taken at low energies, ETRAN might run significantly slower than EGS, but this remains to be verified[*]. The code was initially written for incident energies less than 100 MeV, although recent changes apparently have been made in the program to allow for shower simulation at somewhat higher energies (Berger, 1976). Nevertheless, the code appears to be accurate and available, and we will compare EGS with it.

The results of interest here are for 30 MeV electrons incident upon a thick (3.5 radiation length), semi-infinite tungsten target.

---
[*] Recently, Seltzer (1978) has indicated that EGS runs about nine times faster than ETRAN for the 1 GeV water shower benchmark problem described in Lecture 12, and both codes agree with the experimental data.

Photons that emanate from the tungsten were sorted by energy and by polar angle relative to the direction of the incident beam. When sorting by polar angle, the lateral position of the particle at the target surface was not considered in the analysis. In Fig.1, we plot the bremsstrahlung spectrum on an absolute scale (photons/(MeV-sr-e⁻)) for two angular bins. The agreement between EGS and ETRAN is quite good at the larger of the two angles. In the forward bin, EGS is somewhat higher than ETRAN. The angular distribution of bremsstrahlung is plotted in Fig. 2, again on an absolute basis (MeV/(sr-e⁻)), and the agreement between EGS and ETRAN is excellant everywhere except in the angular region past 90 degrees, where EGS gives results that are 50% lower than ETRAN.

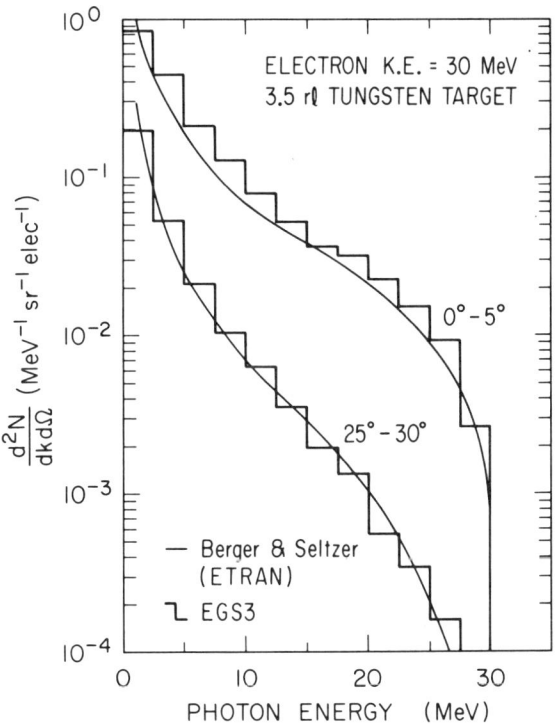

Fig. 1. Comparison of EGS and ETRAN Monte Carlo bremsstrahlung spectrum results for 30 MeV electrons incident on a thick, high-Z target.

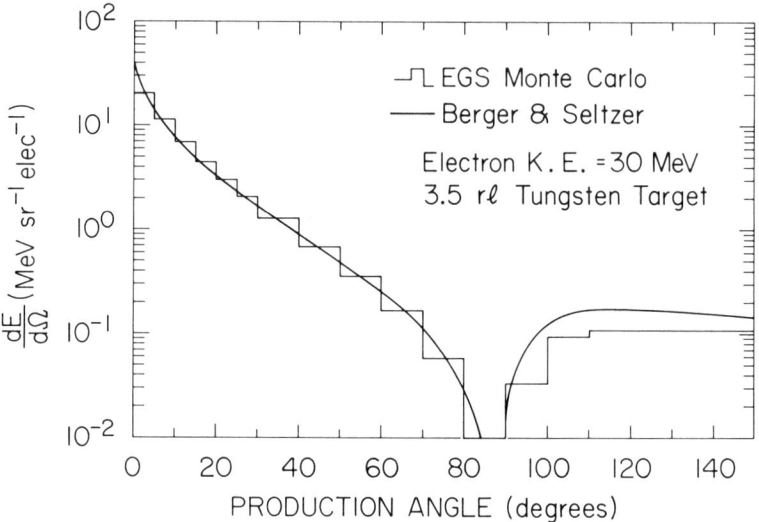

Fig. 2. Comparison of the bremsstrahlung angular distribution results obtained with EGS and ETRAN for 30 MeV electrons incident on a thick, high-Z target.

As stated above, ETRAN uses the Goudsmit-Saunderson multiple scattering formulism as compared to our Bethe-corrected Moliere approach (Bethe, 1953). Whether this is the reason for the difference in the backward direction results is subject for further investigation. The slight increase in the $0°$-$5°$ spectrum over that calculated by ETRAN will be discussed in the next section dealing with experimental results. All in all, the two programs give essentially the same thick-target, high-Z, low-energy bremsstrahlung results.

Comparison of EGS with Experimental Results

Although agreement with another Monte Carlo code is comforting, obviously it is desirable to compare with experimental data in this energy realm as well. A recent experiment by Levy et al (1974) has measured the thick target bremsstrahlung photon spectrum from a 25 MeV electron beam incident on a 6 radiation length lead target.

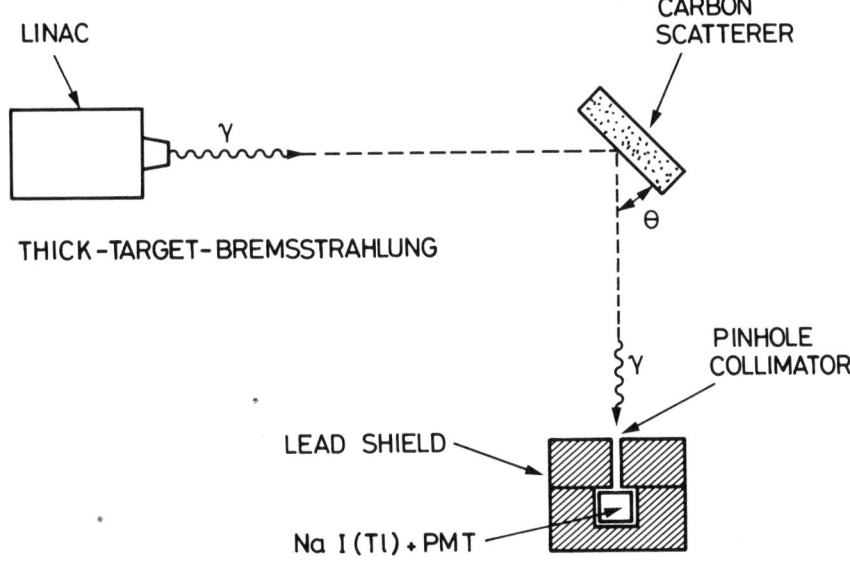

Fig. 3. Experimental arrangement by Levy et al (1974).

The experimental setup (Fig. 3) consisted of a well-shielded NaI(Tl) crystal that looked at the bremsstrahlung after it scattered from a carbon slab, the purpose of the latter being two-fold: 1) to shift the energy of the photons by Compton scattering; 2) to reduce the overall intensity of the bremsstrahlung. By using two-body kinematics and the Klein-Nishina (1929) cross section, the spectrum emanating from the lead target and scattering from the carbon into the angle defined by the pinhole in the detector was thus determined.

The results are shown in Fig. 4 along with the EGS spectrum produced from a 6 radiation length lead target into a 5° cone in the forward direction. The experimental data has been normalized to the EGS spectrum in the high energy region. The agreement is quite good except perhaps in the low energy portion where there seem to be fewer photons calculated than measured. Because Compton scattered electrons could also contribute to the signal processed by the detector, we offer this as a possible explanation for the slight discrepancy that is observed in Fig. 4. In general, the agreement is good.

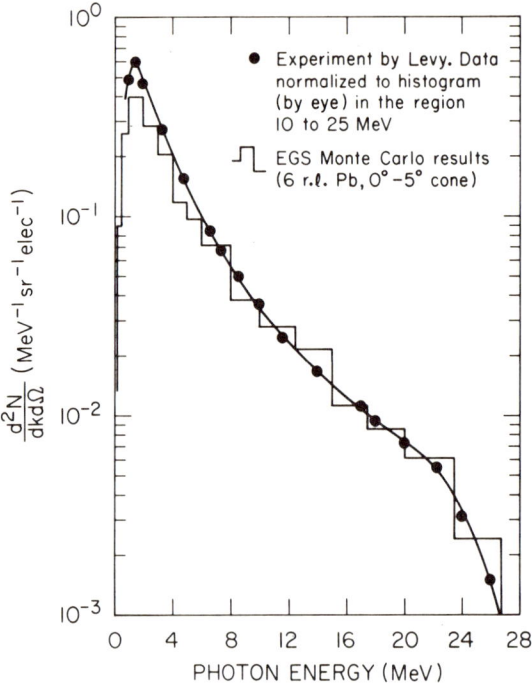

Fig. 4. Comparison of EGS Monte Carlo bremsstrahlung spectrum with experiment by Levy et al (1974).

Another bremsstrahlung spectrum measurement has been performed by O'Dell et al (1968), which is even more useful in that the spectrum was determined on an absolute basis in terms of the incident electron beam current. The experimental arrangement is shown in Fig. 5. Photons strike a deuterium oxide target to produce neutrons by deuteron photodisintegration. The reaction is two-body and a time-of-flight measurement of the neutron energy determines the energy of the photon making the interaction. The neutron "background" from competing photo-reactions with oxygen nuclei was determined by repeating the experiment using an ordinary water target. The neutron detector efficiency was known from previous work and the intensity of the electron beam was monitored. The bremsstrahlung spectrum was determined with the help of the reaction cross section.

APPLICATION OF EGS AND ETRAN 259

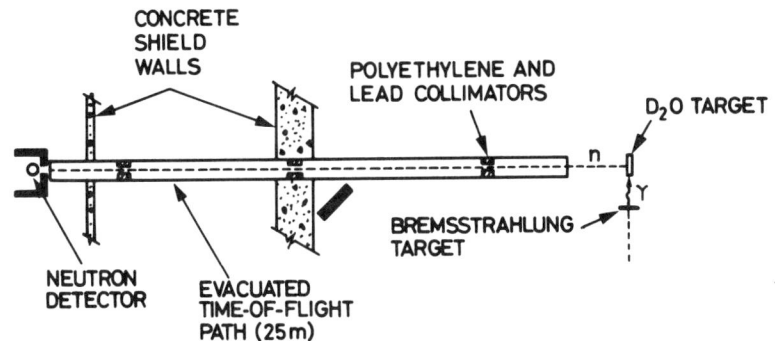

Fig. 5. Experimental arrangement by O'Dell et al (1968).

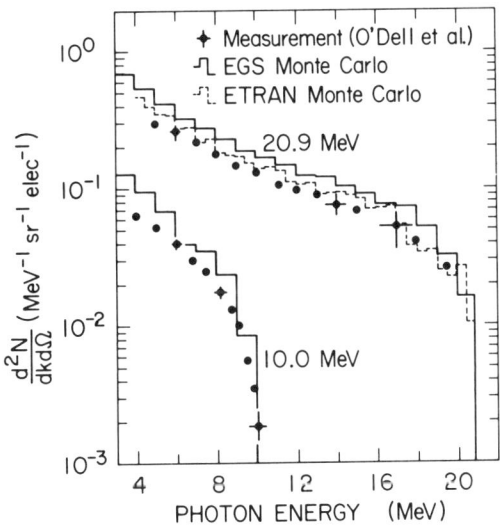

Fig. 6. Comparison of EGS and ETRAN Monte Carlo bremsstrahlung spectrum calculations with the experimental results of O'Dell et al (1968).

The experimental results are compared with Monte Carlo calculations in Fig. 6. The EGS histograms and experimental points shown are for 20.9 and 10.0 MeV electron beams incident on a tungsten-gold target (0.490 and 0.245 g/cm^2, respectively). Also shown are the Monte Carlo results (ETRAN) of Berger and Seltzer (1970). In this comparison one observes that the bremsstrahlung shape is excellant, but the absolute results are about 25% too high in the case of EGS. The experimental error bars are in the range of 10-20%, depending on energy, and ETRAN seems to be able to predict the absolute spectrum better than EGS does.

It was pointed out earlier that EGS, when compared to ETRAN, seems to overestimate the bremsstrahlung yield in the forward direction by as much as 15% (see Fig. 1). This seems to be true in the present comparison as well. Berger (1978) has suggested that both ETRAN and EGS might be using inaccurate bremsstrahlung cross sections for high-Z media and for energies less than 50 MeV, with the choice of cross section used in ETRAN being fortuitously better. The conclusion reached, however, is that EGS and ETRAN can accurately predict the overall shape of low energy bremsstrahlung from high-Z targets, but the absolute yield might not be correct to better than 10-25% using EGS.

PHOTON DOSIMETRY AT MEDIUM TO LOW ENERGIES

The EGS Code System is very useful, as we have seen, in solving electromagnetic cascade problems---particularly those at high energy, for complex geometries, when statistical fluctuations are needed, and when difficult scoring is required. One can use EGS to solve other types of problems, including the one we are about to describe, but we must continue to remind ourselves that we may have better, more efficient methods at our disposal for doing the job. In the case of photon dosimetry in the energy range 1-10 MeV (incident), and where the detector is small, one should first perform some "back-of-the-envelope" calculations to determine whether or not the Monte Carlo approach is even worthwhile. In this next example we will present a situation that is not particularly well suited for EGS. In fact, we are reminded of the statement made by Stevenson in the first lecture of this course: "...one should not use a sledge-hammer to crack a nut." EGS was used in this problem for a number of reasons that we will not go into in this lecture. However, there are two good reasons for discussing it here: 1) it demonstrates another feature that is available with EGS that could prove useful in problems where EGS is more suitable; 2) in a course like this one, we have an obligation to point out the disadvantages as well as the advantages of the various codes and techniques.

Consider the situation (Fig. 7) in which a silicon diode detector is inside a polystyrene phantom that is placed in a radiation field. We will choose a monoenergetic (10 MeV), isotropic source of photons for purposes of discussion without questioning the motivation for doing so. The drawing in Fig. 7 is certainly not to scale, but it is useful for creating the EGS User Code. The problem is to determine the absorbed dose in the silicon detector. The dimensions are shown in this figure and the reader should recognize two things at once:

1.) The silicon detector subtends a very small fraction of the total solid angle (2×10^{-8}) and, for that matter, so does the phantom itself (1.6×10^{-4}).

2.) The probability that a photon will interact within the phantom is not very large (about 8%), and the secondary charged particles must then traverse the silicon region in order to deposit energy by ionization. Therefore, the total probability for the deposition of energy by photons is rather small.

It is obvious from the first item that one will have to bias

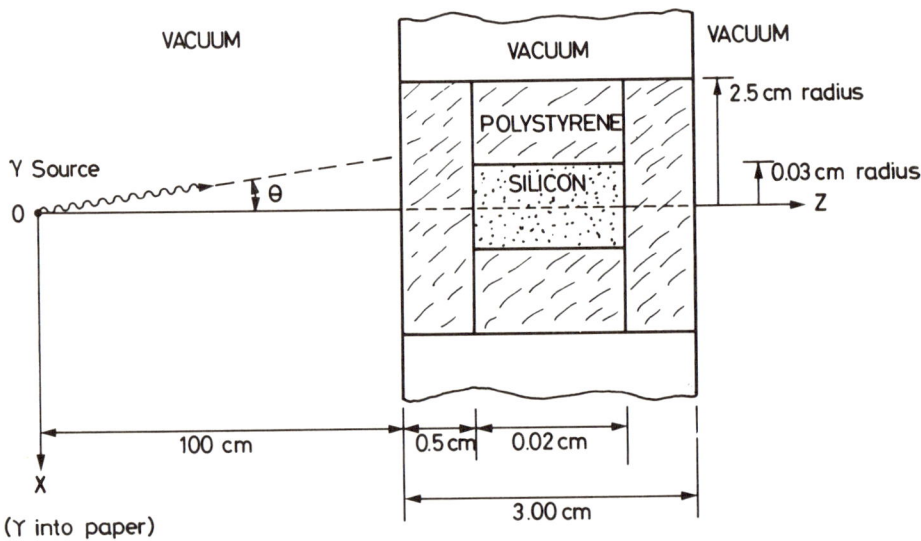

Fig. 7. Diagram showing a silicon diode detector inside a polystyrene phantom that is being irradiated by an isotropic source of monoenergetic photons.

the direction of the incident photons in order to force particles toward the detector. This can be done in a rather straight forward manner in EGS by preferential direction sampling in MAIN with a corresponding weight factor attached to each particle in the SHOWER-call. The method that we actually used, however, is even more direct and obvious. Because of symmetry, shower runs were made for fixed photon directions and the results were integrated and normalized to correspond to a point isotropic source. Figure 8 is a plot of the result of seven such runs. The absorbed dose times the sine of the incident photon angle is the quantity of interest for purposes of integration, and this is plotted as a function of the incident polar angle. Each run consisted of approximately 100 batches of 25000

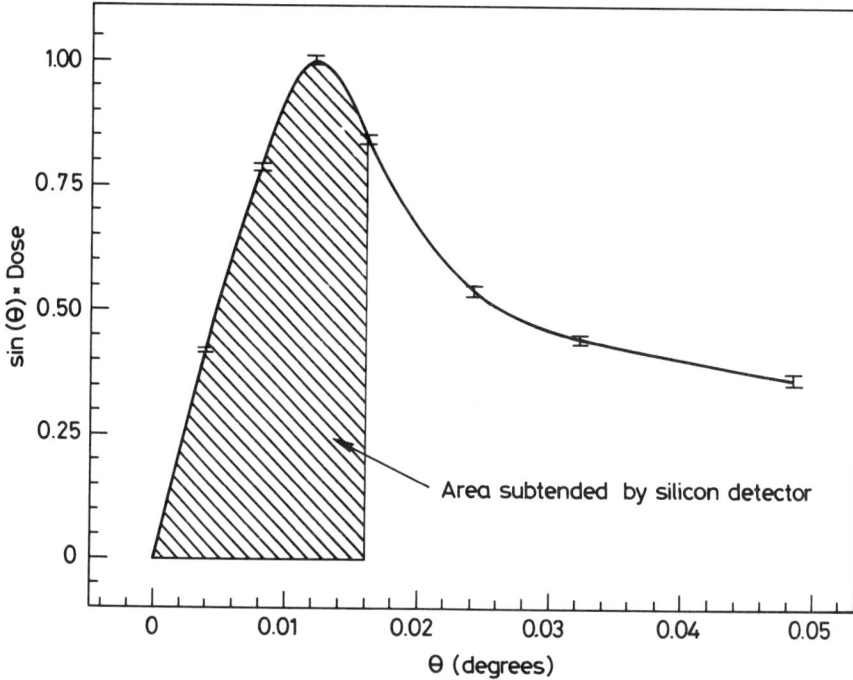

Fig. 8. The absorbed dose (times the sine of the incident angle) in a silicon diode surrounded by a polystyrene cylinder as a function of the incident photon angle.

photon histories. By breaking up the run into batches of 25000 cases, we were able to perform a standard statistical analysis as if each batch were an independent experiment. Even though the photons were "aimed" at the phantom, it obviously was not an easy task to obtain good statistical results (each run took about 60 minutes on the IBM-370/168), but then this was anticipated by our discussion above. Stated simply, the real-life probability of photons interacting in this phantom, such that secondary charged particles traverse the silicon detector and deposit energy by ionization, is small---at least in terms of the number of random numbers that can be generated and particles transported in EGS.

The size of the silicon region is indicated in Fig. 8. We see that a large fraction of the total dose comes from secondaries that are created in the polystyrene at radial distances greater than the radius of the silicon (i.e., the non-shaded area under the curve). One of the objectives of the calculation was to determine just how large to make the polystyrene cylinder in order to measure the effect of all of the secondaries created. Information of this nature would have been difficult to obtain had we used the biasing and weighting approach that we discussed above.

It is fairly easy for the user of EGS to distinguish between effects due to primary photons and those due to secondary ones. In this problem we also wanted to determine the kerma* caused by primary and secondary photons. Furthermore, we required that this information be treated on a region-by-region basis, and we were able to do all of this by means of the User Code interface of the EGS Code System. We will not be able to go into the programming details in this lecture, but the basic technique will be discussed in order to give the reader some idea as to how this is done.

As we have stated in Lecture 12, the AUSGAB subroutine that is required by EGS allows the user the opportunity to "score" many things. AUSGAB has one argument (IARG) that tells the user why the subroutine is currently being called. IARG can take on any one of 23 values, but it is the latter 18 that are of interest to us here. Table 3 of Lecture 12 gives us the situation corresponding to each

---

* The concept of kerma was first introduced by Roesch (1958) to describe the energy transfered from indirectly to directly ionizing radiations per unit mass of the material in which the interactions take place. A formal definition of this and related quantities, useful in radiologic physics, will be found in ICRU Report 19 (1971), and further explanations are given in the text by Kase and Nelson (1978).

of these 18 values of IARG. In scoring kerma one must sum the kinetic energy that is _initially_ released in the form of charged particles, so that one must score differently for each interaction as it occurs. In the case of pair production, for example, the user can write a section of code in AUSGAB that is executed if and only if IARG equals 15 (see Table 3 of Lecture 12), such as

    IF(IARG.EQ.15)   SUMKER=SUMKER + E(NP) - 2.*RM;

where SUMKER = sum of kerma energy (initialized to zero in MAIN),
      E(NP) = energy of photon,
      RM   = rest mass energy of electron.

This is a rather simple example and the actual User Code that was employed in the silicon diode problem was naturally much more complicated. The example, however, demonstrates the multiple-AUSGAB-call feature of EGS3. With it and the MORTRAN-macro facility, the user can do some extremely detailed transport analysis.

CONCLUDING REMARKS

A few examples showing the use of electromagnetic shower codes in treating problems in radiologic physics have been presented. For high-Z bremsstrahlung production, the codes seem to be quite useful. In fact, they probably could be used to improve on the design of clinical accelerators. Using analog Monte Carlo programs such as EGS, however, could result in a costly effort with marginal gain depending on the problem. The user of such codes should be aware of this before selecting such a code for his problem.

REFERENCES

Berger, M. J., and Seltzer, S. M., 1970, Bremsstrahlung and photo-
    neutrons from thick tungsten and tantalum targets, _Phys. Rev._,
    C2:621.

Berger, M. J., 1976, Private communication with W. R. Nelson.

Berger, M. J., 1978, Private communication with W. R. Nelson.

Bethe, H. A., 1953, Molière's theory of multiple scattering, Phys. Rev., 89:1256.

Colbert, H. M., 1973, SANDYL: A computer program for calculating combined photon-electron transport in complex systems, Sandia Laboratories (Livermore) Report Number SCL-DR-720109.

Goudsmit, S., and Saunderson, J. L., 1940a, Multiple scattering of electrons, Phys. Rev., 57:24.

Goudsmit, S., and Saunderson, J. L., 1940b, Multiple scattering of electrons II, Phys. Rev., 58:36.

ICRU Report 19, 1971, Radiation quantities and units, The International Commission on Radiation Units and Measurements.

Kase, K. R., and Nelson, W. R., 1978, Concepts of Radiation Dosimetry, Pergamon Press, Inc., New York.

Klein, O., and Nishina, Y., 1929, Über die Streuung von Strahlung durch freie Electronen nach der neuen relativistischen Quantum Dynamik von Dirac, Z. für Physik, 52:853.

Levy, L. B., Waggener, R. G., McDavid, W. D., and Payne, W. H., 1974, Experimental and calculated bremsstrahlung spectra from a 25-MeV linear accelerator and a 19-MeV betatron, Medical Physics, 1:62.

Molière, G. Z., 1947, Theorie der Streuung schneller geladener Teilchen I. Einzelstreuung am abgeschirmten Coulomb-Feld, Z. Naturforsch, 2a:133.

Molière, G. Z., 1948, Theorie der Streuung schneller geladner Teilchen II. Mehrfach- und Vielfachstreuung, Z. Naturforsch, 3a:78.

O'Dell, Jr., A. A., Sandifer, C. W., Knowlen, R. B., and George, W. D., 1968, Measurement of absolute thick-target bremsstrahlung spectra, Nucl. Instr. Meth., 61:340.

Roesch, W. C., 1958, Dose for nonelectronic equilibrium conditions, Radiation Research, 9:399.

Seltzer, S. M., 1978, Private communication with W. R. Nelson.

HADRONIC CASCADE PROGRAMS
AND THEIR APPLICATIONS

LECTURE 18: INTRODUCTION TO HADRONIC CASCADES

Tony W. Armstrong
Science Applications, Inc.
La Jolla, CA 92038
USA

INTRODUCTION

The purpose here is to give a very brief overview of the basic phenomenology associated with the interaction of hadrons with matter. Hadrons (i.e., strongly interacting particles) of interest here are neutrons, protons, charged pions, and charged kaons.

The prediction of hadronic cascades in thick targets is of interest in several areas - e.g., accelerator shielding (both for the "prompt" radiation, which exists while the beam is on, and the "delayed" radiation from activation, which exists after the beam has been turned off); cosmic ray effects (backgrounds, radiation hazards, etc.); and numerous other application areas (medical applications, dosimetry, etc.).

INTERACTION MECHANISMS

Atomic Interactions

The most important atomic interactions are usually ionization and excitation effects, and, occasionally, coulomb scattering. The physics of such atomic interactions for hadrons is well known,

Nuclear Interactions

Elastic hadron-nucleus interactions are characterized by very small energy losses and small angular deflections. Often they can be safely neglected in calculating the effects of hadronic cascades in thick targets. For problems where scattering by elastic collisions are not negligible (e.g., a narrow beam source with the effects of interest near the edge of the beam), semi-empirical formulas are available for the differential elastic scattering cross section (e.g., Ref. 1).

Nonelastic collisions are always important in predicting high-energy hadronic cascades, and it is this interaction mechanism that is least understood and has the largest physics uncertainties. Some qualitative features of nonelastic hadron-nucleus collisions are discussed later.

Decay

It is usually necessary to take into account hadron decay in predicting thick target cascades. Examples are:

$$\pi^\circ \to 2\gamma,$$
$$\pi^\pm \to \mu^\pm \to e^\pm,$$
$$K^\circ, K^\pm \to \mu^\pm, \pi^\pm, \pi^\circ.$$

The decay product energies and directions can be determined straightforwardly from decay kinematics.

Hadronic/Electromagnetic Cascade Couplings

It should be noted that often both hadronic and electromagnetic cascades are important in the same problem whether the source is hadrons or photons, electrons, and positrons. The "coupling" of these two types of cascades can occur as follows, where N denotes a target nucleus:

INTRODUCTION TO HADRONIC CASCADES

Starting with protons for example:

$$p + N \rightarrow \pi^\circ \rightarrow 2\gamma$$
$$\rightarrow \pi^\pm \rightarrow \mu^\pm \rightarrow e^\pm$$
$$\rightarrow K^\circ, K^\pm \rightarrow \mu^\pm, \pi^\pm, \pi^\circ \rightarrow e^\pm, \gamma$$

Starting with electrons for example:

$$e^- \rightarrow \gamma, e^+, e^-$$

$$\gamma + N \rightarrow \text{hadron cascade via photonuclear interactions}$$

Several types of photonulcear interactions can result in hadrons, namely: giant resonance (for $E_\gamma \sim 10\text{-}20$ MeV), photodisintegration (for $E_\gamma \sim 50\text{-}500$ MeV), and photo-pion absorption (for $E_\gamma \gtrsim 140$ MeV) (see, for example, Ref. 2).

HADRON-NUCLEUS COLLISIONS

Qualitative Features

Some qualitative features of hadron-nucleus collisions which often allow significant simplifications in predicting thick target cascades are:

(a) Above about 100 MeV, the total elastic and nonelastic cross sections are approximately constant.

(b) For the high-energy particles produced in hadron-nucleus collisions, the perpendicular momentum is much smaller than the parallel momentum (i.e., the secondary particle directions are approximately "straightahead" relative to the incident particle direction), and the perpendicular momentum component increases slowly with increasing projectile particle momentum.

(c) The multiplicity of the high-energy particles created increases slowly with increasing kinetic energy of the incident particle, roughly as $\ln(E_{cm})$.

(d) The residual excitation energy (and, hence, low-energy particle multiplicities) left after the high-energy particle production increases slowly with increasing bombarding particle energy.

Examples of these general characteristics can be found in various data reported in the literature (e.g., Refs. 3-6).

Total Cross Sections

Various semi-empirical expressions have been derived for the total elastic and nonelastic cross sections; for example, for protons (from Ref. 1)

$$\sigma_{el} \approx 6A \text{ mb},$$

$$\sigma_{ne} \approx aA^{2/3},$$

where $a \approx 30$ mb for A=1, and $a \approx 50$ mb for A>1. For neutrons and pions, the nonelastic cross section is smaller by about 10% and 20%, respectively. The mean-free-path for nonelastic interactions is not very dependent upon material ($\lambda_{ne} \propto A^{1/3}$), varying only from ~100 to ~200 g/cm² in going from aluminum to lead.

Differential Cross Sections

For nonelastic collisions, we would like to know, for example,

$$\frac{d\sigma_{ne}}{dE \, d\overline{\Omega}} (E_j, \overline{\Omega}_j | i, E_i, A_k),$$

the energy and direction of secondary particles of type j produced when a hadron of type i and energy E_i interacts with a target nucleus of mass number A_k. This information is presently not known accurately over the wide range of parameters of interest, and this is the major physics uncertainty in predicting hadronic cascades.

Basic approaches to estimating $d\sigma_{ne}/dE \, d\overline{\Omega}$ are to use available hadron-hadron data and extend to hadron-nucleus collisions by: (a) theoretical models - e.g., the intranuclear-cascade-evaporation model (Lecture 20), or (b) in conjunction with

INTRODUCTION TO HADRONIC CASCADES

accelerator and cosmic ray data, arrive at semi-empirical formulae (see Lecture 19).

For the elastic differential cross section, semi-empirical formulae are available (e.g., Ref. 1).

THICK-TARGET CASCADES

General Characteristics

The main feature of high-energy hadronic cascades generated in thick targets is an initial increase in particle intensity with depth, due to secondary particle production, followed by a gradual decline as the average energy of the cascade particles decreases, multiplicities decrease, and a greater fraction of individual particle energies are disipated by ionization losses. For example, Fig. 1 shows the depth-dependence of the energy deposition produced in a semi-infinite aluminum target by 250-MeV, 1-GeV, and 10-GeV proton beams, and for comparison, a 1-GeV electron beam. At 250 MeV, the proton range (~ 50 g/cm^2) is less than the mean-free-path for nuclear collision, so most of the protons stop without undergoing collision, and the deposition curve exhibits a "Bragg peak". For the 1 and 10 GeV cases, nuclear collision occurs before range stopping, and the deposition curves are characteristic of "well-developed" hadronic cascades.

Some Qualitative Features

Listed below are some qualitative features of hadronic cascades which sometime allow simplifications to be made in solving practical problems and which can be useful in the interpretation of thick-target cascade results.

Spatial Propagation

For source energies $\gtrsim 10$ GeV, the most penetrating component in the cascade is usually nucleons above a few hundred MeV. High-energy neutrons are particularly important since the protons get

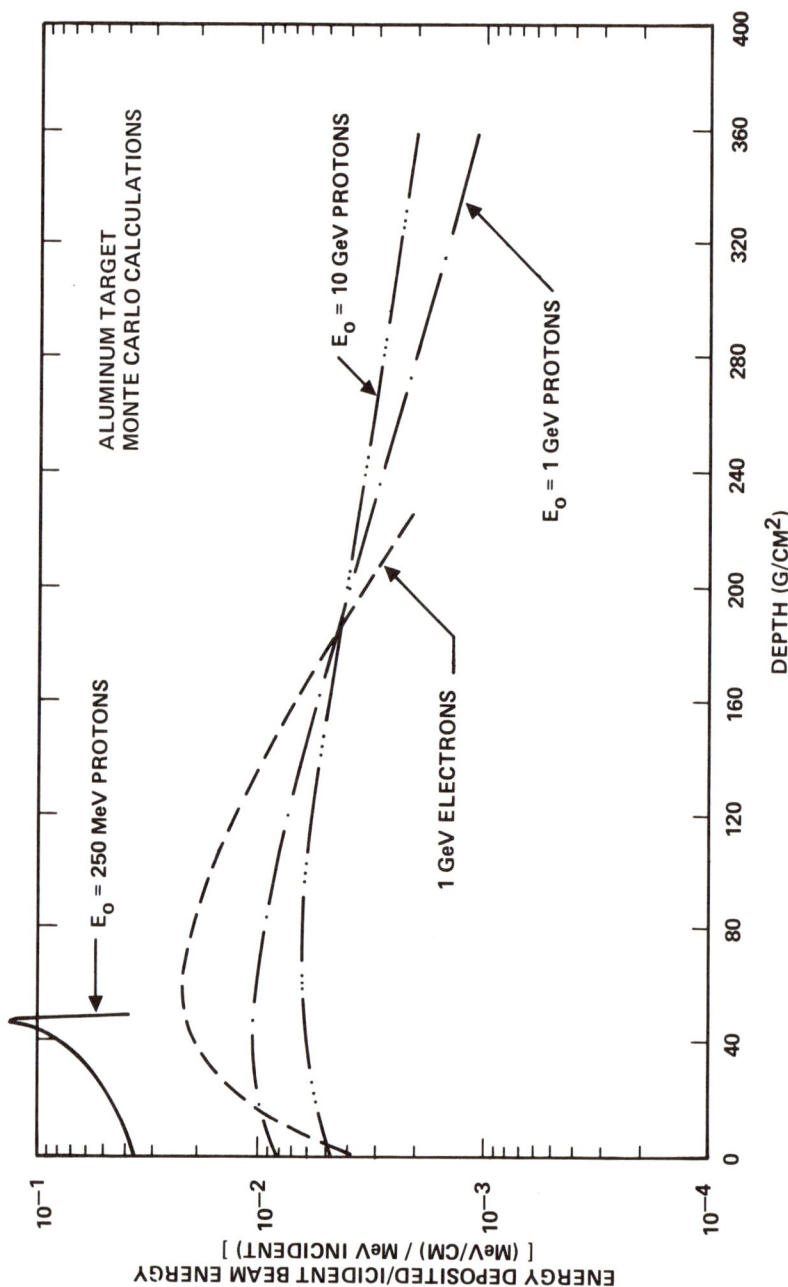

FIGURE 1: Energy deposition (per unit incident beam energy) produced in semi-infinite aluminum target by proton and electron beams.

degraded by ionization losses and lower energy neutrons have smaller mean-free-paths. For depths greater than a few mean-free-paths, the effective cascade attenuation length approaches a relatively constant value, and the shape of the neutron spectrum varies weakly with depth.

For E $\gtrsim$ 10 GeV, muon production becomes appreciable, and, since muons are degraded only by ionization losses, they become the dominate cascade component at very large depths for E $\gtrsim$ 100 GeV.

Pions

For energies $\gtrsim$ 10 GeV, pion production becomes important. The charged pions do not usually have a major effect on the spatial propagation of the cascade because of their soft production spectrum. However, they can be an important contributor to energy deposition and an important muon source. Neutral pions are not usually a major component for spatial propagation because the radiation length for the electromagnetic cascade initiated by the high-energy photons from $\pi°$ decay is typically much less than the hadronic cascade attenuation length. However, neutral pions become important contributors to energy deposition $\gtrsim$ 10 GeV, and provide the dominate deposition mechanism above 100 GeV.

Kaons

The kaon production in nuclear collisions is typically $\sim 1/10$ of the pion production. Where muons are of interest, the kaons can provide an important source because they decay faster than pions.

Muons

For energies above 10 GeV, muons may be abundant, particularly for those cases where (a) the target material is of low density (e.g., air) where pions and kaons predominately decay rather than interact, or (b) in dense materials at large depths after the hadronic cascade has diminished.

Small Angle Effects

There are two interaction mechanisms, multiple coulomb scattering and elastic nuclear scattering, which normally have little effect on the hadronic cascade development. This can usually be neglected because the angular deflections are small and there is very little energy loss. An example of an exception is where the source beam is very narrow and the interest is in effects near the beam axis where small angle scattering of the primary particles may govern the radial distribution of the radiation field.

High-Energy Fission

For very heavy target nuclei (e.g., uranium), the probability that a struck nucleus undergoes fission during de-excitation becomes significant. For example, for protons on uranium the fission/nonelastic (non-fissioning) cross section ratio is $\sigma_f/\sigma_{ne} \sim 0.8$ at a few hundred MeV. The particle production is roughly the same in fission, and non-fission effects (e.g., Ref. 8). However, the "local" energy deposition at the site of the interaction is quite different if fission occurs since the deposition from fission fragments is ≈ 180 MeV per fission, whereas the deposition from heavy, short-range secondary particles in nonelastic, non-fission events is ≈ 20-30 MeV per collision. For some applications (e.g., energy deposition studies) with target compositions containing fissionable nuclei, it is an adequate approximation to treat all nonelastic collisions as nonfissioning but make simple corrections to the local energy deposition for collisions with fissionable nuclei according to the fission probability.

Calculational Approaches

Several calculational methods that have been developed for predicting hadronic cascades are listed in Table 1. All employ

TABLE 1. Calculation Methods for Predicting Hadronic Cascades

	Method	Nuclear Model	Code Name/ Author	Reference
1.	Analytical	Empirical Formulae	LUIN/O'Brien	9
2.	Monte Carlo	"	CASIM/Van Ginneken	10
3.	Monte Carlo	"	FLUKA, KASPRO/Ranft	1
4.	Monte Carlo	Intranuclear Cascade	"Dubna Code"/ Barashenkov, et al.	11
5.	Monte Carlo	"	HETC/Armstrong, et al.	12

three-dimensional Monte Carlo methods except for the work of O'Brien (e.g., Ref. 7), which uses analytical methods and is one-dimensional. O'Brien has been very successful in predicting the general cascade characteristics based on minimal information on the details of hadron-nucleus collisions. He finds that the general behavior (in one spatial dimension) is rather accurately predicted if good estimates of total cross sections and partial inelasticities (i.e., the fraction of the incident particle energy which goes into production of various particle types) are incorporated.

Three of the hadronic cascade computer codes listed in Table 1 will be discussed in subsequent lectures: CASIM (Lecture 21), FLUKA, KASPRO (Lecture 22), and HETC (Lecture 23).

REFERENCES

1. J. Ranft, "Estimation of Radiation Problems Around High Energy Accelerators Using Calculations of the Hadronic Cascade in Matter", Particle Accelerators $\underline{3}$, 129 (1972).

2. R. G. Jaeger (Ed.), Engineering Compendium on Radiation Shielding, Vol. III, Springer-Verlag, New York, 1970.

3. V. A. Konshin and E. S. Matusevich, "Characteristics of the Interaction of High-Energy Nucleons with Nuclei", At. Energy Rev. $\underline{6}$, 3 (1968).

4. Hugo W. Bertini, "Spallation Reactions: Calculations", in Spallation Reactions and Their Applications (B.S.P. Shen and M. Merker, Editors), D. Reidel Publ. Co., Boston, 1976.

5. W. Wade Patterson and Ralph H. Thomas, Accelerator Health Physics, Academic Press, New York, 1973.

6. "Basic Aspects of High Energy Particle Interactions and Radiation Dosimetry", International Commission on Radiation Units and Measurements Report ICRU 28 (1978).

7. K. O'Brien, "Cosmic Ray Propagation in the Atmosphere", Nuovo Cimento $\underline{3A}$, 521 (1971).

8. R. L. Hahn and H. W. Bertini, "Calculations of Spallation-Fission Competition in the Reactions of Protons with Heavy Elements at Energies <3 GeV", Phys. Rev. $\underline{C6}$, 660 (1972).

9. Keran O'Brien, "LUIN, A Code for the Calculation of Cosmic Ray Propagation in the Atmosphere", Health and Safety Lab. Report HASL-275 (1973).

10. A. Van Ginneken, "CASIM, Program to Simulate Hadronic Cascades in Bulk Matter", Fermilab Report FN-272 (1975).

11. V. S. Barashenkov, N. M. Sobolevskii, and V. D. Toneev, "The Penetration of Beams of High-Energy Particles Through Thick Layers of Material", Atomnaya Energiya $\underline{32}$, 217 (1972).

12. T. W. Armstrong and K. C. Chandler, "HETC - A High Energy Transport Code", Nucl. Sci. and Engr. $\underline{49}$, 110 (1972).

LECTURE 19: PARTICLE PRODUCTION MODELS, SAMPLING HIGH-ENERGY MULTIPARTICLE EVENTS FROM INCLUSIVE SINGLE-PARTICLE DISTRIBUTIONS

J. Ranft[*],

Sektion Physik, Karl-Marx-Universität,
Leipzig,
German Democratic Republic

ABSTRACT

The experimental or theoretical knowledge about particle production in high-energy hadron-hadron collisions is limited to inclusive single- (and perhaps two-) particle cross-sections. In hadronic cascade calculations, complete realistic multiparticle production events are sampled which should agree with these single-particle distributions, fulfil energy-momentum conservation and other conservation laws, have a multiplicity distribution which agrees with data, etc. Different strategies for sampling these events are discussed. The present experimental knowledge about particle production in high-energy (10 GeV - 2 TeV) hadron-hadron collisions as well as theoretical models are also discussed. The particle production in hadron-nucleus collisions is extremely important for all hadron cascade calculations, but experimental data in the GeV to TeV energy region are rather incomplete. An extension of the computer program SPUKST is described, which predicts these particle spectra from hadron-hadron data using a recently proposed model for particle production on nuclei.

---

[*] Work supported in part by the HS Division at CERN.

INTRODUCTION: HOW TO SAMPLE HIGH-ENERGY MULTIPARTICLE PRODUCTION EVENTS IN HADRON CASCADE CALCULATIONS

In this paper we will discuss two topics:

i) Methods for sampling inelastic events in high-energy hadron cascade calculations.

ii) General properties of particle production in hadron-hadron and hadron-nucleus collisions in the GeV and TeV energy region.

What is special about high-energy hadron interactions? The interactions of electrons and photons in an electromagnetic cascade or the interactions of neutrons in neutron transport calculations are known and can be described by a limited number of interaction cross-sections, mostly for processes with two or three particles in the final state. Most of the corresponding cross-sections are well known either theoretically or experimentally. The situation regarding hadron interactions in the GeV and TeV region is quite different. The typical multiplicity of secondary particles is around 10. Even if we consider only some ground-state hadrons like p, n, $\pi^{\pm}$, π^0, $K^{\pm}$, $\bar{p}$, the number of possible reaction channels is a very large number. It is impossible to calculate (even if a strong interaction theory would exist) or to measure such a large number of cross-sections.

These cross-sections, where a definite final state is produced, for instance (see Fig. 1)

$$p + p \rightarrow p + n + \bar{p} + n + 4\pi^+ + 2\pi^- + 2\pi^0 \tag{1}$$

$$p + p \rightarrow n_1 c_1 + n_2 c_2 + \ldots + n_i c_i , \tag{2}$$

are referred to as exclusive cross-sections. Only exclusive cross-sections with small numbers of secondaries have ever been measured. If only one or some of the secondary particles are measured we speak about inclusive single-particle cross-sections or inclusive n-particle cross-sections; for instance (see Fig. 1),

$$p + p \rightarrow \pi^+ + \text{anything else} \tag{3}$$

$$p + p \rightarrow c_1 + \ldots + c_m + \text{anything else.} \tag{4}$$

PARTICLE PRODUCTION MODELS

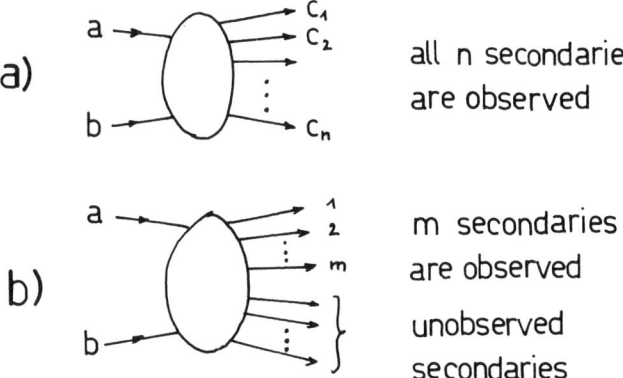

Fig. 1 a) In an exclusive reaction all n-particles in the final state are observed.
b) In an inclusive reaction only m-final-state particles are observed.

In high-energy experiments, only inclusive single- and two-particle cross-sections have been measured systematically. The particle production process can be described completely by all exclusive or all inclusive cross-sections. Let us consider a few details of the theoretical description of exclusive and inclusive cross-sections in order to understand the connection between both descriptions.

In the exclusive reaction (2) the $c_1, \ldots, c_i$ are particles with the multiplicities n_i. The total secondary multiplicity is

$$n = n_1 + n_2 + \ldots + n_i . \qquad (5)$$

To keep the expressions for the cross-sections simple, we restrict ourselves to only one kind of secondary particle. The invariant differential exclusive n-particle production cross-section is defined as

$$g_n(s, p_1, p_2, \ldots, p_n) = \frac{d^{(3n)}\sigma_n^{ex}(s)}{dp_1 dp_2 \ldots dp_n} \tag{6}$$

$$= (2\pi)^{4-3n} \frac{1}{2}\left[\lambda(s, m_a^2, m_b^2)\right]^{-\frac{1}{2}} \times$$

$$\times \frac{1}{n!} \delta^4\left(p_a + p_b - \sum_{i=1}^n p_i\right) \times$$

$$\times M_n(p_a, p_b, p_1, \ldots, p_n) ,$$

where $dp = d^3p/2p_0$, s is the square of the total c.m.s. collision energy $\left[s = (p_a + p_b)^2\right]$, and M_n is the invariant matrix element of the reaction $a + b \to 1 + \ldots + n$. We define the total n-particle production cross-section,

$$\sigma_n(s) = \int g_n(s, p_1, \ldots, p_n) dp_1, \ldots, dp_n , \tag{7}$$

and the total inelastic cross-section,

$$\sigma_{inel}(s) = \sum_n \sigma_n(s) . \tag{8}$$

The great number of variables necessary to describe the exclusive differential cross-sections make experimental measurements extremely difficult.

The possibility of performing experiments is easier if inclusive cross-sections are considered. The inclusive cross-section is obtained from the knowledge of all exclusive cross-sections. We consider the inclusive m-particle cross-section:

$$F_m(s, p_1, \ldots, p_m) = \sigma_{inel} \, f_m(s, p_1, \ldots, p_m)$$

$$= \frac{d^{(3m)}\sigma_m^{in}(s)}{dp_1 dp_2 \ldots dp_m} = \sum_{k=0}^{\infty} \frac{(m+k)!}{k!} \int dq_1, \ldots, dq_k \times$$

$$\times g_{(m+k)}(p_1, \ldots, p_m, q_1, \ldots, q_k, s) .$$

$$\tag{9}$$

This expression can be inverted to obtain the exclusive cross-section in terms of all inclusive cross-sections:

$$g_n(p_1, p_2, \ldots, p_n, s) = \sum_{k=0}^{\infty} \frac{(-1)^k}{k!} \int dq_1, \ldots, dq_k \times$$

$$\times F_{(n+k)}(p_1, \ldots, p_n, q_1, \ldots, q_k, s) \ . \tag{10}$$

If we integrate the inclusive cross-sections (9) over all momenta, we obtain multiplicity moments

$$\phi_m = \int f_m(p_1, p_2, \ldots, p_m, s) \, dp_1, \ldots, dp_m$$

$$= \sum_{n=2}^{\infty} n(n-1) \ldots (n-m+1) \frac{\sigma_n}{\sigma_{inel}} \tag{11}$$

$$= \langle n(n-1) \ldots (n-m+1) \rangle \ .$$

Energy momentum conservation is explicitly introduced into the exclusive cross-sections by the δ-function in (6). In the inclusive case this leads to sum rules. For instance, if p_x^μ is one component of the four-momentum of particle x, we obtain for the inclusive single-particle cross-section

$$\sum_x \int dp_x \, p_x^\mu \frac{d^3 \sigma_1^{in}}{dp_x} = (p_a + p_b)^\mu \, \sigma_{inel} \ , \tag{12}$$

and for the n + 1-particle cross-section

$$\sum_x \int dp_x \, \frac{d^{(n+1)\cdot 3} \sigma_{n+1}^{in}}{dp_1 dp_2 \ldots dp_n dp_x} \, p_x^\mu = (p_a + p_b - p_1 - p_2 \ldots p_n)^\mu \, \frac{d^{(3n)} \sigma_n^{in}}{dp_1 \ldots dp_n} \ . \tag{13}$$

The sum runs over all x-kinds of secondaries.

Let us return to the problem of sampling events for hadron cascade calculations. Clearly what would be needed is i) to select one of the exclusive reaction channels, and ii) to sample momenta of all secondaries according to the corresponding exclusive production cross-section.

Such a program is, however, extremely unpractical owing to the large number of exclusive channels and the difficulties of measuring or calculating the exclusive cross-sections. Experimental data exist almost only for inclusive single- and two-particle distributions. Therefore we cannot use Eq. (10) to generate the exclusive cross-sections. The only practical way remaining is to sample exclusive multiparticle production events directly from inclusive cross-sections. In the next section we will discuss strategies for doing this. Following that, we describe the particle production in hadron-hadron collisions, and we describe general properties and one specific model for particle production in hadron-nucleus collisions.

SAMPLING HIGH-ENERGY MULTIPARTICLE EVENTS FROM INCLUSIVE SINGLE-PARTICLE DISTRIBUTIONS

We will list four methods:

- The method used in the hadron cascade program FLUKA and related programs, which is rather efficient but uses approximations.

- Two "exact" methods. In the first of these the well-known techniques for sampling events from phase space are used to sample exclusive events from inclusive distributions. In the second, which so far is only practical at relatively low energies, no inclusive cross-sections are used. The inelastic reactions are described by quasi-two-particle production of resonances which subsequently decay into stable hadrons.

- A method used in CASIM or KASPRO where no complete exclusive events are created and where general properties of the interactions, such as energy-momentum conservation, are only fulfilled on the average.

The Method Used in FLUKA and Related Programs[1-3]

The secondary particle production is described by inclusive single-particle distributions defined in the c.m.s.:

PARTICLE PRODUCTION MODELS 285

$$\frac{d^2 N(a + b \to i)}{dp_\| dp_\perp} = \frac{d^2\sigma^{in}}{\sigma_{inel} dp_\| dp_\perp} = f_i(p_\|, p_\perp, E_{cm}), \quad (14)$$

where E_{cm} is the total energy in a hadron-hadron centre-of-mass frame, $p_\|$ and $p_\perp$ are the longitudinal and transverse momenta of the produced particle i, and $f_i(p_\|, p_\perp, E_{cm})$ is an empirical expression with free parameters fitted to experimental data. The reason for defining those distributions in the c.m.s. is the simple and well-known form of the energy dependence of the inclusive single-particle distributions (Feynman scaling). The reason for using $p_\|$ and $p_\perp$ as variables is the approximate factorization of the $p_\|$ and $p_\perp$ dependence. Random longitudinal and transverse momenta according to Eq. (14) can be sampled using factorization property and standard random selection techniques. The laboratory frame particle momenta and production angles are obtained from $p_\|$ and $p_\perp$ by a Lorentz transformation.

The average number of secondaries produced, $\langle n_i \rangle$, is obtained from Eq. (14):

$$\langle n_i \rangle = \int f_i(p_\|, p_\perp, E_{cm}) \, dp_\| \, dp_\perp . \quad (15)$$

In the calculation, however, we use the inelasticities defined as

$$K_{ij} = \frac{1}{E_i} \int E_j \frac{d^2 N}{dp d\Omega} \, dp \, d\Omega , \quad (16)$$

where K_{ij} gives the fraction of the incident energy in the laboratory frame E_i carried away by secondary particles of the kind j. The sum rule [equivalent to Eq. (12)]

$$\sum_j K_{ij} = 1 , \quad (17)$$

follows from energy conservation.

To sample complete events we start from the requirements: i) that energy is conserved in each collision; ii) that, averaged over many collisions, the inelasticities for various secondary particles must agree with the prescribed values; iii) that all secondary particles are sampled from the distribution functions and that each sampled particle must be used in the calculation in order to avoid bias in the distributions.

Sampling one event, we randomly choose at the beginning the order in which the different kinds of secondary particles are to be selected. The energy conservation counter, C, referring to the event, is set to zero. Other counters are used to keep the inelasticities K_{ji} near to the values required when averaging over many collisions. The counters D_{jik} and G_{jik} refer to a collision where a primary particle j and secondary particle i are involved with the momentum of the primary particle p_j inside the momentum bin k. At the beginning of each event all K_{ji} are added to the counters D_{jik}. Each time a particle is sampled, its energy fraction E_i/E_j is added to C and to G_{jik}. The production of a particle is only allowed as long as the resulting C remains smaller than 1 and the resulting D_{jik} remains larger than G_{jik}. If $D_{jik} < G_{jik}$ and/or $C > 1$, the corresponding particle cannot be created in that event and it has to be subtracted from G_{jik} and from C. The momentum p_i and angles θ_i of this particle are stored in arrays P_{jik} and θ_{jik} to be used in the next event of this kind. No new particle of this kind (jik) is sampled as long as this particle has not been used. After storing this particle the program continues with the selection of secondary particles of the next kind.

A complication arises when selecting the last kind of secondaries if $D_{jik} < G_{jik}$ and/or $C > 1$. In this situation a particle of just the required kinetic energy E_i and momentum p_i, necessary to fulfil energy conservation (C = 1) has to be invented, with only the production angle being sampled from the distribution. By inventing this particle, we violate the momentum distribution and introduce bias. This bias is subsequently removed by the following procedure. Every time a particle is invented we increase the element I_{jiks} of an array by unity. The index s defines the secondary momentum bin to which p_i belongs. Each time a secondary particle momemtum p_i is sampled, the program checks whether the corresponding I_{jiks} is different from zero. If so, the momentum p_i is thrown away and I_{jiks} is decreased correspondingly.

The procedure described works rather efficiently; however, one could ask, "Are there essential features of multiparticle production which are missed by using only inclusive single-particle distributions as input?" This problem can be studied, at least partly, by comparing the generated inclusive two-particle distributions with experimental data. It is useful to decompose the inclusive two-particle distribution as follows:

$$f_2(p_1, p_2, E_{cm}) = f_1(p_1, E_{cm})f_1(p_2, E_{cm}) + C(p_1, p_2, E_{cm}) \,. \quad (18)$$

The correlation function C is well known experimentally and can be understood from two sources[4]: i) kinematical correlations which follow from energy momentum conservation; ii) dynamical short-range correlations, which are well understood to be due to the production of clusters or resonances.

It turns out that the kinematical correlations are well reproduced by the particle production technique described. The short-range correlations could be incorporated by first sampling the cluster production with subsequent decay. However, it is believed that the short-range correlations of the produced particles are not essential for most applications of hadron cascade calculations.

Exact Methods

A method using techniques to select events from phase space. We have seen above that the problem of selecting events from inclusive distributions is the difficulty of ensuring energy-momentum conservation. To describe an n-body event we need 3n variables; the three-momentum components for each particle. Owing to the energy-momentum conservation constraints, only 3n-4 of these variables can be chosen independently. Energy-momentum conservation constrains physical events to lie on 3n-4-dimensional hyper-surfaces of the

3n-dimensional momentum space. If we first select the particles independently in the 3n-dimensional momentum space, we can subsequently ensure energy momentum conservation by projecting the event on this 3n-4-dimensional hyper-surface. The use of such a method for the Monte Carlo calculation of phase-space integrals was described by Chen and Peierls[5]. To explain the method, we first describe the calculation of an n-dimensional integral

$$R(a) = \int_{R_x} f(x) G(x, a) \, d^n x , \tag{19}$$

where R_x is a region in n-dimensional space and $G(x, a)$ is a condition expressed as a δ-function

$$G(x, a) = \delta[\phi(x) - a] . \tag{20}$$

We transform the integral into the region R_y using the transformation of variables

$$x = g(y, b) , \tag{21}$$

where b is a parameter of the transformation, so that

$$R(a) = \int_{R_y} f[g(y, b)] \delta\{\phi[g(y, b)] - a\} J(y, b) \, d^n y , \tag{22}$$

where $J(y, b)$ is the Jacobian.

Owing to the δ-function in Eqs. (19) and (22), we cannot calculate the integral by a Monte Carlo method. In order to remove the δ-function, we introduce in Eq. (22) an integration over the transformation parameter b using the normalized weight function

$$\int_\beta \sigma(b) \, db = 1 . \tag{23}$$

We obtain from Eq. (22)

$$R(a) = \int_\beta \int_{R_y} f[g(y, b)] \delta\{\phi[g(y, b)] - a\} \times J(y, b) \sigma(b) \, d^n y \, db . \tag{24}$$

We perform the b-integral with the help of the δ-function and obtain

$$R(a) = \int_{R_y} f[g(y, b_0)] \, \sigma(b_0) J(y, b_0) \times \left[\frac{\partial \phi[g(y, b_0)]}{\partial b}\right]^{-1} d^n y \, . \tag{25}$$

This is an n-dimensional integral which corresponds to Eq. (19), but no longer contains the δ-function and can be calculated in a straightforward manner using a Monte Carlo method.

The sampling of multiparticle events corresponds to the Monte Carlo calculation of a phase-space integral weighted with the square of the matrix element. We consider the n-particle phase-space integral

$$R(P) = \int_{\Omega^n} \prod_{i=1}^{n} \frac{d^3 p_i}{2 p_i^0} \, \delta\!\left(P^0 - \sum_i p_i^0\right) \delta\!\left(\vec{P} - \sum_i \vec{p}_i\right) |M|^2 \, . \tag{26}$$

One n-particle event is described by 3n-momentum components subject to four energy-momentum constraints. The method consists in choosing one event in R^{3n} and projecting with the help of a transformation on the proper R^{3n-4}. We introduce these transformations in the form

$$\vec{p}_i = \eta(\vec{q}_i + \vec{\lambda}) \, , \quad p_i^0 = [\eta^2(\vec{q}_i + \vec{\lambda})^2 + m_i^2]^{1/2} \, , \tag{27}$$

where η and $\vec{\lambda}$ are parameters of the transformation which are determined from energy and momentum conservation:

$$\vec{P} = \sum_i \vec{p}_i = \eta \sum_i \vec{q}_i + \eta n \vec{\lambda} \, , \tag{28}$$

$$P^0 = \sum_i [\eta^2(\vec{q}_i + \vec{\lambda})^2 + m_i^2]^{1/2} \, . \tag{29}$$

From Eq. (28) we obtain

$$\vec{\lambda}_0 = \frac{\vec{P}}{n \eta_0} - \frac{1}{n} \sum_i \vec{q}_i \, . \tag{30}$$

Inserting into Eq. (29) we obtain an equation which can be used to determine the root η_0:

$$\sum_i \left[(n\vec{q}_i + \frac{\vec{P}}{n} - \frac{n}{n} \sum \vec{q}_i)^2 + m_i^2 \right]^{\frac{1}{2}} = E ; \qquad (31)$$

after this, Eq. (30) gives the solution for $\vec{\lambda}_0$. We transform the phase-space integral (26) to the new variables $\vec{q}_i$, and at the same time transform the energy momentum δ-functions, so that

$$R(P) = \int_{\Omega^n} \prod_{i=1}^{n} \left(\frac{n^3 d^3 \vec{q}_i}{2[n^2(\vec{q}_i + \vec{\lambda})^2 + m_i^2]^{\frac{1}{2}}} \right) \times$$

$$\times \frac{|M|^2 \delta(\lambda - \lambda_0)\delta(n - n_0)}{n^3 n^3 \sum_i \frac{n(\vec{q}_i + \vec{\lambda}) [\vec{q}_i + \vec{\lambda} - (\vec{P}/nn)]}{[n^2(\vec{q}_i + \vec{\lambda})^2 + m_i^2]^{\frac{1}{2}}}} . \qquad (32)$$

As already discussed [Eqs. (23) and (25)], we introduce the integral over the transformation parameters η and $\vec{\lambda}$ using a normalized distribution function:

$$\int \rho(\eta) \tau(\vec{\lambda}_\perp) w(\lambda_\parallel) \, d\eta \, d\vec{\lambda}_\perp \, d\lambda_\parallel = 1 , \qquad (33)$$

where $\vec{\lambda}_\perp$ and $\lambda_\parallel$ are the transverse and longitudinal components of $\vec{\lambda}$. The η and $\vec{\lambda}$ integrations can be performed using the δ-functions, and we finally obtain

$$R(P) = \int_{\Omega^n} \prod_i \left(\frac{n_0^3 d^3 \vec{q}_i}{2[n_0^2(\vec{q}_i + \vec{\lambda}_0)^2 + m_i^2]^{\frac{1}{2}}} \right) \times$$

$$\times \frac{|M|^2 \delta(n_0) \tau(\vec{\lambda}_{\perp 0}) w(\lambda_{\parallel 0})}{n_0^3 n^3 \sum_i \frac{n_0(\vec{q}_i + \vec{\lambda}_0) [\vec{q}_i + \vec{\lambda}_0 - (\vec{P}/nn)]}{[n_0^2(\vec{q}_i + \vec{\lambda}_0)^2 + m_i^2]^{\frac{1}{2}}}} . \qquad (34)$$

This n-dimensional integral no longer contains δ-functions. If we use a Monte Carlo method to complete this integral -- preferably using a method with importance sampling -- each sampled term corresponds to one generated multiparticle event. First, we select the n-momenta $\vec{q}_i$ without considering energy-momentum conservations. The transformed momenta $\vec{p}_i$ describe the event with energy-momentum conservation. Note that these events will in general have weights W different from unity.

Based on this method, we have developed a program at Leipzig[6] to sample events from inclusive distributions. The events with the momenta $\vec{q}_i$ are sampled as in FLUKA, and the selection of secondaries is stopped as soon as the sum of the secondary energies exceeds the primary energy. The corrected momenta, $\vec{p}_i$, and weights of the event W, are then calculated. Besides energy-momentum conservation, the conservation of additive quantum numbers, such as charge, baryon number, and strangeness, is also considered.

Generating inelastic events via quasi-two-body production of hadron resonances which decay subsequently into two or three particles or resonances. At low energies near to the threshold of particle production, the methods discussed above, starting from inclusive distributions, will not give the best results. At energies below 3 to 5 GeV, inclusive distributions are not well known. Furthermore, the number of exclusive channels is not too large and quite a few of the exclusive cross-sections have been measured. The correct Monte Carlo treatment in this energy region is not very important if radiation effects induced by particles with tens or hundreds of GeV are the object of the study. However, if the interaction of particles in this energy range is the reason for the cascade calculation, it is preferable to use all available exclusive data to construct the inelastic events. While this approach has been used extensively elsewhere[7], we have only recently started to develop such a program at Leipzig[8].

We are interested in the interactions of $\pi^{\pm}$, $K^{\pm}$, p, n, and $\bar{p}$ with target protons and neutrons. Extensive compilations of cross-sections have been compiled in the last decade[9]. Studying these cross-sections in the energy range under consideration, one finds that some 15-30 exclusive channels dominate the cross-section in each channel. Many of these channels have been studied in detail and are known to be dominated by quasi-two-body production of excited mesons and baryons which subsequently decay. For each reaction channel we were able to isolate some 10 to 25 quasi-two-body reactions and compile their energy-dependent cross-sections. All dominant exclusive reaction channels are well represented by means of the subsequent two- and three-body decay of these hadron resonances, according to the table of particle properties[10].

Methods with Variance Reduction and Energy-Momentum Conservation in the Average over many Collision Events

For some applications of hadron cascade calculations, detailed energy-momentum conservation in each single event is not essential if energy is conserved in the average over many collisions. An example is the calculation of the particle flux from an extended target done by the program KASPRO[11]. This is an average quantity which is not influenced by fluctuations of single collisions. Otherwise, in order to sample the particle flux efficiently for momenta and secondary particles which occur very infrequently, it is advisable to use methods with variance reduction. Such methods are used in CASIM[12] and also in KASPRO[11]. Next we describe the method employed in KASPRO.

In KASPRO and related programs the production of particles of kind i (i = p, n, π^0, $\pi^\pm$, $K^\pm$, $\bar{p}$) in collisions of incoming particles of kind j on nuclei is described by normalized inclusive single-particle distributions $\hat{f}$:

$$\frac{d^3N}{d^3p} = f_{ij}(\vec{p}_i) , \qquad (35)$$

$$\hat{f}_{ij}(\vec{p}) = \frac{f_{ij}(\vec{p}_i)}{\int f_{ij}(\vec{p}_i) \, d^3p_i} . \qquad (36)$$

It has been noted that the inelasticities

$$K_{ij} = \frac{1}{E_{tot}} \int E_i \, f_{ij}(\vec{p}_i) \, d^3p_i \qquad (37)$$

are, in a wide range, rather independent of the total collision energy E_{tot}. If we use the inelasticities and the average energies $\langle E_{ij} \rangle$ of particles i according to the inclusive distributions

$$\langle E_{ij} \rangle = K_{ij} \cdot E_{tot} = \int E_i \, \hat{f}_{ij}(\vec{p}_i) \, d^3p_i \,, \qquad (38)$$

we obtain the average multiplicities for the particles i

$$n_{ij} = \frac{K_{ij} \cdot E_{tot}}{\langle E_{ij} \rangle} = \frac{K_{ij}}{k_{ij}} \,, \qquad (39)$$

which normalize the inclusive distributions

$$f_{ij}(\vec{p}_i) = n_{ij} \, \hat{f}_{ij}(\vec{p}_i) \,. \qquad (40)$$

The total secondary particle multiplicity is

$$n_j = \sum_i n_{ij} \,. \qquad (41)$$

Similarly, as in Eqs. (35) and (36), we introduce a selection function

$$\hat{s}_{ij}(\vec{p}_i) = \frac{s_{ij}(\vec{p}_i)}{\int s_{ij}(\vec{p}_i) \, d^3p_i} \,. \qquad (42)$$

This function corresponds to f_{ij} but is chosen in such a way that the efficient selection of random momenta is possible and furthermore should allow enhancement of the production in certain regions of momentum space (large $p_\perp$, large $p_\parallel$, large angles, backward region, etc.). Similarly, as in Eqs. (37) to (41), we define quantities G_{ij}, g_{ij}, m_{ij} and m_j. The inelasticity for selection is

$$G_{ij} = \frac{1}{E_{tot}} \int E_i \, s_{ij}(\vec{p}_i) \, d^3p_i \,, \qquad (43)$$

which can be arbitrarily chosen subject to the sum rule

$$\sum_i G_{ij} = 1 \,.$$

The average energy of the selected particle i is

$$g_{ij} E_{tot} = \int E_i \hat{s}(\vec{p}_i) d^3 p_i , \qquad (44)$$

and the selection multiplicities are

$$m_{ij} = \frac{G_{ij}}{g_{ij}} , \qquad (45)$$

$$m_j = \sum_i m_{ij} . \qquad (46)$$

In this method only a fixed number of secondaries, n, is selected in each inelastic event. Let us choose n = 1. The selected particle is characterized by: i) the label i determining its kind (i is selected according to the multiplicities m_{ij}); ii) the momentum $\vec{p}_i$ selected from $\hat{s}_{ij}(\vec{p}_i)$; and iii) a weight w_{ij} determined by

$$w_{ij} = \frac{\hat{f}_{ij}(\vec{p}_i)}{\hat{s}_{ij}(\vec{p}_i)} \frac{n_{ij}}{\frac{m_{ij}}{m_j}} = \frac{m_j f_{ij}(\vec{p}_i)}{s_{ij}(\vec{p}_i)} . \qquad (47)$$

HADRON PRODUCTION IN HIGH-ENERGY HADRON-HADRON COLLISIONS

General Properties

We restrict ourselves to particle production at low transverse momenta ($p_\perp$ < 1-2 GeV/c). Large transverse momentum particle production is not important for hadron cascade calculations. There are a few general empirical rules which apply to all particle production processes:

i) The average transverse momentum ($p_\perp$ = p sin θ) of the produced hadrons is rather small; that is,

$$<p_\perp> \approx 0.3 \text{ to } 0.4 \text{ GeV/c} . \qquad (48)$$

It is nearly independent of the collision energy E_{tot} and the kind of primary particles, but it rises slowly with the mass of the produced particle i. The $p_\perp$ and m_i dependence is rather well described by the distribution

$$\frac{d\sigma}{dp_\perp^2} \approx \exp\left(-\frac{\sqrt{p_\perp^2 + m_i^2}}{T}\right). \qquad (49)$$

ii) The average multiplicity of the produced particles is relatively small and rises only slowly with the collision energy. If a finite, energy-independent fraction of the collision energy would be used to create the rest mass of the produced particles, we would observe multiplicities rising proportionally to the c.m.s. energy, E_{cm}. Instead the observed multiplicities (see Fig. 2) rise like

$$\langle n \rangle \sim \ln E_{cm} \qquad (50)$$

or

$$\langle n \rangle \sim E_{cm}^B \; ; \; B \approx 1/4 \, . \qquad (51)$$

iii) As seen in Fig. 2, the largest contribution to the secondaries comes from π mesons.

iv) In the inelastic collisions the primary particles lose only about 50% of their energy by particle production processes. The through-going primary (leading) particles have, on the average, much larger momenta than do the newly created particles. Typical momentum distributions of different kinds of secondaries are plotted in Fig. 3. The dominance of leading particles appears rather strongly in the secondary proton momentum distribution.

v) For the energies of interest, the multiplicity distribution is approximately Poisson. The energy dependence of the charged prong cross-sections in pp collisions is plotted in Fig. 4.

vi) According to the scaling hypothesis of Feynman, the invariant inclusive distributions in the c.m.s.,

$$E^* \frac{d^3\sigma^{in}}{d^3p^*} = F_1(p_\parallel^*, \vec{p}_\perp, s) \, , \qquad (52)$$

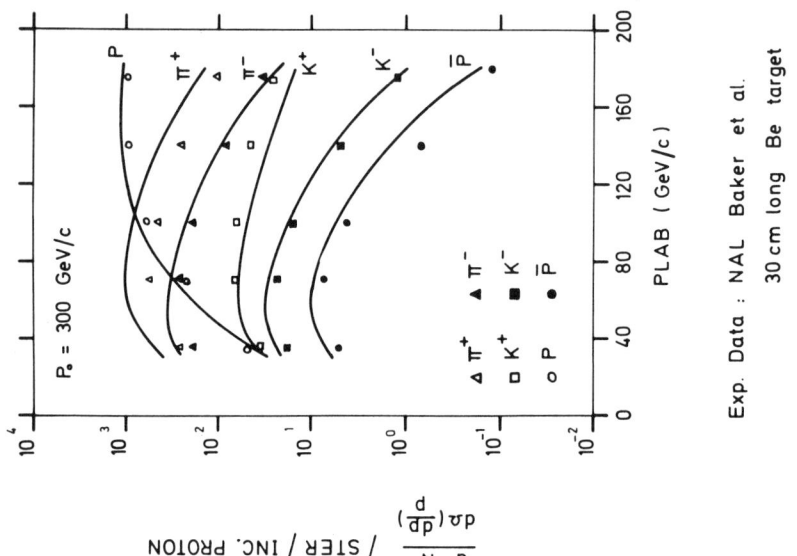

Fig. 3 Typical momentum distributions of secondary particles. The particles were produced in pp collisions at 300 GeV/c.

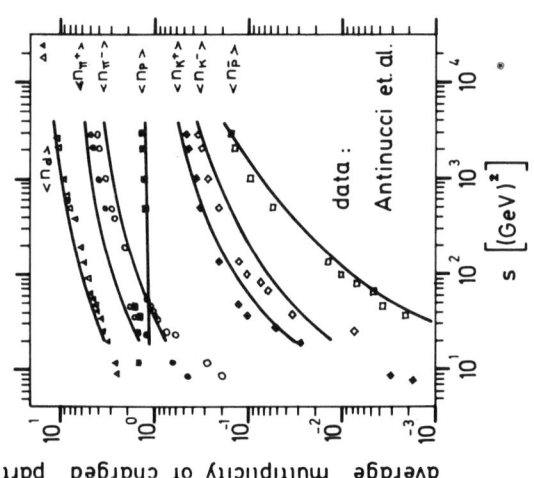

Fig. 2 Average multiplicities of charged secondary particles in pp collisions, $\langle n_{ch} \rangle$, and of π^+, π^-, K^+, K^-, p, and $\bar{p}$ separately, as function of the square of the c.m.s. energy, $s = E_{cm}^2$.

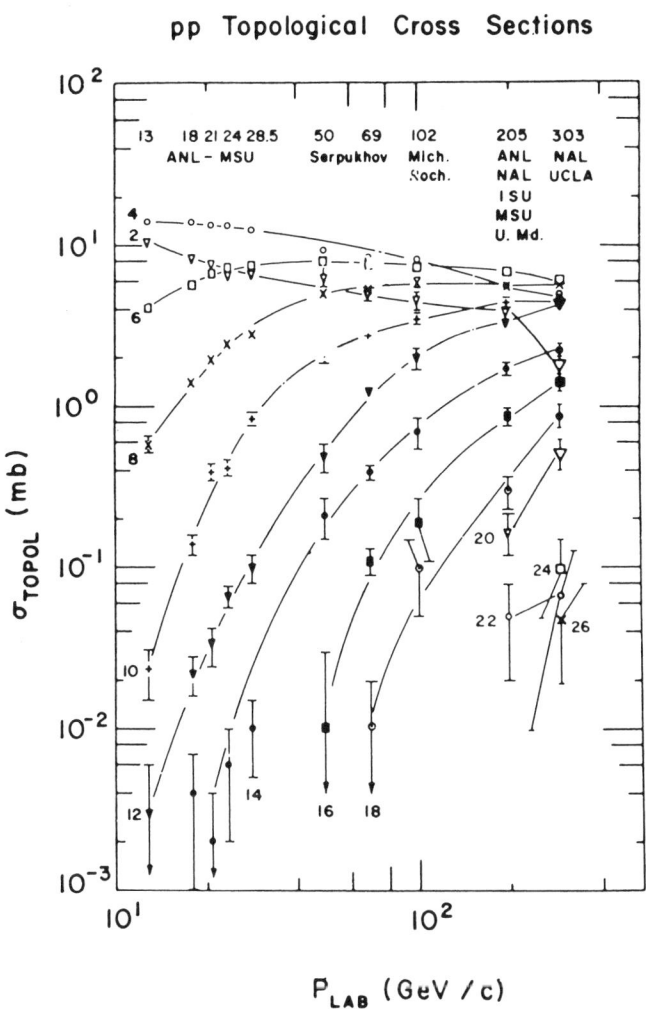

Fig. 4 Topological cross-sections for n-charged particle production in pp collisions

become asymptotically independent of s, the square of the c.m.s. energy; that is,

$$F_1(p_\parallel^*, p_\perp, s) \xrightarrow[s \to \infty]{} f\left(x = \frac{2p_\parallel}{\sqrt{s}}, p_\perp\right). \tag{53}$$

There is, at present, considerable doubt about the validity of this hypothesis, but in the energy region of interest here the data show approximate Feynman scaling. Therefore, this hypothesis is very useful and allows us to write down empirical expressions for the inclusive distributions, whereby data can be extrapolated from one energy to another.

Theoretical Models for Multiparticle Production in Hadron-Hadron Collisions

Theoretical models for multiparticle production should reproduce the general properties (i) to (vi) of multiparticle production, as well as other properties which are not important for our purposes. Furthermore, they should start from a theory or model of the strong interaction of elementary particles and allow for an understanding of the validity of this model. Finally, they should be worked out to the extent that detailed predictions for the inclusive and exclusive cross-sections emerge, which agree in detail with data. Not all models which have been discussed in the last decade have been carried out to such detail.

One of the first models that was used to obtain inclusive cross-sections in considerable detail was the thermodynamic model[13], which is based on the statistical bootstrap model of strong interactions[14]. In particular, this model was very successful in predicting particle production at the new generation of accelerators which came into being in the seventies: the 400 GeV proton synchrotrons at Fermilab and CERN and the CERN Intersecting Storage Rings. Computer programs and detailed descriptions are available. Since the middle of the seventies the underlying theoretical model based on statistical concepts has lost some of its attraction.

Currently, a theory of strong interactions, based on gauge field theories and quark and gluon constituents, is in the centre of experimental and theoretical research in strong interaction physics. As yet, no description of multiparticle distributions as detailed as the thermodynamic or cluster models has evolved from this kind of approach. But quark recombination models[15] seem to be very promising and are actively being developed[16]. To characterize these kinds of models, we treat only the production of mesons in proton-proton collisions. We start from the quark momentum distributions in hadrons which are measurable in deep inelastic lepton-hadron collisions. Let

$$u(x), \quad d(x), \quad \bar{u}(x), \quad \text{and} \quad \bar{d}(x)$$

be the distributions of u, d, $\bar{u}$, and $\bar{d}$ quarks in the proton, respectively, where x gives the fraction of the proton momentum carried by the quark. In the quark recombination model the produced mesons, which are quark-antiquark bound states, are formed by recombination of one quark and one antiquark out of the original hadrons. For example, if we consider the inclusive Feynman x-distribution of produced π^+ mesons, we write

$$\frac{dN}{dx} = \int F_2(x_u, x_{\bar{d}}) R(x_u, x_{\bar{d}}, x) \, dx_u \, dx_{\bar{d}}, \qquad (54)$$

where $F_2(x_u, x_{\bar{d}})$ is the two-particle distribution for the presence of one u quark with momentum fraction x_u and one $\bar{d}$ quark with momentum fraction $x_{\bar{d}}$ in the proton, and $R(x_u, x_{\bar{d}}, x)$ is a quark recombination function which should be related to the quark wave functions of the produced π^+ meson. Introducing the momentum δ-function and a scaling hypothesis into the recombination function, we can write

$$R(x_u, x_{\bar{d}}, x) = \delta(1 - \xi_u - \xi_{\bar{d}}) \rho(\xi_u, \xi_{\bar{d}}), \qquad (55)$$

where

$$\xi_u = \frac{x_u}{x}, \quad \xi_{\bar{d}} = \frac{x_{\bar{d}}}{x}. \qquad (56)$$

Quite good results are obtained using a very simple form

$$\rho(\xi_u, \xi_{\bar{d}}) \approx \xi_u \cdot \xi_{\bar{d}} .\qquad(57)$$

This is encouraging and I would expect that a very successful and detailed description of multiparticle production will emerge from this scheme[16].

HADRON PRODUCTION IN HADRON-NUCLEUS COLLISIONS

Empirical Parametrization of Inclusive Single-Particle Distributions from Hadron-Nucleus Collisions

The inclusive production of p, n, $\pi^\pm$, π^0, $K^\pm$, and $\bar{p}$ in collisions of p, n, $\pi^\pm$, $K^\pm$, and $\bar{p}$ with nuclei is described in the FLUKA and KASPRO codes by empirical formulae, fitted to experimental data as far as available. The formulae, and parameters in the formulae, are given in Tables 1 and 2. All formulae are defined in a projectile nucleus c.m.s. where the parameters are fitted to data from different nuclei. We interpolate the parameters if the production from different nuclei is needed. This procedure is not too reliable, especially for light nuclei; no problems have been found for heavy nuclei. Most of the data which were used in the fits are for primary energies in the region of 10 to 30 GeV. The extrapolation to higher energies is done using the Feynman scaling hypothesis. For $\bar{p}$ production the deviations from Feynman scaling are too large and we use energy-dependent inelasticities as described in Ref. 11. Regarding the present situation with experimental data on particle production in hadron-nucleus collisions, I think it is preferable to use a theoretical model which relates the particle production in hadron-nucleus collisions to the particle production in hadron-hadron collisions. In the next section I describe an attempt to do this. This procedure seems to me more capable of describing the nucleus dependence of particle production in a systematic way.

Table 1

Empirical formulae describing inclusive particle production in hadron-nucleus collisions

$\dfrac{d^2N}{dp_\| dp_\perp}$ single particle distribution in the projectile-nucleon c.m.s.

$p_\|$ c.m.s. longitudinal momentum

$p_\perp$ transverse momentum

$E_{cm} = \sqrt{s}$ c.m.s. total energy

$y = \sinh^{-1}\left(\dfrac{p_\|}{m_\perp}\right)$ c.m.s. rapidity

$m_{\perp i} = \sqrt{p_\perp^2 + m_i^2}$ m_i : mass of the particle of kind i

a_i parameters in particle production formulae, see Table 2.

Reaction	Formula
$\left.\begin{array}{l}p\\n\end{array}\right\}+A \to \left\{\begin{array}{l}p\\n\end{array}\right.$	$\dfrac{d^2N}{dp_\| dp_\perp} = \dfrac{a_1}{E_{cm}}\left(1 + \dfrac{a_2}{E_{cm}}p_\| + \dfrac{a_3}{E_{cm}^2}p_\|^2\right) p_\perp \left[\exp(-a_4 p_\perp^2) + a_5 \exp(-a_6 p_\perp)\right]$
$\left.\begin{array}{l}p\\n\\p\end{array}\right\}+A \to \left\{\begin{array}{l}\pi^+\\ \pi^0\\ \pi^-\end{array}\right.$	$\dfrac{d^2N}{dp_\| dp_\perp} = \dfrac{a_1 \exp\left(-\dfrac{a_2}{E_{cm}^2}p_\|^2\right) p_\perp \left[\exp(-a_3 p_\perp^2) + a_4 \exp(-a_5 p_\perp)\right]}{(p_\|^2 + p_\perp^2 + m_\pi^2)^{1/2}}$
$\left.\begin{array}{l}p\\n\end{array}\right\}+A \to \{\bar{p}$	$\dfrac{d^2N}{dp_\| dp_\perp} = a_1 E_{cm}^{a_2} \exp\left[-a_3 \left(\dfrac{p_\|}{E_{cm}}\right)^2 - a_4 p_\perp^2\right] \dfrac{p_\perp}{(p_\|^2 + p_\perp^2 + m_p^2)^{1/2}}$
$\left.\begin{array}{l}\pi^+\\ \pi^-\end{array}\right\}+A \to \left\{\begin{array}{l}p\\n\end{array}\right.$	$\dfrac{d^2N}{dp_\| dp_\perp} = \begin{cases}\dfrac{a_1}{E_{cm}}\left(1 + \dfrac{a_2}{E_{cm}}p_\| + \dfrac{a_3}{E_{cm}^2}p_\|^2\right) p_\perp [\exp(-a_4 p_\perp^2) + a_5 \exp(-a_6 p_\perp)], & p_\| < 0 \\ 0, & p_\| > 0\end{cases}$
$\left.\begin{array}{l}\pi^+\\ \pi^-\end{array}\right\}+A \to \left\{\begin{array}{l}\pi^+\\ \pi^0\\ \pi^-\end{array}\right.$	$\dfrac{d^2N}{dp_\| dp_\perp} = \begin{cases}\left[a_5 \exp\left(-4\dfrac{a_6}{E_{cm}^2}p_\|^2\right) + a_7\right] a_8^2 p_\perp \exp(-a_8 p_\perp)(p_\|^2 + p_\perp^2 + m_\pi^2)^{-1/2}, & p_\| > 0 \\ \left[a_1 \exp\left(-4\dfrac{a_2}{E_{cm}^2}p_\|^2\right) + a_3\right] a_4^2 p_\perp \exp(-a_4 p_\perp)(p_\|^2 + p_\perp^2 + m_\pi^2)^{-1/2}, & p_\| > 0\end{cases}$
$\left.\begin{array}{l}\pi^+\\ \pi^-\end{array}\right\}+A \to \{\bar{p}$	$\dfrac{d^2N}{dp_\| dp_\perp} = \exp\left[-a_1 m_\perp - a_2\left(\dfrac{|y^*|}{\frac{1}{2}\ln E_{cm}}\right)^2\right] \dfrac{p_\perp}{(p_\|^2 + p_\perp^2 + m_p^2)^{1/2}}$
$\bar{p} + A \to \left\{\begin{array}{l}p\\n\end{array}\right.$	$\dfrac{d^2N}{dp_\| dp_\perp} = \begin{cases}\dfrac{a_1}{E_{cm}}\left(1 + \dfrac{a_2}{E_{cm}}p_\| + \dfrac{a_3}{E_{cm}^2}p_\|^2\right) p_\perp [\exp(-a_4 p_\perp^2) + a_5 \exp(-a_6 p_\perp)], & p < 0 \\ 0, & p > 0\end{cases}$
$\bar{p} + A \to \bar{p}$	$\dfrac{d^2N}{dp_\| dp_\perp} = \begin{cases}0, & p < 0 \\ \dfrac{a_1}{E_{cm}}\left(1 + \dfrac{a_2}{E_{cm}}p_\| + \dfrac{a_3}{E_{cm}^2}p_\|^2\right) p_\| [\exp(-a_4 p_\perp^2) + a_5 \exp(-a_6 p_\perp)], & p > 0\end{cases}$
$\left.\begin{array}{l}p\\n\\ \pi^+\\ \pi^-\end{array}\right\}+A \to K^\pm$	$10^{a_1}\left(\dfrac{\sqrt{s}}{6}\right)^{a_2} p_\perp \exp\left(-a_3\left(\dfrac{p_\|}{\sqrt{s}}\right)^2 - a_4 p_\perp^2\right) \dfrac{1}{(p_\|^2 + p_\perp^2 + m_K^2)^{1/2}}$

Table 2

Parameters of the particle production formulae given in Table 1 as determined by fits from data on different materials

Reaction	A	a_1	a_2	a_3	a_4	a_5	a_6	a_7	a_8
$\left.\begin{array}{l}p\\n\\\pi^+\\\pi^-\\\bar{p}\\\bar{p}\end{array}\right\} + A \to \left\{\begin{array}{l}p\\n\end{array}\right.$ $\bar{p} + A \to \bar{p}$	H Be Al Cu Pb	0.94 0.42 0.45 0.44 0.44	0.86 1.03 -1.78 -2.99 -2.99	-3.37 -3.85 0.3 4.9 4.9	3.78 5.63 5.38 3.91 3.91	0.47 3.49 3.8 5.82 5.82	3.6 2.89 2.8 2.99 2.99		
$\left.\begin{array}{l}p\\n\\\bar{p}\end{array}\right\} + A \to \left\{\begin{array}{l}\pi^+\\\pi^0\end{array}\right.$ $\bar{p} + A \to \pi^-$	H Be Al Cu Pb	4.94 1.81 1.54 2.36 2.36	33.83 33.39 35.54 37.21 37.21	6.11 3.01 3.7 5.83 5.83	0.69 5.12 3.03 0.76 0.76	4.12 7.34 4.94 3.22 3.22			
$\left.\begin{array}{l}p\\n\end{array}\right\} + A \to \pi^-$	H Be Al Cu Pb	2.81 1.52 1.54 1.60 1.60	44.08 42.47 44.62 46.52 46.52	5.17 5.33 5.67 6.47 6.47	0.81 0.82 0.83 0.93 0.93	4.34 3.53 3.17 3.05 3.05			
$\pi^+ + A \to \pi^+$	H	0.22	7	0	5.7	0.22	5	0.115	5.7
$\pi^+ + A \to \pi^0$ $\pi^+ + A \to \pi^-$ $\pi^- + A \to \pi^0$ $\pi^- + A \to \pi^+$	H	$0.14 - \frac{0.14}{\sqrt{s}}$	13	0	5.7	$0.14 - \frac{0.14}{\sqrt{s}}$	9	0	5.7
$\pi^- + A \to \pi^-$	H	0.2	13	0	5.7	0.2	5	0.115	5.7
$\left.\begin{array}{l}p\\n\end{array}\right\} + A \to \bar{p}$	H Be Al Cu Pb	10^{-8} 1.0×10^{-8} 1.4×10^{-8} 1.7×10^{-8} 1.4×10^{-8}	7.75 7.75 7.75 5.02 11.21	85.2 85.2 91.86 94.25 87.43	4.26 4.26 4.25 3.65 4.48				
$\left.\begin{array}{l}\pi^+\\\pi^-\end{array}\right\} + A \to \bar{p}$	H	8.6	2.2						
$\left.\begin{array}{l}p\\n\\\pi^+\\\pi^-\end{array}\right\} + A \to K^+$	Be Al Cu Pb	0.7 0.67 0.67 0.71	2.00 2.16 2.97 3.84	29.1 30.3 30.9 33.0	2.97 2.95 3.33 3.32				
$\left.\begin{array}{l}p\\n\\\pi^+\\\pi^-\end{array}\right\} + A \to K^-$	Be Al Cu Pb	-1.02 -1.01 -1.06 -1.09	1.91 2.18 2.97 8.03	60.1 61.8 62.7 65.2	3.98 4.11 4.01 3.88				

General Properties of Particle Production in Hadron-Nucleus Collisions[17]

Particle production in hadron-nucleus collisions is characterized by a few general properties which appear, in part, rather surprising. To describe these properties we first introduce a few concepts. The density of nuclear matter inside the nucleus is characterized by a density distribution $\rho(r)$, most often parametrized in the Woods-Saxon form

$$\rho(r) = \frac{\rho_0}{\exp\left[(r-c)/z_1\right] + 1} ,$$

$$c = \left(0.978 + 0.0206\, A^{1/3}\right) A^{1/3} , \qquad (58)$$

$$z_1 = 0.54 .$$

The impact parameter b measures the distance of the incident particle trajectory from the centre of the nucleus. The nuclear thickness is measured usually by the parameter $\bar{\nu}$ which gives the average number of collisions of a projectile inside the nucleus. It can be defined as

$$\bar{\nu} = \frac{A\sigma_{hN}^{inel}}{\sigma_{hA}^{inel}} \approx \frac{A}{A^{2/3}} \approx A^{1/3} , \qquad (59)$$

where A is the mass number of the nucleus. According to this measure, the amount of nuclear matter depends on the target as well as on the incident projectile.

Before measurements at present-day high-energy accelerators became available, an attempt was made to predict the properties of particle production on nuclei by intranuclear cascade calculations, a technique that was found to be very successful at energies around 1 GeV. It belongs now to the folklore of high-energy physics that the intranuclear cascade picture breaks down completely at high energies. The secondary particle multiplicities predicted from this picture are much too large when compared to experimental data.

A rather surprising property of high-energy particle production on nuclei is the fact that many features of particle production are, to a first approximation, independent of the nuclear target. It is not an impossible approximation to perform hadron cascade calculations in Pb by using secondary particle distributions as measured in pp collisions.

We come to a few general properties of inelastic collisions on nuclei:

i) The total inelastic cross-section rises approximately as $A^{2/3}$; for instance,

$$\sigma_{pA} = 46 \, A^{0.69}, \tag{60}$$

$$\sigma_{\pi^{\pm}A} = 28.5 \, A^{0.73}. \tag{61}$$

ii) The multiplicity ratio, defined as

$$R_A = \frac{\langle n_{hA} \rangle}{\langle n_{hh} \rangle}, \tag{62}$$

is rather small. At large energies R_A rises about linearly with $\bar{\nu}$ from $R_A(\bar{\nu} = 1) = 1$ to $R_A(\bar{\nu} = 4) \approx 2\text{-}3$. For fixed A, R_A shows a threshold behaviour with the collision energy; saturation is reached around E = 100 GeV. To present the secondary particle momentum distributions, the rapidity variable is most often used

$$y_i = \ln\left(\frac{p_{\|i} + \sqrt{p_{\|i}^2 + \mu_i^2}}{\mu_i}\right), \tag{63}$$

where

$$\mu_i = \sqrt{p_{\perp i}^2 + m_i^2}.$$

In hadron-hadron collisions the rapidity distribution is expected to show a plateau behaviour. With increasing collision energy the rapidity distribution becomes only wider with higher energy:

PARTICLE PRODUCTION MODELS

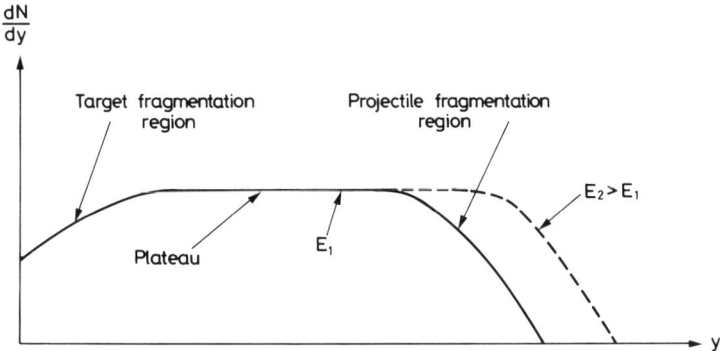

In nuclei the rapidity distribution depends also on $\bar{\nu}$.

iii) At low rapidity, in the target fragmentation region and part of the central region, a $\bar{\nu}$-dependent enhancement of the rapidity distribution is observed.

iv) In the projectile fragmentation region, one observes a suppression of fast secondaries. This suppression is not only observed for leading particles, but for all kinds of secondaries.

The properties (iii) and (iv) can be well described by defining the exponent $\alpha(y, p_\perp)$ by the expression

$$\frac{d\sigma_{hA}}{dy} = \frac{d\sigma_{hh}}{dy} A^{\alpha(y,p_\perp)} . \qquad (64)$$

For the total inelastic cross-section, $\alpha = 2/3$. The experimental data indicate that α is larger than $2/3$ at small rapidities, crosses through $2/3$ with rising rapidity, and decreases in the projectile fragmentation region down to $\alpha = 0.5$.

The interpretation of these general properties is the following: Particle production is a process in space and time. It follows from rather general considerations that the slow secondaries are produced first and the fast secondaries last. The fast secondaries in the hadron-nucleus collision are produced at a time when the excited hadronic matter, which gives rise to the secondaries, has already left the nucleus; no secondary interactions inside the nucleus are therefore possible. Slow secondaries might be produced inside the nucleus, and secondary interactions explain the slow particle enhancement (iii) in hadron-nucleus collisions.

Predictions of a Model for Particle Production on Nuclei

Most models considered for particle production in hadron-hadron collisions have been applied to hadron-nucleus collisions. Up to now, however, almost all of these models have only dealt with charged particle multiplicities, total cross-sections, and general properties of charged particle distributions in their energy and A-dependence[18]. Inclusive distributions of identified secondaries are usually not considered. Having in mind the application for hadron cascade calculations we[19] have extended the model of Capella and Krzywicki[20], which describes well the dependence of the total charged particle rapidity distribution on A and energy. As input to describe the single-particle distributions, we use the thermodynamic model.

First, let us describe the Capella-Krzywicki model[20], which is founded on the Reggeon theory and includes energy-momentum conservation exactly. Within this model the hadron-nucleus interaction is assumed to proceed via a certain number of universal subcollisions of hadron constituents with nucleons of the nucleus. The number of subcollisions is defined by the probability distribution

$$w_n = \frac{\sigma_n}{\sigma_{inel}^{hA}}, \tag{65}$$

with

$$\sigma_n = \binom{A}{n} \int d^2b \left[\sigma^{hN} T(b)\right]^n \left[1 - \sigma^{hN} T(b)\right]^{A-n} \tag{66}$$

and

$$\sigma_{inel}^{hA} = \sum_{n=1}^{A} \sigma_n = \int d^2b \left\{1 - \left[1 - \sigma^{hN} T(b)\right]^A\right\}. \tag{67}$$

For calculating the nuclear profile function,

$$T(b) = \int dz\, \rho(b, z), \tag{68}$$

we use the nuclear density (58) normalized to one. Neglecting cascading of secondaries, Capella and Krzywicki arrive at the following formula for the rapidity distribution in the laboratory frame:

$$\frac{dN^{hA}}{dy}\left(y, E_{lab}\right) = \sum_n n\, w_n \int dE'\, P\left(E', E_{lab}, n\right) \times$$
$$\times \frac{dn}{dy}\left[y - y_0(E'), E'\right]. \tag{69}$$

Here the normalized probability $P(E', E_{lab}, n)$ that the secondary particle is produced in an event with n subcollisions by a constituent with laboratory energy E' is given by

$$P\left(E', E_{lab}, n\right) = \begin{cases} \dfrac{(n-1)\left[E_{lab} - E' - (n-1)E_0\right]^{n-2}}{(E - nE_0)^{n-1}}, & n > 1 \\ (E_{lab} - E'), & n = 1 \end{cases} \tag{70}$$

for $E_{lab} - (n-1)E_0 > E' > E_0 = 5$ GeV. The rapidity distribution $dn[y - y_0(E'), E']/dy$ for the constituent-nucleon interaction is defined in the constituent-nucleon c.m.s. Therefore, the rapidity shift, $y_0(E')$, is given by

$$y_0(E') = \ln\left(\frac{\sqrt{s'}}{m_N}\right), \tag{71}$$

where

$$s' = 2E'm_N + 2m_N^2. \tag{72}$$

To obtain results on the production of different kinds of secondaries in proton-nucleus collisions, we treat the constituent nucleon interactions as p-nucleon ones using the thermodynamic model. A good description of the low rapidity enhancement and fast particle suppression effects can be obtained this way, as can be seen in Fig. 5 where we compare the results with experimental data.

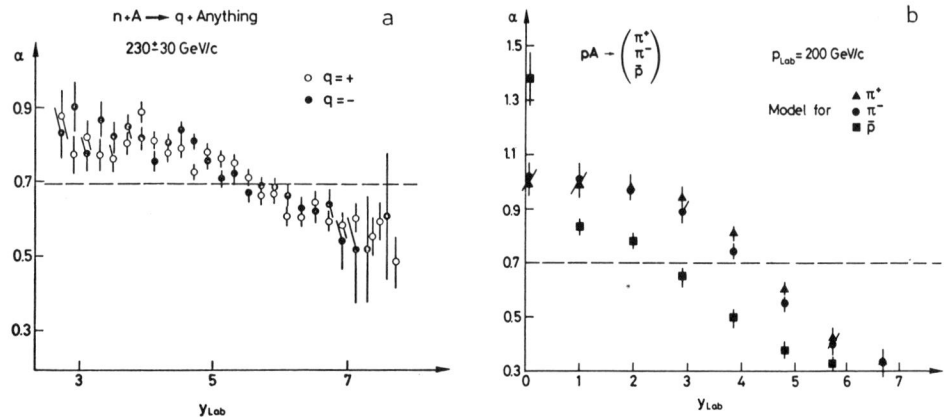

Fig. 5 a) The rapidity dependence of the exponent α as measured by Bleser et al.[23], for positive and negative charged secondary particles compared to b) the calculated exponent α for π^+ and π^-.

REFERENCES

1. J. Ranft, Nucl. Instrum. Methods 48, 133 (1967); 48, 261 (1967); 81, 29 (1970).
2. J. Ranft and J.T. Routti, Comput. Phys. Commun. 7, 327 (1974).
3. J. Ranft, Particle Accel. 3, 129 (1972).
4. For a review see, for instance, J. Ranft, Fortschr. Phys. 23, 468 (1975).
5. M.S. Chen and R.F. Peierls, A mapping technique for efficient random event generation with constraints, Brookhaven National Laboratory Report BNL 18656 (1973).
6. H.J. Discher and J. Ranft, unpublished results (1978).

7. W.A. Coleman and T.W. Armstrong, The nucleon-meson transport code NMTC, Oak Ridge National Laboratory Report ORNL-4606 (1970).
   H.E. Bertini, Phys. Rev. 188, 1711 (1969).
8. H.J. Discher, K. Hänssgen, R. Kirschner, J. Ranft, and J. Wetzig, unpublished results (1978).
9. Compilations of cross-sections by the CERN-HERA group. For further information, write to: Technical Information Division, Lawrence Berkeley Laboratory, Berkeley, California 94720; or CERN-Scientific Information Service, CH-1211 Geneva 23, Switzerland.
10. Particle Properties Data Booklet, from Phys. Lett. 75B, No. 1 (1978).
11. J. Ranft, KASPRO, a Monte Carlo program to calculate: a) particle fluxes as modified by the hadron cascade in targets, and b) energy deposition and star densities in the material, CERN Report LAB II-RA/75-1 (1975).
    B.V. Chirikov, V.A. Tayurski, H.-J. Möhring, J. Ranft, and V. Schirrmeister, Nucl. Instrum. Methods 144, 129 (1977).
12. A. Van Ginneken, Weighted Monte Carlo calculation of hadronic cascades in thick targets, Fermilab Report NAL-FN-250 (1972).
13. R. Hagedorn and J. Ranft, Suppl. Nuovo Cimento 6, 169 (1968).
    H. Grote, R. Hagedorn, and J. Ranft, Particle Spectra (CERN, Geneva, 1970).
    R. Hagedorn and J. Ranft, Nucl. Phys. B48, 157 (1972).
14. R. Hagedorn, Suppl. Nuovo Cimento 3, 147 (1965).
15. K.P. Das and R.C. Hwa, Phys. Lett. 68B, 459 (1977).
    J. Ranft, Phys. Rev. D 18, 1491 (1978).
    D.W. Duke and F.E. Taylor, A determination of the sea quark distributions in the proton by single particle inclusive reactions, Fermilab Report, Fermilab-PUB-77/96-THY (1977).
    T.A. De Grand and H.I. Miettinen, Hadronic production with a Drell-Yan trigger, Stanford Linear Accelerator Center Report SLAC-PUB-2036 (1977).

16. T.A. De Grand, Hadronic fragmentation initiated by pointlike probes, Stanford Linear Accelerator Center Report SLAC-PUB-2812 (1978).

    J. Ranft, Contribution to Erice Workshop on Theoretical Physics, Hadronic Matter at Extreme Energy Density: Particle production in soft and hard hadronic collisions --- is there evidence for hadronic constituents?, E. Majorana Centre, October 1978.

17. For reviews see, for instance, D.E. Nagel (editor), Proc. 5th Conf. on High-Energy Physics and Nuclear Structure, Santa Fe and Los Alamos (American Institute of Physics, New York, 1975).

18. For reviews, see L. Bertocchi in Ref. 17.

19. H.-J. Möhring and J. Ranft, Particle production in hadron-nucleus collisions, CERN Report HS-RP/019 (1977).

20. A. Capella and A. Krzywicki, Phys. Lett. $\underline{67B}$, 84 (1977).

21. M. Antinucci, A. Bertin, P. Capiluppi, M. D'Agostino-Bruno, A.M. Rossi, G. Vannini, G. Giacomelli and A. Bussière, Nuovo Cimento Lett. $\underline{6}$, 121 (1973).

22. W.F. Baker, A.S. Carroll, I.-H. Chiang, D.P. Eartly, O. Fackler, G. Giacomelli, P.F.M. Koehler, T.F. Kycia, K.K. Li, P.O. Mazur, P.M. Mockett, K.P. Pretzl, S.M. Pruss, D.C. Rahm, R. Rubinstein, and A.A. Wehmann, Phys. Lett. $\underline{51B}$, 303 (1974).

23. FNAL-Northwestern-Rochester Collaboration, presented by T. Ferbel at the 8th Internat. Colloquium on Multiparticle Production, Kayserberg (1977).

LECTURE 20:   THE INTRANUCLEAR-CASCADE-EVAPORATION MODEL

Tony W. Armstrong
Science Applications, Inc.
La Jolla, CA 92038
USA

INTRODUCTION

The most important physics uncertainty associated with predicting the interaction of high-energy hadrons with matter lies in determining the description (multiplicities and energy and angular distributions) of particles produced in nonelastic nuclear collisions. A theoretical model, the intranuclear-cascade-evaporation model, has been developed and applied to such problems over the past twenty years or so, with considerable success, and this model has been incorporated in thick-target hadronic transport codes (e.g., Ref. 1 and Lecture 23). The purpose here is to give a brief introduction to this method for treating the nonelastic interaction of high-energy hadrons with complex nuclei.

A key feature of this model is that, at sufficiently high energies, the initial phase of the reaction can be treated in terms of collisions of the incident particle with individual nucleons inside the nucleus. The struck nucleons can cause further collisions, giving rise to a particle "cascade" _inside_ the nucleus; hence the term _intra_nuclear cascade to describe this process. (In thick targets, such nuclear collision products can proceed to produce further collisions with other nuclei, giving rise to an "extranuclear", or <u>inter</u>nuclear cascade.) After the intranuclear cascade, the nucleus is left in an excited state, and the subsequent de-excitation is determined using an evaporation model where the excitation energy is dissipated by the successive "boiling off" of low-energy ($\sim$ few MeV) particles. Calculationally, the model is well-suited to a statistical approach using Monte Carlo techniques, and this has been the method used in all implementations of the model.

Historically, the intranuclear cascade model was first proposed by Serber in 1947 (Ref. 2). He recognized that at high energies (few tens of MeV), the deBroglie wave length of the incident particle becomes comparable to, or shorter than, the average internucleon distance within the nucleus ($\sim 10^{-13}$ cm). That is, the collision time between the incident particle and a nucleon inside the nucleus becomes short compared to the time between nucleon collisions inside the nucleus. Hence, the justification for describing the interaction in terms of particle-particle collisions. Goldberger, in 1948 (Ref. 3), was the first to perform calculations using Serber's model. (It is reported that two people working for two weeks were able to generate one hundred Monte Carlo particle histories by hand calculations.) Metropolis et al. (Ref. 4), in 1958, were the first to report extensive calculations using computers, and generated large amounts of data for low-energy incident particles ($\lesssim$ 400 MeV, below the pion production threshold). Since then, numerous refinements and extensions of the model have been made over a number of years, primarily by three groups: Barashenkov and colleagues at Dubna, U.S.S.R. (e.g., Ref. 5), the Brookhaven National Laboratory/Columbia University Group (e.g., Ref. 6), and by Bertini and co-workers at the Oak Ridge National Laboratory (e.g., Ref. 7). (A numerical comparison of results from the model as implemented by these three groups for low-energy protons (150 and 300 MeV) has been made, Ref. 8.) The discussion in this paper is primarily based on the model as implemented by Bertini.

THE INTRANUCLEAR-CASCADE-EVAPORATION MODEL

BASIC CONCEPT

The basic idea then is that the nonelastic interaction of hadrons with nuclei takes place in two stages. In the first stage, the intranuclear cascade (or "knock-on" or "fast" phase), the incident particle interacts with a single nucleon inside the nucleus, much as if the nucleon were in free space. The collision is not exactly analogous to one in free space because the Pauli principle will exclude those encounters with certain momentum transfers. Also, at high energies the bombarding particle can traverse the nucleus without experiencing an interaction, resulting in a "nuclear transparency". The collision products may undergo subsequent collisions, giving rise to a particle cascade inside the nucleus. For incident particle energies above the pion production threshold ($\sim$ 300 MeV), inelastic pion-production collisions have a significant effect on the propagation of the cascade. The locations of the collisions, the type of collision, the momentum of the struck nucleon, scattering angles, etc., can be determined by using statistical sampling techniques together with free-particle (p-p, p-n, π-p, etc.) cross section data.

After the completion of the intranuclear cascade, the kinetic energy possessed by those nucleons which remain inside the nucleus is assumed to be equilibrated among all of the nucleons. This residual excitation energy, as well as the mass and charge of the residual nucleus, can take on a range of values because of the variety of different outcomes possible for reactions at high energies. Subsequent de-excitation is determined by applying evaporation theory (e.g., Ref. 9), which is a statistical treatment of compound nuclei. Thus, the second stage of the reaction is the evaporation phase. Particles emitted during evaporation are of low energies ($\sim$ 1-10 MeV) and have isotropic angular distributions. Nucleon emission, especially neutrons, is more probable than the evaporation of positively charged clusters such as deuterons, tritons, ^{3}He, or alpha or even heavier, particles. After particle emission is no longer energetically possible, the nucleus can still be left with a small amount of excitation energy, which is assumed to be released by photon emission (e.g., Ref. 10). For very heavy target nuclei, fission can occur during the evaporation phase of the reaction (e.g., Ref. 11).

BASIC FEATURES OF THE MODEL

Intranuclear Cascade

Summarized below are some of the assumptions and representations that have been used in implementing the intranuclear cascade model. These specific descriptions are those used by Bertini et al. (Ref. 7); other implementations are similar but sometimes differ in detail.

Nucleon Density. The nucleon density distribution inside the nucleus is represented as a three-region model (a central sphere and two surrounding spherical annuli). The neutron and proton densities are constant in each region (Fig. 1) with magnitudes chosen to approximate the continuous distribution as determined from Hofstadter's experiments on electron scattering by nuclei (Ref.12). The outer radii of the three regions is determined such that the distances are at 90, 20, and 10% of the central value of the continuous distribution. The region boundaries are the same for neutrons and protons with the neutron-to-proton ratios in each region the same as for the nucleus as a whole.

Nucleon Momentum Distribution. The nucleons inside the nucleus are assumed to have a zero-temperature Fermi momentum distribution in each region - i.e., $f(p) = c\, p^2$ with c determined such that $\int_o^{p_f} f(p)\, dp$ is the total number of neutrons (protons) in a region and p_f is the momentum of a nucleon corresponding to the Fermi energy. Since this energy is dependent upon the nucleon density and, therefore, different for each type of nucleon in each region, the composite momentum distribution for the entire nucleus is not a zero-temperature Fermi distribution (Fig. 2).

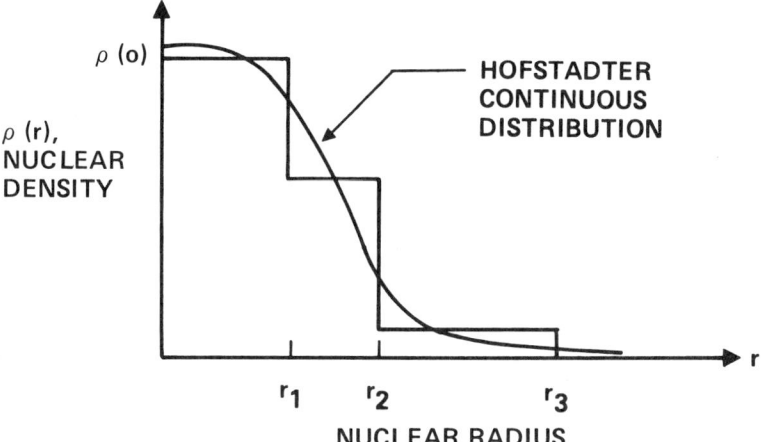

Fig. 1. Assumed three-step nucleon density representation and actual continuous distribution.

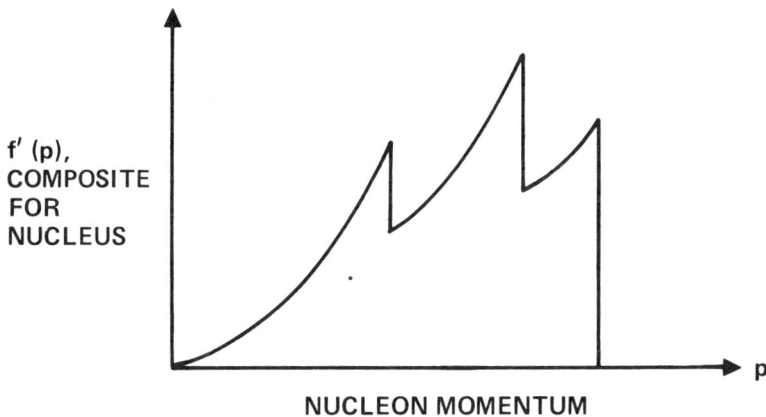

Fig. 2. Assumed momentum distribution of nucleons inside the nucleus.

Potential Energy. The nucleon potential energy is the sum of the zero-temperature of the nucleons in each region plus the binding energy of the most loosely bound nucleon, and this binding energy is taken to be constant at 7 MeV for all regions and all nuclei (Fig. 3).

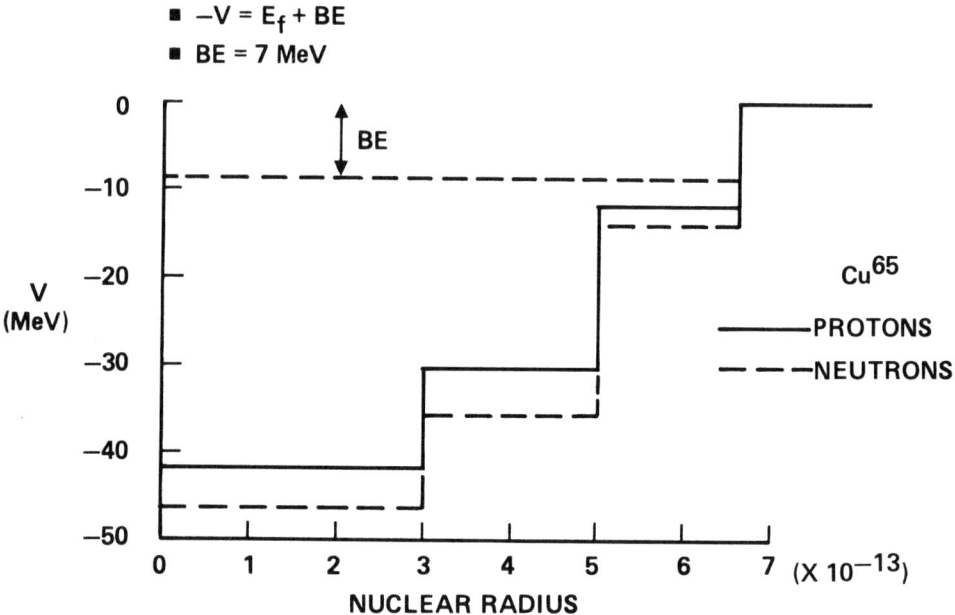

Fig. 3. Assumed potential energy distribution of nucleons inside nucleus. Example for Cu^{65}.

Cross Sections. The cross sections needed in computing the cascade are the nucleon-nucleon elastic and inelastic (pion production) and the pion-nucleon elastic scattering, charge-exchange, and pion production cross sections (Fig. 4). For the most part, these cross sections are rather well known from experimental data (e.g., Ref. 13). In the Bertini codes, single and double pion production in nucleon-nucleon collisions and single pion production in pion-nucleon collisions are taken into account. This restricts the upper energy limit for applications to about 3-GeV since at higher energies higher order pion production becomes important.

Pion Production. For those collisions that lead to pion production, the Lindenbaum-Sternheimer isobar model (Ref. 14) is used to determine the final states. This model assumes that pions are produced through the decay of an isobar that is formed when a nucleon is excited in a collision. That is, $N + N \to N^* + N$, or $\to N^* + N^*$, and then $N^* \to \pi + N$ for each N^*. To determine the final-state momentum distributions, the angular distributions of the isobars and decay pions must be specified, and these are arrived at phenomenologically through comparisons with experimental data (Ref. 15).

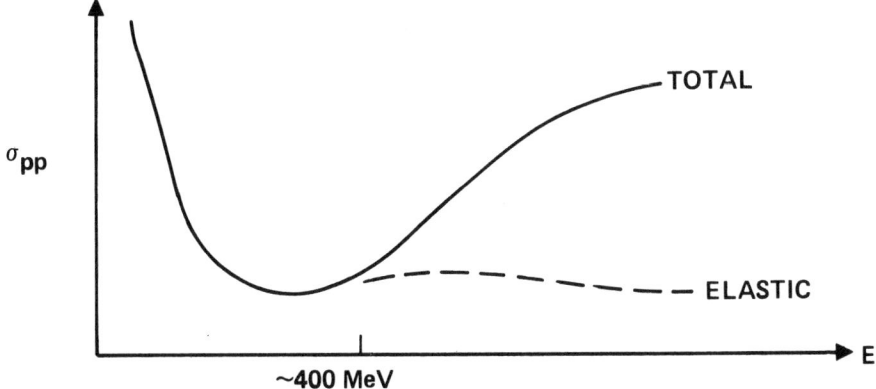

Fig. 4. Schematic of particle-particle cross section energy dependence needed as input for the intranuclear cascade calculation. Example is for proton-proton collisions, showing the onset of pion production at ∿ 400 MeV.

Pion Absorption. Pion absorption is considered to occur on a two-nucleon cluster. The type of cluster (p-p, p-n, or n-n) for absorption is determined from the number of each type of particle pairs within the nucleus, excluding those pair types which would violate charge conservation.

Exclusion Principle. Since we are considering collisions inside nuclear matter, the Pauli exclusion principle forbids interaction where the collision products would be in occupied states. For a completely degenerate Fermi gas, the levels are filled starting from the lowest level; therefore, collisions are forbidden in which either very large or very small amounts of energy are transferred. The minimum energy allowed for the low-energy product of a collision corresponds to the lowest unfilled level of the system, which is the Fermi energy in the region. Thus, in the Monte Carlo calculation, the exclusion principle can be taken into account by neglecting any collision where the energy of any of the collision products falls below the Fermi energy.

Evaporation

The de-excitation, and associated particle emission, of the highly excited nucleus remaining after the intranuclear cascade is determined by applying the statistical theory of evaporation, which was originally developed by Weisskopf (Ref. 16) for treating relatively low excitations from compound nucleus formation by low energy particles. Inherent assumptions in this model are 1) complete energy equilibration before the first particle emission, and 2) reequilibration of excitation energies between successive evaporations.

The probability of emission of a particular particle type (neutron, proton, or clusters - such as deuterons, tritons, ^{3}He, or alpha particles) can be formulated in terms of: a) the inverse cross section (the cross section for the formation of a compound nucleus by the interaction of the emitted particle and the residual nucleus), b) the level densities of the final and original nucleus, c) the Coulomb barrier, separation and binding energies, d) the spin, mass, and charge of the nucleus, and e) the excitation energy (Ref. 16 and 17). The Monte Carlo method, as developed by Dostrovsky et al. (Ref. 9) for such applications, is the computational procedure most often used. A Maxwellian energy distribution, based on the excitation energy and mass of the nucleus with modifications to take into account the Coulomb barrier, is used to select the kinetic energy of the emitted particle. The angular distribution of emitted particles is isotropic in the center of mass.

The emission of particles is computed sequentially until the excitation energy falls below some specified cutoff, usually $\sim$ 7 MeV. Further de-excitation is assumed to occur by photon emission. Normally the photon spectrum is not provided, although some studies on such photon emission and spectra have been made (e.g., Ref. 10,18).

COMPUTATION

The model can be implemented by carrying out Monte Carlo calculations in essentially an analog fashion. The basic steps of the calculation are straightforward, as summarized below:

1. The spatial point where the incident particle enters the nucleus is determined by selecting uniformly over a circular area representing the projected area of the nucleus.

2. A path length is selected for the distance the projectile particle travels before having a collision, using the total particle-particle cross sections and region-dependent nucleon densities.

3. If the particle escapes the nucleus without having a collision, this incident particle contributes to the "nuclear transparency". Otherwise, the momentum of the struck nucleon, the type of reaction, and the energy and direction of the reaction products are determined.

4. If the collision is allowed according to the exclusion principle, and if the kinetic energy of the product is above a predefined cutoff energy, then go to step (2) to transport the products. (A cutoff energy is necessary because at very low energies it is not valid to represent subsequent reactions as particle-particle collisions; instead, the particle energy contributes to the excitation energy of the residual nucleus.)

5. After completion of the cascade calculation for an incident particle, the mass and charge, A' and Z', of the residual nucleus is determined from a particle balance, and the residual excitation energy, E^*, is determined from an energy balance.

6. The resulting A', Z' and E^* are then used as input for the evaporation calculation to determine the energy of the nucleons and heavier particles (deuterons, tritons, ^{3}He, and alpha particles) produced, assuming isotropic emission. The remaining excitation energy is assumed to dissipate by photon emission, but the photon spectrum is not determined. The A and Z of the final residual nucleus is then determined.

ADVANTAGES OF METHOD

The main advantages of the intranuclear-cascade-evaporation model are as follows:

1. The model is essentially parameter free and provides results with absolute normalization.

2. For input the model requires mainly only particle-particle cross section data, which is relatively well known.

3. The model has been applied rather extensively and shown to be in reasonably good agreement with a wide range of experimental data.

4. The model has rather general applicability to: a) all target nuclei with $A \gtrsim 5$; b) different projectile particles (neutrons, protons, and charged pions, with recent extensions to heavy-ions, Ref. 19), and c) a wide energy range. The valid energy range is not well defined. The lower energy limit is $\sim$ 15-100 MeV. The Bertini implementation of the model as described here is limited to energies below $\sim$ 3 GeV because only single and double pion production are included. Rather ad hoc extensions to tens, even hundreds, of GeV have been made by using approximate scaling relations (Ref. 20). Barashenkov et al. (Ref. 5), and recently Bertini et al., (Ref. 21), have made modifications to the basic model (e.g., to incorporate nuclear depletion, which takes into account the time dependence of the changing nucleon density in the path of the developing cascade) to extend the upper energy limit to $\sim$ 1000 GeV.

5. The model is capable of providing very detailed output results, including a) the type, energy, and direction of each emitted particle, b) the type and recoil energy of residual nuclei, c) the photon source from π° decay and nuclear de-excitation, and d) the low-energy ($\lesssim$ 15 MeV) neutron production, which is of interest in many practical problems.

COMPUTER CODES

The most current and available computer code of the intranuclear-cascade-evaporation model as implemented by Bertini et al., is designated MECC-7/EVAP-IV.*+¹ It is this version that is used in the HETC transport code (Lecture 23). There have been several earlier versions, but they are now considered obsolete.

There have been two recent extensions of the model by Bertini et al., that are important. First, major modifications have been made to allow particle energies up to $\sim$ 1000 GeV; the cascade portion of this now code is called HECC*+²(Ref. 21). Secondly, MECC-7/EVAP-IV have been extended to treat collisions of projectile particles heavier than nucleons with nuclei; this code system is called HIC‡ (Ref. 22).

REFERENCES

1. T.W. Armstrong and K.C. Chandler, Nucl. Sci. Engr. $\underline{49}$, 110 (1972
2. R. Serber, Phys. Ref. $\underline{72}$, 1114 (1947).
3. M.L. Goldberger, Phys. Rev. $\underline{74}$, 1268 (1948).
4. N. Metropolis, R. Bivins, M. Storm, A. Turkevich, J.M. Miller, and G. Friedlander, Phys. Rev. $\underline{110}$, 185 (1958).
5. V.S. Barashenkov and V.D. Toneyev, Vzaimodeystviya Vysokoenergeticheskikh Chastits i Atomnykh Yader 5 Yadrami, Moscow (1972).
6. K. Chen, Z. Fraenkel, G. Friedlander, J.R. Grouer, J.M. Miller, and Y. Shimamoto, Phys. Rev. $\underline{166}$, 949 (1969).
7. Hugo W. Bertini, Phys. Rev. $\underline{188}$, 1711 (1969).
8. V.S. Barashenkov, H.W. Bertini, K. Chen, G. Friedlander, G.D. Harp, A.S. Iljinov, J.M. Miller, and V.D. Toneev, Nucl. Phys. $\underline{A187}$, 531 (1972).
9. I. Dostrovsky, P. Rabinowitz, and G. Friedlander, Phys. Rev. $\underline{111}$, 1659 (1958).

---

*¹ MECC-7 = Medium Energy Cascade Code, version 7.
EVAP-IV = EVAPoration code, version IV.

+¹ Available from the Radiation Shielding Information Center (see Lecture 9).

*² HECC - High Energy Cascade Code

+² Barashenkov et al. have also extended their cascade code to treat energies up to $\sim$ 1000 GeV (Ref. 5).

‡ HIC = Heavy Ion Code

10. Y. Shima and R. G. Alsmiller, Jr., "Calculation of the Photon-Production Spectrum from Proton-Nucleus Collisions in the Energy Range 15 to 150 MeV and Comparison with Experiment", Oak Ridge National Laboratory Report ORLN-TM-2908, 1969.

11. R.L. Hahn and H.W. Bertini, Phys. Rev. C$\underline{6}$. 660 (1972).

12. R. Hofstadter, Rev. Mod. Phys. $\underline{28}$, 214 (1956).

13. V.S. Barashenkov, "Interaction Cross Sections of Elementary Particles", IPST Press, Jerusalem, 1968.

14. R.M. Sternheimer and S.J. Lindenbaum, Phys. Rev. $\underline{123}$, 333 (1961).

15. H.W. Bertini, M.P. Guthrie, and A.H. Culkowski, "Phenomenologically Determined Isobar Angular Distributions for Nucleon-Nucleon and Pion-Nucleon Reactions Below 3 GeV", Oak Ridge National Laboratory Report, ORNL-TM-3132 (1970).

16. V. Weisskopf, Phys. Rev. $\underline{52}$, 295 (1937).

17. K.J. Le Couteur, Proc. Phys. Soc. (London), $\underline{A63}$, 259 (1950).

18. C.W. Hill and K.M. Simpson, Jr., "Calculation of Proton-Induced Gamma-Ray Spectrum and Comparison with Experiment", Second Symposium on Protection Against Radiations in Space, NASA SP-71, 1964.

19. H.W. Bertini, et al., "HIC-1, A First Approach to the Calculation of Heavy-Ion Reactions at Energies $\gtrsim$ 50 MeV/Nucleon", Oak Ridge National Laboratory Report ORNL-TM-4134, 1974.

20. T.A. Gabriel, R.T. Santoro and J. Barish, " A Calculational Method for Predicting Particle Spectra From High Energy Nucleon-Pion Collisions ($\geq$ 3 GeV) with Protons", Oak Ridge National Laboratory Report ORNL-TN-3615, 1971.

21. H.W. Bertini, Private communication, 1979.

22. R.T. Santoro, " Operating Instructions for the Heavy-Ion Code HIC-1", Oak Ridge National Laboratory Report ORNL-TM-4791, March 1975.

LECTURE 21: CALCULATION OF THE AVERAGE PROPERTIES OF HADRONIC CASCADES AT HIGH ENERGIES (CASIM)

A. Van Ginneken

Fermi National Accelerator Laboratory
Batavia, Illinois 60510
USA

INTRODUCTION

CASIM[1] is a Monte Carlo program which simulates the average development of internuclear cascades when high energy hadrons (10-1000 GeV) are incident on large targets. The program computes such desiderata as star densities, momentum spectra of interacting particles and energy deposited by the cascade as a function of location throughout the target. From these quantities, effects of interest in radiobiological protection, radiation damage, accelerator design, detector development, etc., can be estimated. CASIM is sufficiently general to apply to cascades in heterogeneous targets, arbitrary geometries and the presence of magnetic fields. Only the less conventional features of the program will be discussed here.

The properties of hadronic cascades are largely determined by the numbers and distributions of particles produced in nuclear collisions. These basic features of hadron-nucleus interactions are, at present, not well understood and for the most part are rather difficult to describe. There are several reasons for this: (a) an acceptable and sufficiently detailed fundamental theory of strong interactions is still lacking, (b) the nuclear many-body problem introduces added complexities and (c) the high average multiplicities and the variety of particle types produced in high energy collisions. One might hope that once (a) is solved, then (b) and (c) will also cease to be a problem and all calculation can be simply carried out at the newly understood level.

In a companion lecture (Lecture 14 of this volume), another Monte Carlo code, AEGIS, is described. This program calculates the average properties of electromagnetic (EM) showers. The techniques used there are basically the same as the ones used in CASIM. Most of the discussion in Lecture 14 on the motivation and basic properties of biased calculations applies equally well to the case of hadronic showers. For this reason, the discussion on biasing below is limited to a summary of the main points, and the reader is referred to Lecture 14 for a somewhat more detailed treatment. The basic concepts are much easier stated as well as understood in the EM case for several reasons: fewer types of participating particles, fewer basic processes, only two outgoing particles per interaction (neglecting recoil), the cross sections are well known and they can be expressed in closed form.

CASIM is a biased Monte Carlo program which aims at broad applicability. The program structure is considerably simplified by tracing one and only one particle from each interaction. Execution time is decreased and grows only logarithmically with energy. (It grows linearly for analogue calculations.) The presence of a general bias facilitates the introduction by simple substitution of special biases into the code. Such special biases are, in certain cases, the only manner in which valid results can be obtained. The disadvantage of this approach is that one no longer keeps track of the correlations among the members of the cascade. This makes biased calculations less suited to study shower-to-shower fluctuations.

Fig. 1 shows a symbolic representation of a hadronic shower. The solid lines indicate nucleons, the dashed lines are pions. In the familiar analogue method, every particle participating in the shower is traced and its effects (if any) on the problem being analyzed are noted. This scheme has definite advantages such as in studying fluctuation problems. However for most applications and almost all those encountered in dosimetry and radiation damage, a great deal of the information generated is actually of little use but makes the calculation slow to execute, especially at high energies.

A possible simplification is shown in Fig. 2 where one particle is selected at each vertex to represent its entire generation. If all particles have equal probability of being selected, then the weight of the representative particle increases by a factor equal to the multiplicity with each generation. These weights are indicated besides each trajectory in Fig. 2. For most applications, such a scheme will lead to a very poorly converging calculation. Generally the more energetic particles have either by themselves or through their progeny the largest effects on the problem. To treat all particles indiscriminately is therefore inefficient.

The solution is in large part similar to the EM case, i.e. using a biasing factor linear in energy in the selection of the representative particle. In fact, such a scheme is already sufficient as the basis for a general purpose energy deposition calculation. It is still inefficient for star density estimates. The needed modifications will be described below.

The basic processes considered in CASIM are (1) particle production, (2) the slowing down and (3) multiple Coulomb scattering of charged particles. The treatment of (2) and (3) is along quite familiar lines and will not be discussed here. The program traces only particles which are above a lower momentum limit. Typically, this is set at 0.3 GeV/c but the program allows this to

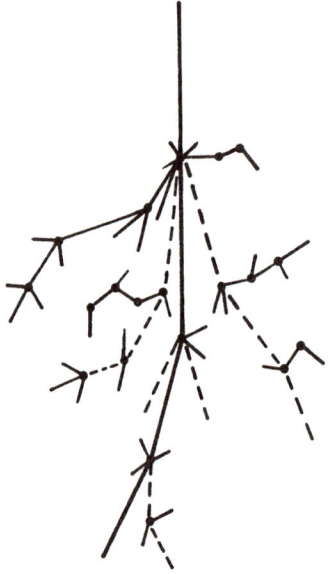

Fig. 1 Schematic representation of a hadron shower. Solid lines indicate nucleons. Dashed lines represent pions.

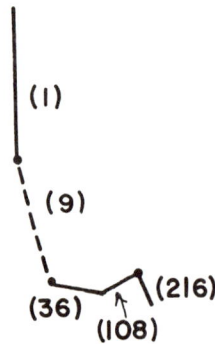

Fig. 2 A subcascade taken from Fig. 1. The numbers in parenthesis next to each trajectory indicate the weights if particles are selected with equal probability at each vertex.

be as low as 0.1 GeV/c. This restriction considerably simplifies
the model since at the higher energies, the cross sections for
various nuclear processes tend to vary slowly with energy, atomic
mass and particle type. This is in contrast with the behavior of
cross sections in the low energy regime (from a few MeV down to
thermal energies). The price that is paid is that only sketchy
information can be obtained about these low energy particles. An
extension of the code into the low energy domain would be very
desirable. At present, the upper momentum limit of the program is
1000 GeV/c. It would be a simple matter to extend this to higher
momenta, aside from questions about the validity of the model.

PARTICLE PRODUCTION MODEL

The only particles considered in the calculation are protons,
neutrons and pions. Muons are easily treated as an appendage to
the hadronic cascade. While other particles which are known to be
created in the shower are not explicitly included, their effects
are not outright neglected. This is achieved by forcing the model
to be energy conserving. In this manner, it can be claimed that
e.g. effects of kaons are mainly absorbed into the effects of pions.

Since only one outgoing particle is traced from each vertex,
the basic particle production model can be in the form of a set of
inclusive distributions, i.e. distributions of the type

$$a + b \rightarrow c + X \tag{1}$$

where only the yield of particle c (e.g. as a function of angle
and momentum) is measured or predicted and all the rest (X) is
ignored. In CASIM, a and c are protons, neutrons and charged pions
in any combination and b is a nucleus. There are then (for each
nuclear species b) a total of 16 different reactions to consider.
(Neutral pion yields are calculated from charged pion yields.)
Inclusive distributions have enjoyed considerable attention both
experimentally and theoretically. Hence reliable estimates for

them are readily available. The same cannot be said for reactions where two or more particles in the final state are measured or predicted. Indeed, even for a modest number of final state particles, a complete measurement or theoretical description becomes prohibitively long and complicated. This is somewhat of a difficulty in analogue calculations which, out of necessity, typically also start from inclusive distributions with at most a few added correlations.

Specifically, the inclusive distributions adopted in CASIM are those predicted by the Hagedorn-Ranft (HR) model.[2] A correction to this model based on an empirical formula[3] is applied to describe particle production with large transverse momenta ($p_t \geq 1$ GeV/c). This model applies only to the energetic products of elementary particle collisons. An additional distribution represents low energy constituent nucleons knocked out of the nucleus. This is taken from a parametrization of Ranft and Routti[4] of Bertini's intranuclear cascade calculations.[5]

At least until relatively recently, the HR model has been formulated only for pp collisons. Subsequent to its introduction, Ranft[6] adjusted the remaining free parameters of the model to fit 19.2 GeV/c p-nucleus distributions.[7] The HR model distinguishes between leading particles and produced particles i.e. whether the particle c in (1) can be identified with either a or b or with neither. Using a few general assumptions, the 16 different types of reactions can be sorted into four basic types: (a) leading particle forward and (b) leading particle backward in the pp center of mass, (c) produced π^+ and (d) produced π^- in pp collisions. Each such distribution is a sum of either two or four terms, each of which is in turn expressed as a one-dimensional integral. Even with the free parameter fit to nuclear collisions, the HR spectra are forward-backward symmetric. Since the selection of particles and calculation of the weights are generally the most time consuming parts of CASIM, this forward-backward symmetry is exploited to arrive at a convenient way for particle splitting in star density calculations (see below).

In addition to particles from the five distributions mentioned above (four HR distributions plus slow nucleons), each inelastic interaction creates a certain amount of nuclear excitation energy which is assumed to be a simple function of the atomic mass and the incident energy. A binding energy correction of 8 MeV per nucleon emitted is applied.

The HR nucleon spectra are normalized to two outgoing nucleons (for incident nucleons) and one outgoing nucleon (for incident pions). If the energy balance does not permit this, the nucleon spectra are normalized so as to force energy conservation and π production is forbidden. If π production is kinematically forbidden, then the nucleon spectra are normalized to force energy conservation. The energy remaining after subtracting from the incident energy, the fraction carried off by nucleons (HR plus knock-out nucleons), as well as the excitation and binding energy, is assumed to be carried off by pions. Their spectra are accordingly normalized. This procedure ensures that energy is conserved (on the average) in nuclear interactions. The normalization factors are never far from unity except at rather low energies ($\leq$ 3 GeV). Even down to the low momentum limit of the calculation (0.1 GeV/c), the above assumptions do not appear to have any apparent non-physical consequences. A comparison of the HR model with a limited sample of data from high energy p-nucleus experiments shows no gross discrepancies except as noted above for particles with high transverse momenta.

STAR DENSITY CALCULATION

The scheme which uses a selection function proportional to $E(d\sigma/dpd\Omega)$, where E is the energy of the outgoing particle representing the cascade, generally converges too slowly for a star density calculation. This means that to estimate with reasonable accuracy the star (nuclear interaction) density, or equivalently the omnidirectional particle flux from counting the number of simulated interactions per unit volume, would generally require long computer times. Briefly stated, the reason for this is that such a calculation follows the "energy flow", while the star density measures the "particle flow". Yet this does not mean that one should adopt something like the scheme of Fig. 2, because to select with equal probability from among all outgoing particles, or equivalently to sample according to the unweighted differential cross section, correctly follows the particle flow only for that generation. The ability to produce stars in all subsequent generations is actually roughly proportional to the particle's energy (a well known rule-of-thumb often invoked in shielding calculations). A satisfactory compromise is the following scheme: at each vertex, the program selects a cascade <u>propagating particle</u> from a selection function proportional to the product of the lab energy of the particle and its differential production cross section, as well as one or more <u>recording particles</u> selected according to their differential production cross section. By separating the purpose of the representative particles, one applies to each the selection scheme best suited: the propagating particles are selected for their ability to produce stars in all subsequent generations, the recording particles to produce stars in the present generation.

A symbolic representation is shown in Fig. 3 where the dotted lines indicate the recording particles. These recording particles are traced through the medium while the expected average number of stars along their trajectory is recorded. This continues until the particle is no longer of interest to the calculation either because (a) it is beyond the geometric bounds of the problem (b) it falls below the low energy cutoff or (c) its contribution to

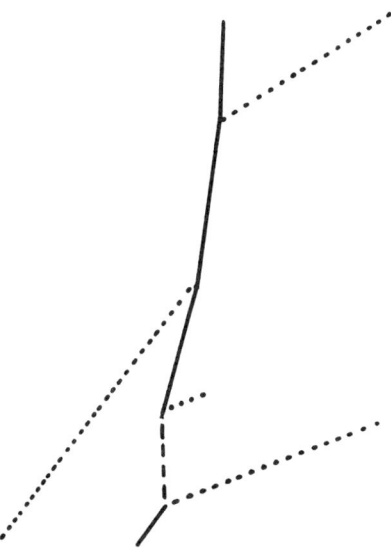

Fig. 3 A subcascade taken from Fig. 1 supplemented by recording particles (dotted lines). Propagating particles (solid and dashed lines) are selected from among all outgoing particles with probability proportional to their energy. Recording particles are selected with equal probability.

the average number of stars is insignificant. The propagating particle is traced until it undergoes a nuclear interaction or until it is of no further interest for the reasons given above. In the latter event, this particular chain comes to an end and a new incident particle is introduced.

One can readily apply the well known technique of particle splitting to the recording particles i. e. to divide, in certain instances, the weight among two or more particles. This might be motivated by calculational convenience. For example, the differential cross section for particle production is independent of the azimuthal angle of the emitted particle. Therefore, a particle can be split into several different particles with different azimuthal angles. (It is best to choose them evenly spaced over the available

2π radians.) Such splitting is generally only worthwhile if (a) the weight of the particles is substantial, (b) the polar angle is sufficiently large, since otherwise the trajectories at different azimuthal angles nearly coincide and (c) the parent particle's trajectory is not along an axis of cylindrical symmetry of the problem, or nearly so. Another instance of recording particle splitting motivated by calculational convenience and somewhat peculiar to CASIM has already been referred to above. This applies to HR spectra and proceeds by tracing, in addition to the recording particle selected, a second particle with equal and opposite momentum in the pp center of mass, if this particle is above calculational threshold. The weights of these two particles are identical.

Another motivation for splitting of recording particles is to improve selectively the statistical significance of certain parts of the calculation e.g. pertaining to some geometric region of the shield or to a given type of particle. The rules for such splitting could even be varied as the calculation proceeds; for example, increase of splitting in regions of poor statistics. Most of these problem-dependent types of particle splitting are not included in CASIM, though they may be easily inserted.

ENERGY DEPOSITION CALCULATION

The program divides this calculation into two main parts: (a) energy deposition due to hadrons, (b) due to EM showers initiated by photons from π^o decay.

For the part due to hadrons, the calculation uses the propagating particle and estimates the expected average energy deposition along its trajectory. As in the star density calculation, the trajectory is extended until the particle is no longer of interest to the calculation. The mechanisms of energy deposition considered are (a) ionization losses by charged particles, (b) de-excitation by charged particle emission of struck nuclei and (c) low energy charged particles emitted below the threshold of the calculation. At the same time, an estimate is made of energy deposited in the vicinity of the trajectory by (d) low energy neutrons and photons emitted from struck nuclei and (e) low energy neutrons emitted below calculational threshold.

For reasons of calculational efficiency, the program does not generate π^o among the propagating (or recording) particles. However, a simple algorithm relates π^o emission to $\pi^\pm$ emission. The energy deposition due to the π^o initiated EM showers is derived from an algorithm based on data and calculated results predominantly in the few GeV range. This algorithm has been extensively tested against the program AEGIS and the results reveal a close comparison. Alternatively, but at the expense of computer time, one can run AEGIS as a subprogram of CASIM. This is generally necessary where magnetic fields are to be included in the calculation, since the algorithm is not applicable under those conditions.

SELECTION OF PARTICLE TYPE AND DISTRIBUTION

In contrast with the program AEGIS for EM showers, most of the selection functions in CASIM appear in numerical form. To select a particle type for recording particles, a table is consulted which contains the fractional multiplicities above threshold, i.e.,

$$S_i^m = \int_{p_{min}}^{p_{max}} \int_{4\pi} \left(\frac{d\sigma}{dpd\Omega}\right)_i dpd\Omega \Big/ \sum_j \int_{p_{min}}^{p_{max}} \int_{4\pi} \left(\frac{d\sigma}{dpd\Omega}\right)_j dpd\Omega . \tag{2}$$

In this expression i and j indicate particle type as well as distribution. The limits of integration p_{min}, p_{max} are the minimum and maximum allowed momenta including effects of calculation threshold. For propagating particles, the table contains instead the fractional inelasticities above threshold:

$$S_i^i = \int_{p_{min}^i}^{p_{max}^i} \int_{4\pi} E_{lab}^i \left(\frac{d\sigma}{dpd\Omega}\right)_i dpd\Omega / \Sigma \int_{p_{min}^j}^{p_{max}^j} \int_{4\pi} E_{lab}^j \left(\frac{d\sigma}{dpd\Omega}\right)_j dpd\Omega \qquad (3)$$

Note that this procedure introduces a weight factor equal to $(S_i^m)^{-1}$ or $(S_i^i)^{-1}$ respectively for recording and propagating particles.

For ease of computation, the values of S_i^m and S_i^i are tabulated for eleven different incident momenta which range from 0.1 GeV/c to 1000 GeV/c in logarithmic intervals. There is such a table for each material present in the calculation.

SELECTION OF MOMENTUM

In case a "knock-out" nucleon is selected by (2) or (3), the differential cross section or inelasticity from which to select the momentum are rather simple analytical expressions. Though the expressions are entirely different, the procedure rather strongly. resembles the ones discussed in the AEGIS lecture and are omitted here to avoid repetition.

To describe high energy processes, the (lab) momenta and angles appearing in the integrations of (2) and (3) are somewhat poorly suited. It is more convenient to select a set of variables in terms of which differential cross sections and inelasticities vary rather slowly over the range of physical interest. One such set is:

$$x = \exp(-kp_t^2), \qquad (4)$$

where $k \simeq 4$ (GeV/c)$^{-2}$ is a parameter which might be varied, and p_t is the transverse momentum,

$$y = p_z^{cm}/p_{z,max}^{cm}, \qquad (5)$$

where p_z^{cm} and $p_{z,max}^{cm}$ are the longitudinal momentum in the center of mass and its maximum value, respectively. For the present purpose, there is no fundamental significance to this choice of variables and other sets may serve equally well or better.

Again, the selection functions for x and y are stored in tabular form: a set of tables to select x given the incident momentum, P_{inc}, and type of distribution, D_i, and a set to select y given x, P_{inc} and D_i. A considerable amount of both computation time and storage space is required to prepare the tables of the selection functions. To minimize this, these tables are not created separately for each material present in the calculation. Advantage is taken of the fact that the HR model predictions are not very sensitive to nuclear mass, and only one set of tables is computed. This slow dependence upon nuclear mass has been extensively verified by experiment. However, it must be emphasized that when calculating the weight factor

$$w = S^{-1}(x,y) \, (dN/dxdy) \,, \tag{6}$$

the cross section is calculated for the actual material in which the collision takes place. Thus the use of single set of selection function tables may affect the efficiency of the calculation, but it will not introduce systematic errors. In the case of a multi-medium calculation, the tables are based on a material with atomic weight equal to the average atomic weight of all materials present. This, along with the low sensitivity to atomic weight, means that the loss in efficiency is expected to be small. The x selection tables contain the integral probabilities

$$P_1(x_i) = k_1 \int_{x_{min}}^{x_i} dx' \int_{y_{min}}^{y_{max}} (dN/dx'dy) dy \tag{7}$$

for recording particles, where k_1 provides normalization such that $P_1(x_{max})$ is unity and $x_{min} \equiv \exp(-kp_{max}^{cm\,2})$, $y_{min} = -1$ or 0 and $y_{max} = 0$ or 1, depending on the type of distribution. For each of the 11 incident momenta for which the tabulations are made, there are 11 entries of $P(x_i)$ equally spaced in the variable x_i. There are corresponding tabulated probabilities for propagating particles

$$P_1'(x_i) = k_1' \int_{x_{min}}^{x_i} dx' \int_{y_{min}}^{y_{max}} E_{lab}(x',y)(dN/dx'dy)dy \qquad (8)$$

To each x_i entry of the x-selection tables, there corresponds a table containing 26 entries of the y selection probability:

$$P_2(x_i,y_j) = k_2 \int_{y_{min}}^{y_j} (dN/dx_i dy')dy' \qquad (9)$$

$$P_2'(x_i,y_j) = k_2' \int_{y_{min}}^{y_j} (dN/dx_i dy')E_{lab}(x_i,y')dy' \;. \qquad (10)$$

The selection of a given particle type and distribution proceeds by comparing a uniform random number ($0<r_1<1$) with the entries $P_1(x_i)$ in the appropriate table. If $P_1(x_{i-1})<r_1<P_1(x_i)$ then x is chosen uniformly between x_{i-1} and x_i. A new random number, r_2, is generated and, for x closer to x_i than to x_{i-1}, compared with the entries $P_2(x_i,y_j)$. Given that $P_2(x_i,y_{j-1})<r_2<P_2(x_i,y_j)$, then y is chosen uniformly between y_{j-1} and y_j. To calculate the weight (6), the (momentum) selection function is expressed as

$$S(x,y) = [P_1(x_i)-P_1(x_{i-1})][P_2(x_i,j_m)-P_2(x_i,y_{j-1})]/\Delta x \Delta y, \qquad (11)$$

where $\Delta x \equiv x_i - x_{i-1}$ and $\Delta y \equiv y_j - y_{j-1}$.

The factors in square brackets in (11) represent the probabilities for x and y to be bound by the indicated limits. The factors Δx^{-1} and Δy^{-1} represent the probabilities for uniform selection of x and y between the indicated limits. The efficiency is enhanced by using non-uniform intra-bin selection rules (a) for the low x (high p_t) region and (b) for the small y region in the case of pion production. The details of this are omitted. Likewise for reasons of efficiency, the above procedures are modified to avoid selecting particles below the calculational thresholds. Ref. 1 describes these and other details about the code.

MISCELLANEOUS OTHER FEATURES

CASIM has provision for collision length biasing. This is useful in very large (and sometimes very small geometries). This entails selecting the distance to the next collision according to an exponential distribution with a mean free path $c\lambda$ where λ is the true mean free path and c a collision length multiplier. If $c \neq 1$ this introduces an extra weight factor.

Another useful feature is that of correlated sampling between different Monte Carlo runs. This ensures that each incident particle in two or more different but related runs starts with the same random number and thereby provides a certain coherence among the random number chains. This is very useful to bring out small changes in geometry, presence of magnetic fields, etc. which in two independent runs might easily be masked by the randomness of the "background". While the same goal can sometimes be reached by performing different problems in one and the same run, this requires usually a rather substantial extra coding effort and often is not practical for other reasons as well.

ACCURACY

Comparison of CASIM results with experiment has been quite gratifying. Results of such comparisons have been published for two experiments.[9,10] The first concerned energy deposition as measured by temperature rise as a function of depth in a series of targets several interaction lengths long of rather slender dimensions. The second experiment measured various kinds of radionuclides produced in foils placed around a Fermilab dipole magnet when the beam struck a thick aluminum target which blocked the aperture of the magnet. Other types of radiation detectors were also present. CASIM was modified specifically to include radionuclide production in the different foils and to estimate the response of the other various detectors. The details may be found in Refs. 9 and 10. A good comparison with experiment for very large shields is still lacking. Crude measurements on this scale, such as those with hand held instruments, generally appear to confirm CASIM predictions quite well.

REFERENCES

1. A. Van Ginneken, "CASIM. Program to Simulate Hadronic Cascades in Bulk Matter", Fermilab, FN-272 (1975).
2. R. Hagedorn, Suppl. Nuovo Cim. $\underline{3}$, 147 (1965); R. Hagedorn and J. Ranft, Suppl. Nuovo Cim. $\underline{6}$, 169 (1968); H. Grote, R. Hagedorn and J. Ranft, "Atlas of Particle Spectra", CERN, Geneva, Switzerland (1970).
3. D. C. Carey et al., Phys. Rev. Lett. $\underline{33}$, 327 (1974).
4. J. Ranft and J. T. Routti, "Hadron Cascade Calculations of Angular Distributions of Secondary Particle Fluxes from External Targets and Description of Program FLUKU", CERN-II-RA/72-8 (1972).
5. H. W. Bertini, Phys. Rev. $\underline{188}$, 1711 (1969).
6. J. Ranft, "Secondary Particle Spectra According to the Thermodynamical Model. A Fit to Data Measured in p-Nucleus Collisions", TUL-36, Karl Marx Univ., DDR (1970).
7. J. V. Allaby et al., "High Energy Particle Spectra from Proton Interactions at 19.2 GeV/c", CERN-70-12 (1970).
8. A. Van Ginneken, "Comparison of Some Recent Data on p-Nucleus Interactions with the Hagedorn-Ranft Model Predictions", Fermilab, FN-260 (1974).
9. M. Awschalom et al., Nucl. Inst. Meth. $\underline{131}$, 235 (1975).
10. M. Awschalom et al., Nucl. Inst. Meth. $\underline{138}$, 521 (1976).

LECTURE 22:   THE FLUKA AND KASPRO HADRONIC CASCADE CODES

J. Ranft[*]
Sektion Physik, Karl-Marx-Universität
Leipzig,
German Democratic Republic

ABSTRACT

FLUKA, and some closely related programs using the same method, are Monte Carlo programs to calculate the high-energy extranuclear hadronic cascade. Only particles with energies above 50 MeV are treated. The particle production in hadron-nucleus collisions is described by empirical formulae giving the inclusive single-particle distributions of nucleons and mesons. The results of the program should be reliable at energies where the Feynman scaling of inclusive distributions is approximately valid. So far, applications to and comparisons with data are up to energies of 500 to 1000 GeV. The program mainly calculates star and energy deposition densities. Applications are to radiation heating, direct radiation doses, induced radioactivity, punch-through of charged hadrons, hadron calorimeters, etc. This lecture will describe the physical approximations, the Monte Carlo methods used, and typical applications.

KASPRO, and some closely related programs, are Monte Carlo programs to calculate the high-energy extranuclear cascade. It uses an importance sampling method and does not sample complete collision events. The application is mainly to calculate particle production from extended targets and to study the efficiency of schemes to decrease the emittance of the produced particle beam.

---

[*] Work supported in part by the HS Division at CERN.

INTRODUCTION

Hadron cascade calculations and their applications to radiation problems around high-energy accelerators have been described in detail by Ranft[1]. Here we report on the properties and approximations in the hadron cascade codes FLUKA and KASPRO (and related programs), and on recent applications of these calculations.

The cascades considered are initiated by primary hadrons in the energy range of tens to hundreds of GeV. The characteristic feature of collisions in this energy range is the abundant production of new hadrons, a process which is rather independent of whether hydrogen or a heavy material such as Fe or Cu is used as target material[2]. The number of newly produced secondary particles rises logarithmically or with a small power of the laboratory energy of the primary particle:

$$n_s \approx \log E_p \quad \text{or} \quad n_s \approx E_p^{1/4} . \tag{1}$$

In extended matter these secondaries in turn produce more particles in their collisions, and so on. This process finally comes to an end only when the energies of all particles are small enough so that particle production is no longer possible.

In most materials the cascade of high-energy hadrons is the dominant mechanism of energy transport. For many applications, nuclear excitation processes such as the intranuclear cascade and nuclear evaporations, which give rise to large numbers of low-energy secondaries (mostly neutrons), are relatively unimportant and can be treated approximately. In a well-developed cascade the fluxes of these low-energy particles are related to the flux of high-energy particles; collisions of high-energy particles are the main source of the low-energy component.

There are three dominant mechanisms of energy deposition by the hadron cascade, namely:

i) ionization energy loss of high-energy charged hadrons;

ii) electromagnetic cascades initiated by γ quanta from π^0 decay; and

iii) low-energy nuclear fragments depositing the energy component which we will call nuclear excitation energy.

The three mechanisms are roughly of equal importance. The proportion of energy deposited by the electromagnetic cascade rises with energy and the proportion of nuclear excitation energy decreases with primary energy.

For most applications the three-dimensional development of the cascade initiated by a narrow beam of primary particles is of interest. The elementary events in the cascade are rather complicated and particle production is strongly anisotropic. There are different kinds of processes to be considered: elastic and inelastic collisions, ionization energy loss, electromagnetic cascades, etc. Different kinds of particles are involved. We consider protons, neutrons, and charged and neutral pions, as well as kaons and antiprotons. In this situation three-dimensional analytic calculations are extremely difficult and no such calculations have been performed. One-dimensional calculations are valuable for the general understanding of the cascade process but are of only limited use for practical applications.

Physical input data for the cascade calculations include the following: elastic and inelastic cross-sections of hadrons on nuclei; inclusive particle production cross-sections for hadrons colliding with nuclei; ionization energy losses; multiple Coulomb scattering; nuclear excitation energies; and energy deposition by the electromagnetic cascade. All of these input data are discussed elsewhere[1-3]. The design criteria for the hadron cascade calculations in the computer programs to be described are the following:

i) The programs should be easy to use for standard applications. This requires that the input data are simple, that no user-supplied subroutines are necessary, that standard analysis routines are included in the program, and that the results of these analysis routines are printed and plotted in an easily understandable way.

ii) Similar programs are available for a few standard geometries. These standard geometries should usually be used to approximate the exact, more complicated, geometry of the problem. Therefore no geometry routines have to be written by the user. Usually the programs use a cylindrically symmetric geometry, which allows for a significant decrease in the statistical errors of sampled distributions.

iii) Only the physically dominant processes are taken into account in the calculation. This implies that the programs run fast. Since the emphasis is on the high-energy cascade, and the high-energy hadrons are the dominant component of energy transport in the cascade, it is possible to treat the low-energy components with approximations. Usually no particles with energies below 50 MeV are followed, the electromagnetic cascades are not followed by a Monte Carlo method, and only the energy deposition by the electromagnetic cascade is used by sampling from energy deposition distributions.

The uncertainties connected with these approximations are usually smaller than a factor of 2 (and can be decreased further with some special study). The systematic uncertainties of many experimental methods for measuring radiation doses, induced radioactivity, and radiation heating are very often equally large; therefore such an uncertainty is tolerable.

iv) The programs are mostly oriented towards the calculation of radiation effects around high-energy accelerators, including the design and properties of detectors and the calculation of particle fluxes from extended targets.

THE FLUKA CODE[4]

FLUKA is the basic program, but the method is also used in the programs FLUKOO[5,6], MAGKA[7], MAGKO[7,8], CYLKAZ[9] and CYLKOZ[10], which differ only in the arrangement of the geometry, the materials, and in the energy spectrum of particles taken into account. Only particles which dominate the star production in the cascade and the energy deposition are taken into account in FLUKA; namely, protons, neutrons, and charged and neutral pions.

The absorption cross-sections of protons, neutrons, and charged pions on nuclei with atomic weight A for energies between tens of MeV and hundreds of GeV are needed for the calculations. The dependence of the cross-sections on A can usually be represented in the form

$$\sigma_{abs}(hA) = \sigma_0 A^B. \qquad (2)$$

For pA collisions above 20 GeV one might use $\sigma_0 = 50.4$ mb and $B = 0.67$. The cross-sections compiled by Barashenkov[11] for different nuclei are used for pA and πA collisions in the energy range from 50 MeV to 20 GeV. For materials where no measurements are available, FLUKA uses the interpolation

$$\sigma(A) = \sigma_1 \left(\frac{A}{A_1}\right)^B, \qquad (3)$$

$$B = \log\left(\frac{\sigma_2}{\sigma_1}\right) \Big/ \log\left(\frac{A_2}{A_1}\right), \qquad (4)$$

where σ_1 and σ_2 are the measured cross-sections for nuclei with atomic weights A_1 and A_2. Above 20 GeV the cross-sections are assumed to remain constant, which is in good agreement with measurements at FNAL. Not enough cross-sections of neutrons at high energy have been measured, so FLUKA uses

$$\sigma_{nA} = \sigma_{pA}/1.07. \qquad (5)$$

The absorption lengths are calculated from the cross-sections according to

$$\lambda_{abs}(A) = \frac{A}{\sigma_{abs} N} \quad (g/cm^2) , \qquad (6)$$

where N is the Avogadro number (6.022×10^{23}). Elastic scattering is far less important for the hadron cascade. The elastic scattering total and differential cross-sections are taken into account as described in Ref. 1. The production of secondary particles in high-energy hadron-nucleus collisions is described using empirical formulae given in Ref. 2.

Electromagnetic cascades are a very efficient mechanism for depositing energy. The source of electromagnetic cascades are the photons from the decay of neutral pions. There are two possible procedures for taking the electromagnetic cascades into account: 1) to combine a code to calculate the electromagnetic cascade with the hadron cascade; and 2) to use the results of experiments and electromagnetic shower calculations in order to present this energy deposition properly in the hadron cascade calculation. The second choice is used in FLUKA. FLUKA samples the energy to be deposited from tables which give the longitudinal and transverse energy deposition densities due to the electromagnetic cascade. This picture contributes significantly to speeding up the program.

The method of sampling complete events from inclusive single-particle cross-sections that is used in FLUKA is described in Ref. 2. This method is implemented in the routine DISEK. The inelasticities determine the relative number of secondaries of different kinds of particles produced, as well as the energy fraction carried away by these particles. FLUKA uses the following basic inelasticities (K^0_{ij}):

Incoming particle	Outgoing particle			
	p	n	$\pi^{\pm}$	π^0
p	0.3	0.3	0.25	0.15
n	0.3	0.3	0.25	0.15
$\pi^{\pm}$	0.2	0.2	0.4	0.2

Particles not taken into account in FLUKA have to be represented by the kinds of particle components present. A certain part of the energy is used as the nuclear excitation energy E_{ex}, which is considered to include the energies of all nuclear fragments such as cascade and evaporation protons and neutrons, as well as recoiling heavy fragments.

The excitation energy E_{ex} depends on the energy E of the primary hadron and the atomic weight A of the target nuclei. At low energy, below the threshold E_{thr} in the cascade calculation, we put $E_{ex} = E$. For $E > E_{thr}$ we use

$$E_{ex} = \frac{A}{100} \qquad \text{for } E > 3 \text{ GeV},$$

$$E_{ex} = E_{thr} + \frac{(E - E_{thr})(B - E_{thr})}{(3 - E_{thr})} \qquad \text{for } E_{thr} \leq E \leq 3 \text{ GeV}, \qquad (7)$$

where $B = \max(E_{thr}, A/100)$.

The inelasticities given above have to be corrected for the excitation energy, so that FLUKA uses

$$K_{ij} = K_{ij}^0 \left(1 - \frac{E_{ex}}{E_0}\right). \qquad (8)$$

One third of the excitation energy is deposited within the bin where the interaction occurs and two thirds are carried away by low-energy particles (mainly neutrons). This part is deposited isotropically around the position of the hadron interaction using the absorption lengths of 50 MeV neutrons. While this procedure is certainly very approximate, it is reassuring that the importance of the nuclear excitation as an energy deposition mechanism decreases with primary energy.

According to the above design criteria, FLUKA samples various standard results. These are:

i) Three-dimensional densities of hadronic interactions (star densities):

$$F_i(r, z) = \iiint d\left(\cos \theta_x\right) d\left(\cos \theta_y\right) dp \, \phi_i(p, r, z, \cos \theta_x, \cos \theta_y) , \tag{9}$$

where i stands for the interacting hadron and p, $\cos \theta_x$, and $\cos \theta_y$ are momentum and direction cosines of the interacting hadron. $F_i(r, z)$ is also integrated over either r or z.

ii) Three-dimensional energy deposition densities

$$E(r, z) \left[\frac{\text{GeV}}{\text{cm}^3, \text{ incoming hadron}}\right]. \tag{10}$$

Three important mechanisms contribute to the energy deposition density:

- ionization energy losses of charged hadrons (20-25% of total energy deposited);
- energy deposited by electromagnetic cascades (rising from about 20% at low energy to 50% at high energy);
- nuclear excitation energy (decreasing from about 50% at low energy to 20% at high energy).

iii) The probability for charged particle transmission through a shield as a function of the thickness.

iv) Histograms of the total energy of each incoming hadron, deposited in blocks of certain sizes. These histograms are related to the performance of total absorption detectors.

Some words about the use of the programs. Short descriptions of the use, input data, and printed results are available for all programs: FLUKA[4], MAGKA[7], FLUKOO[5], MAGKO[7], CYLKAZ[9], and CYLKOZ[10]. The compilation of the most important parts of these descriptions by Schönbacher[12], which is updated from time to time, is very useful.

THE KASPRO CODE[13,14]

The basic program is KASPRO, but the same method is also used in the programs KAPRYZ[15] and KAPRYM[16], as well as in a version of the program containing rather special magnetic fields. KAPRYM and

KAPRYZ allow the use of a more complicated geometry, as well as of different materials. They differ mainly in the kinds of hadrons taken into account.

The following secondary particles are considered:

$$\begin{aligned}&\text{KASPRO, KAPRYZ:} \quad p, n, \pi^+, \pi^0, \pi^-, \bar{p} \\ &\text{KAPRYM:} \quad p, n, \pi^+, \pi^0, \pi^-, K^+, K^-, \mu \ . \end{aligned} \quad (11)$$

The particle production formulae used in the programs and the method of sampling secondaries are described in detail in Ref. 14. The formulae are defined in one routine only, which makes it very convenient to change them. Integrals over the particle production formulae have to be calculated in order to use the selection method. These integrals are usually calculated only once numerically, and are used as input data in subsequent runs for the same production formulae, selection formulae, and inelasticities.

Because of the method used, no complete inelastic events are sampled. It is not practical to calculate quantities that depend on single events and their fluctuations with KASPRO (e.g. quantities related to the performance of hadron calorimeters, the transmission of hadrons through absorbers, etc.). Owing to the importance sampling method used in the selection of secondaries it is possible to calculate quantities which could not be calculated using the straightforward methods used in FLUKA. This applies especially to the momentum and angular distributions of secondary particles which are emitted from a large target. Distributions for rarely produced particles such as antiprotons, and in regions of phase space where the production is much suppressed, could not be efficiently sampled by any other method.

Secondary particle distributions (in the particle flux option of KASPRO) are sampled in the form

$$\frac{d^2N}{dpd\Omega} = F(p_i, \theta) \quad (12)$$

as a histogram with chosen bins in p and θ, and

$$\frac{dN}{dp} \quad \text{and} \quad \frac{dN}{d\theta}, \tag{13}$$

are obtained by integrating one of the variables in Eq. (12). External target efficiencies are defined by

$$E(p, \theta) = \frac{(d^2N/dpd\Omega)_{\text{extended target}}}{(d^2N/dpd\Omega)_{\substack{\text{production in first} \\ \text{generation of the cascade}}}}. \tag{14}$$

For small targets, all secondaries produced in the primary interaction leave the target, and secondary interactions are unimportant. Therefore $E(p, \theta) = 1$. For large targets, the interaction of secondaries causes $E(p, \theta)$ to decrease at large momenta and increase at small momenta and large angles owing to the particles created by the interaction of these secondaries.

DIFFERENT VERSIONS OF THE PROGRAMS

The Geometry Used in the Programs

The programs FLUKA and FLUKOO calculate the cascade within a solid cylinder of one material. The material is characterized by the atomic weight A, the nuclear charge Z, and the density ρ. If composite materials are used, these three quantities should be averaged to reflect the composition.

The programs CYLKAZ, KAPRYM, and KAPRYZ calculate the hadron cascade in a cylindrical array made up of different materials. The materials M_{ij} are in cylindrical shells between radii R_i and R_{i-1} and the longitudinal positions Z_j and Z_{j-1} ($R_0 = Z_0 = 0$). R_i and Z_j should preferentially be multiples of the bin sizes ΔR and ΔZ. The programs assume that the bins are filled with material and calculate star and energy densities correspondingly. The R_i and Z_j can be chosen in such a way that bins only partially filled with material result. In this case the printed densities should be increased correspondingly.

THE FLUKA AND KASPRO HADRONIC CASCADE CODES

The limits for the number of different radii and lengths are $i \leq 10$ and $j \leq 10$.

The primary hadron beam, usually a Gaussian distribution in the transverse coordinates x and y and directions x' and y', enters the block at z = 0 and with a chosen radial position R. The densities, however, are calculated assuming cylindrical symmetry. The azimuthal asymmetry resulting from a primary beam entering at R > 0 can be studied by restricting the azimuthal angular range from which the calculated stars and energy deposition are accepted into the histograms (however, the histograms are normalized as if the range in azimuthal angle would be $0 \leq \phi \leq 2\pi$). The resulting star and energy deposition densities are given, as in FLUKA, in the form of longitudinal and transverse density curves and in the form of contour plots.

All materials with densities $\rho < 0.05$ g/cm^3 are treated as vacuum. The user should specify, at z = 0 and R = 0, a material different from vacuum; if possible the dominant material in the study (a very thin slab of material which will not influence the cascade is permitted).

FLUKA, FLUKOO, CYLKAZ, KASPRO, KAPRYM, and KAPRYZ normalize the star and energy deposition densities per incoming particle. The assumption is made that the block of material is sufficiently large so that this corresponds, most of the time, to the number of interacting particles.

In MAGKA and MAGKO the geometry is different than in the above codes. The cascade always starts in one or several targets, all at R = 0 but at different longitudinal positions. Since these targets might be small, all normalization is done with respect to the number of interacting hadrons. The target or targets at R = 0 are surrounded by a hollow cylinder starting at R = 0. The star and energy deposition densities are calculated within the material of this cylinder. Only one material is allowed in MAGKA and MAGKO!

Low-Energy Cascade Particles in FLUKOO and Related Programs

FLUKOO (and related programs CYLKOZ and MAGKO) treat, in addition to the particles considered in FLUKA, low-energy secondary protons and neutrons emitted by intranuclear cascade processes in hadron-nucleus collisions. These particles are determined by an empirical formula[17], described as follows. The intranuclear cascade initiated by the collision of high-energy hadrons with nuclei has been calculated by Bertini[18] and Barashenkov et al.[19] using the Monte Carlo method. These cascade protons and neutrons have been ignored in FLUKA and it is argued that, for a cascade that is already well developed, the fluxes of high-energy particles are proportional to the total fluxes and other characteristics of the cascade. The assumption, however, is not justified if the cascade is weakly developed as in the case of a proton hitting an external target, etc. In such cases the fluxes of low-energy particles are certainly important at large angles to the original beam. There are two ways of introducing the cascade protons and neutrons into extranuclear cascade calculations:

i) The program that is used to calculate the intranuclear cascade might be used as a subroutine each time a hadron-nucleus collision is considered in the extranuclear cascade code. This procedure has been used at Oak Ridge[20].

ii) The fluxes of cascade particles might be represented by empirical formulae, which are used to sample particles each time a hadron-nucleus collision is considered in the extranuclear cascade calculation.

The second procedure will be described here. The formulae are extracted from a large amount of published results on intranuclear cascade calculations[18,19]. The aim is not to represent all details of the fluxes of cascade protons and neutrons, and no fit to the data is performed. For our purposes it is sufficient to represent the most important features of the data and their dependence on several parameters. Some of these features include the following:

i) The number of cascade particles grows with the atomic weight of the target nucleus.

ii) The number of cascade particles grows initially with the kinetic energy of the projectile particles, but eventually becomes constant for energies above about 5 GeV.

iii) Low-energy particles are emitted more isotropically than high-energy particles.

iv) There are slightly more neutrons than protons.

v) The average energies of cascade protons are slightly higher than the average energies of cascade neutrons.

The formulae must be in a form suitable for Monte Carlo sampling, as given below.

The distributions of intranuclear cascade particles are represented as functions of the following parameters:

- the kinetic energy, T_0 (GeV), of the projectile particle in the rest frame of the target nucleus;

- the kinetic energy, T (GeV), of the emitted particle in the rest frame of the target nucleus;

- the solid angle Ω (rad);

- the production angle θ (rad); and

- the atomic weight A of the target nucleus.

The double differential particle spectrum is written in a factorized form as

$$\frac{d^2N}{dTd\Omega} = f(A, T_0, T) \times g(A, T, \theta) , \qquad (15)$$

where the function $g(A, T, \theta)$ is a normalized angular distribution that also depends on the kinetic energy of the produced particle. The function $f(A, T_0, T)$ gives the T-dependence and does not depend on the production angle θ. It is normalized to the total multiplicity of emitted intranuclear cascade protons and neutrons. The kinetic energy-dependent function f(A, T₀, T) is described by a superposition of two exponentials whose parameters are given separately for protons (i = p) and neutrons (i = n) and which depend on T_0 and A as follows:

$$\left(\frac{dN}{dT}\right)_i = f_i(A, T_0, T) = \frac{n_{1i} \exp\left(-T/\alpha_{1i}\right)}{\alpha_{1i}\left[1 - \exp\left(-T_0/\alpha_{1i}\right)\right]} + \frac{n_{2i} \exp\left(-T/\alpha_{2i}\right)}{\alpha_{2i}\left[1 - \exp\left(-T_0/\alpha_{2i}\right)\right]} \quad (16)$$

Here n_{1i} and n_{2i} are normalization parameters which are expressed as functions of A and T_0 (GeV) as follows:

$$n_{1p} = \begin{cases} 0.03 \sqrt{A} & \text{for } T_0 \leq 0.1 \\ 0.06 \sqrt{A} \left[0.5 + (1 + \log_{10} T_0)^2\right] & 0.1 < T_0 < 5 \\ 0.21 \sqrt{A} & T_0 \geq 5 \end{cases}$$

$$n_{2p} = \begin{cases} 0.0035 \sqrt{A} & \text{for } T_0 \leq 0.1 \\ 0.007 \sqrt{A} \left[0.5 + (1 + \log_{10} T_0)^2\right] & 0.1 < T_0 < 5 \\ 0.0245 \sqrt{A} & T_0 \geq 5 \end{cases}$$

$$n_{1n} = \begin{cases} 0.036 \sqrt{A} & \text{for } T_0 \leq 0.1 \\ 0.06 \sqrt{A} \left[0.6 + 1.3 (1 + \log_{10} T_0)^2\right] & 0.1 < T_0 < 5 \\ 0.27 \sqrt{A} & T_0 \geq 5 \end{cases}$$

$$n_{2n} = \begin{cases} 0.0042 \sqrt{A} & \text{for } T_0 \leq 0.1 \\ 0.007 \sqrt{A} \left[0.6 + 1.3 (1 + \log_{10} T_0)^2\right] & 0.1 < T_0 < 5 \\ 0.032 \sqrt{A} & T_0 \geq 5 \end{cases}$$

(17)

The total proton and neutron multiplicities are expressed by

$$\langle n_p \rangle = n_{1p} + n_{2p},$$
$$\langle n_n \rangle = n_{1n} + n_{2n}. \quad (18)$$

The parameters α_{1i} and α_{2i} also depend on T_0 and A according to

$$\alpha_{1p} = \begin{cases} (0.019 + 0.0017\, T_0)(1 - 0.001A) & \text{for} \quad T_0 < 5 \\ 0.027\,(1 - 0.001A) & T_0 \geq 5 \end{cases}$$

$$\alpha_{1n} = \begin{cases} (0.017 + 0.0017\, T_0)(1 - 0.001A) & \text{for} \quad T_0 < 5 \\ 0.023\,(1 - 0.001A) & T_0 \geq 5 \end{cases}$$

$$\alpha_{2p} = \begin{cases} (0.11 + 0.01\, T_0)(1 - 0.001A) & \text{for} \quad T_0 < 5 \\ 0.16\,(1 - 0.001A) & T_0 \geq 5 \end{cases}$$

$$\alpha_{2n} = \begin{cases} (0.1 + 0.01\, T_0)(1 - 0.001A) & \text{for} \quad T_0 < 5 \\ 0.15\,(1 - 0.001A) & T_0 \geq 5 \end{cases}$$

(19)

The average kinetic energy of one intranuclear cascade particle emitted is calculated from Eq. (16) to be

$$\langle T_i \rangle = \frac{\int_0^{T_0} T (dN/dT)_i\, dT}{\int_0^{T_0} (dN/dT)_i\, dT} =$$

$$= 0.9\, \frac{\alpha_{1i}\left[1 - \left(T_0/\alpha_{1i} + 1\right)\exp\left(-T_0/\alpha_{1i}\right)\right]}{1 - \exp(-T_0/\alpha_{1i})} +$$

$$+ 0.1\, \frac{\alpha_{2i}\left[1 - \left(T_0/\alpha_{2i} + 1\right)\exp\left(-T_0/\alpha_{2i}\right)\right]}{1 - \exp\left(-T_0/\alpha_{2i}\right)} =$$

$$= 0.9\,\left(T_{i1}\right) + 0.1\,\left(T_{i2}\right).$$

(20)

The total energy spent in producing intranuclear cascade protons and neutrons is then equal to

$$T_{i,tot} = n_{1i}\langle T_{1i} \rangle + n_{2i}\langle T_{2i} \rangle.$$

(21)

The proton and neutron multiplicities $\langle n_p \rangle$ and $\langle n_n \rangle$, as expressed by Eqs. (17) and (18), are plotted in Fig. 1 for several materials in terms of the primary proton energy T_0. The total energies, according to Eq. (21), are plotted in a similar manner in Fig. 2.

The angular distribution function is represented by

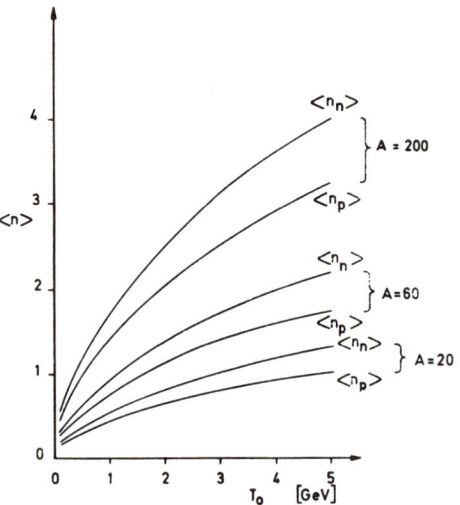

Fig. 1 The average number of intranuclear cascade protons and neutrons [Eq. (18)] as a function of the kinetic energy T_0, for nuclei with atomic weights A = 20, 60, and 200.

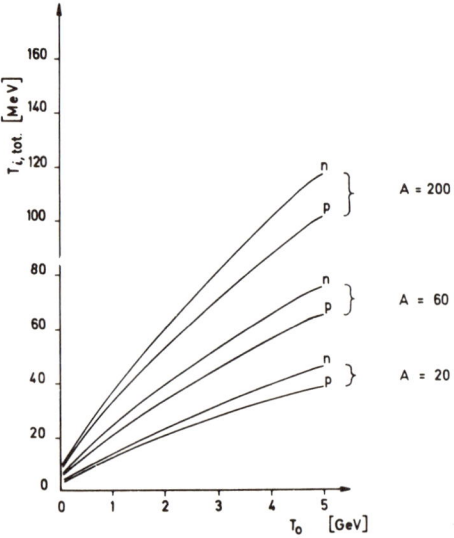

Fig. 2 The total kinetic energy spent for the production of intranuclear cascade protons and neutrons [Eq. (21)] as a function of the kinetic energy T_0, for nuclei with atomic weights A = 20, 60, and 200.

$$g(A, T, \theta) = \begin{cases} N \exp(-\theta^2/\lambda) & \text{for } 0 \leq \theta < \pi/2 \\ N \exp(-\pi^2/4\lambda) = \text{const} & \text{for } \pi/2 \leq \theta \leq \pi. \end{cases} \quad (22)$$

The parameter λ depends on the target material and the kinetic energy T of the produced particle according to

$$\lambda = (0.12 + 0.00036A)/T. \quad (23)$$

N normalizes the integrated angular distribution to unity.

COUPLING KASPRO WITH THE ELECTROMAGNETIC CASCADE CODE EGS

Since the construction of the 400 GeV proton synchrotrons at Fermilab and CERN it has been recognized that such machines are capable of producing γ, e^+ and e^- beams, as well as hadron beams. In fact, such secondary beams are in operation at present. In spite of this, however, the yields are not as systematically well understood as they are in the hadron beam case. The following describes the coupling of KASPRO with EGS[21] for the purpose of calculating γ, e^+ or e^- yields[22].

High-energy photons, mainly from π^0 decay, convert into e^+e^-. KASPRO calls EGS each time a photon has been created by the decay of a π^0 produced in a hadron-nucleus collision. EGS follows the electromagnetic cascade initiated by this photon until all particles are below an energy cut-off (taken to be 1 GeV) or until they leave the target. In the latter case, the KASPRO analysis routine is called and the required histograms are constructed. As one would expect, the results of such calculations show that tungsten targets are very efficient for converting photons into e^+e^-. However, a much higher yield can be obtained by a judicious choice of beryllium target followed by a tungsten converter. The beryllium makes the neutral pions, and the tungsten converts the photons into electrons and positrons.

Electron or positron production from a proton target has been proposed[23] as an elegant method of filling electron or positron storage rings operating together with a proton synchrotron. Such a scheme could be an interesting alternative to the construction of an electron linac or synchrotron as an injector for the storage ring. The KASPRO/EGS method should, therefore, be of use in such projects as CHEEP[24], ISABELLE[25], and the Serpukhov UNK study[26]. In addition, heat deposition (and other) calculations for e^+e^- targets can be done with the hadron cascade codes discussed in this lecture.

REPRESENTATIVE APPLICATIONS OF THE PROGRAMS

Hadronic cascade star and energy deposition densities belong to the most important quantities calculated. Energy deposition densities are directly related to radiation dose and to heat deposition in the material, as well as to the signal size in scintillation detectors. For these applications the following definitions are often needed

$$1 \text{ GeV} = 1.6 \times 10^{-3} \text{ erg} = 1.6 \times 10^{-10} \text{ J} = 0.368 \times 10^{-10} \text{ cal}$$
$$1 \text{ rad} = 10^2 \text{ erg/g}; \quad 1 \text{ GeV/g} = 1.6 \times 10^{-5} \text{ rad} .$$
(24)

Star densities are of use in estimating the production of radioactive isotopes in the material, and from this the remnant dose rates from induced radioactivity after the end of irradiation by the high-energy particles. For iron, the following empirically relates the star densities calculated by the programs to the induced activity after the irradiation time (t_i = 30 d) and the cooling time (t_c = 1 d) after the end of irradiation:

$$\omega(t_i = 30 \text{ d}; \ t_c = 1 \text{ d}) = \begin{cases} 6 \times 10^{-7} \ \frac{\text{rem/h}}{\text{stars/cm}^3 \text{ s}} & \text{FLUKOO} \\ 1.2 \times 10^{-6} \ \frac{\text{rem/h}}{\text{stars/cm}^3 \text{ s}} & \text{FLUKA} \end{cases}$$
(25)

Comparisons of Results of Hadron Cascade Calculations with Experimental Data

The confidence in the prediction power of the rather complex hadron cascade calculations is justified by comparison with experimental data obtained in a wide variety of situations. Such a study has already been done[1], and in this lecture we will concentrate on work performed since that time (1971).

The computer program FLUKA[4] was used by Henny and Potier[27] to calculate the heat deposition in external targets. The results of the calculation were compared with experimental data obtained at the CERN Proton Synchrotron (PS) with a proton beam of 24 GeV/c incident on external targets (length = 180 mm and diameter = 12 mm). In Table 1 we compare the measured heat deposition with the hadron cascade results. It is to be noted that the heat deposition in heavy materials is up to five times larger than that due to the ionization energy loss of primary particles alone. The calculated heat depositions agree rather well with the experimental data.

Table 1

Heat deposition (cal/10^{12} protons) by a proton beam of 10^{12} particles incident on various targets (length = 180 mm, diameter = 12 mm)[27].

Target material	Al	Ti	Cu	Mo	W
Primary ionization alone	3.5	5.5	10.5	11.2	19.0
Experiment[27]	4.7	10.8	38.0	47.0	93.5
Hadron cascade calculation (FLUKA)	7.0	13.3	28.6	38.8	89.0

In a Fermilab experiment[28] the energy deposited by 300 GeV protons in long targets (Be, Al, Cu, and W) was measured. Each target consisted of six cylinders, separated by 3.76 cm, mounted coaxially with the incident proton beam. The lengths of the target cylinders were $\lambda/4$, $\lambda/2$, $\lambda/2$, λ, λ, and λ, where λ is the absorption length. The results of this experiment were compared with the energy deposition calculated by CYLKAZ[29], which allows for the exact experimental geometry. Good agreement was found in the comparison.

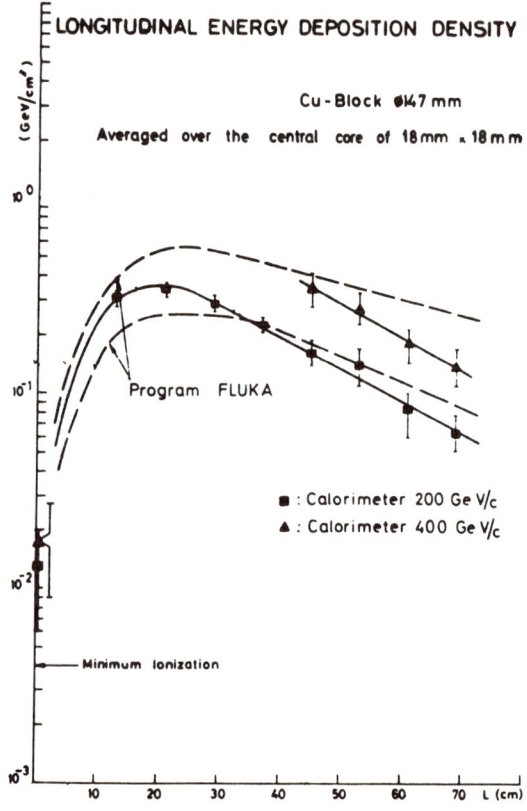

Fig. 3 Experimental and computed energy deposition densities along the axis of a Cu cylinder irradiated with protons of 200 and 400 GeV/c [30].

Measurements of the energy deposition by 200 and 400 GeV/c protons in aluminium and copper targets were reported by Sievers[30], who compared the results with hadron cascade calculations using FLUKA. An example of this is presented in Fig. 3.

FLUKA can be used to simulate the action of total absorption detectors[1]. A modified version of FLUKA was used by Engler et al.[31] in order to optimize the energy resolution of sampling total absorption counters (STAC) for incident neutrons (or other strongly interacting particles). In Fig. 4 we compare the measured energy resolution[20] with their modified version of FLUKA.

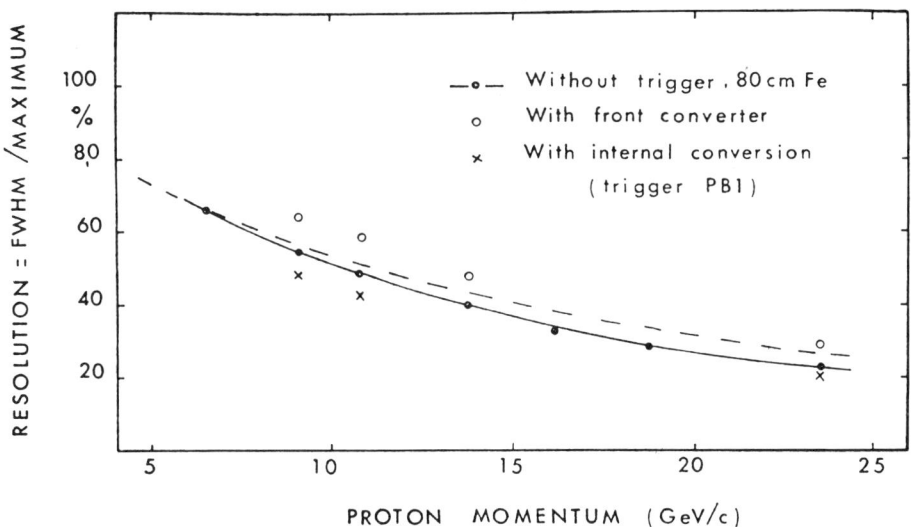

Fig. 4 The energy resolution of a STAC counter as a function of the proton momentum for different trigger conditions compared with the results of the hadronic cascade calculation[31].

Particle fluxes measured at large angles around massive targets in external proton beam lines were compared by Ranft and Routti[22] with results obtained with the computer program FLUK00, and good agreement was obtained. Hadron fluxes in a side shield near a target in an external proton beam line were measured at BNL[32] for incident protons of 15.5 and 28 GeV/c. The results have been compared with MAGKO calculations by Ranft[33]. In situations where particle fluxes and doses perpendicular to the incident proton beam are of interest, it is advantageous to use the programs FLUK00[5], CYLKOZ[10], and MAGKO[7]. In these programs the protons and neutrons resulting from the intranuclear cascade are considered in more detail than in other computer programs.

These programs should also be used if cascades initiated by particles around 1 GeV are studied. FLUK00 has been compared with data in this energy range by Schirrmeister and Ranft[6]. Energy deposition due to 1 GeV protons incident on copper has been studied experimentally by a group from Leningrad using thermoluminescent detectors at various depths in the absorber[34]. The proton beam had a diameter of 0.5 cm and the copper absorber was a cylinder of 20 cm diameter and about 80 cm length. The mean range of 1 GeV protons was found to be Z_R = 468.1 ± 2.3 g/cm^2. By integrating radial distributions at several depths, longitudinal energy deposition points, especially in the region around the Bragg peak, were obtained. FLUKOO results are compared with the data in Fig. 5. The agreement is good, and the Bragg peak is well reproduced. At depths beyond the Bragg peak the calculated energy deposition drops by about one order of magnitude.

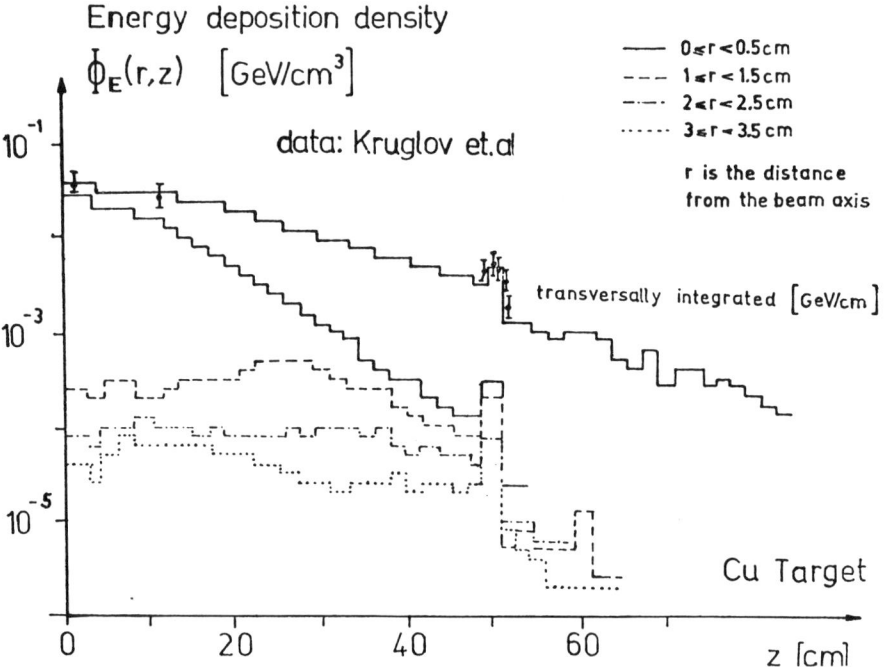

Fig. 5 Longitudinal energy deposition density in a copper absorber by 1 GeV protons as a function of the depth in the absorber. The curves are for four different radial bins, and the transversally integrated density curve is compared with the experimental data[34].

A charged particle detector placed behind a sufficiently thick hadron absorber can be used to sort muons from hadrons. Calculation or measurement of the hadron transmission probability is necessary in order to estimate the hadronic background of such a range detector. We define the hadron transmission probability P(z) as the probability of finding one or more charged particles (p, $\pi^{\pm}$, $e^{\pm}$, etc.) behind an absorber of thickness z for one incident hadron. The transmission probability P(z) has been measured[35] for primary pions with momenta between 0.6 and 4 GeV/c, and for primary protons with momenta between 1 and 4 GeV/c. The data indicate significant differences between incident protons and pions. At large depths P(z) decreases exponentially with a λ equal to the nuclear absorption length. Comparison of FLUK00 results with the experimental data is shown in Fig. 6 for incident protons and in Fig. 7 for incident pions. The agreement is good. The stronger absorption of protons is well reproduced.

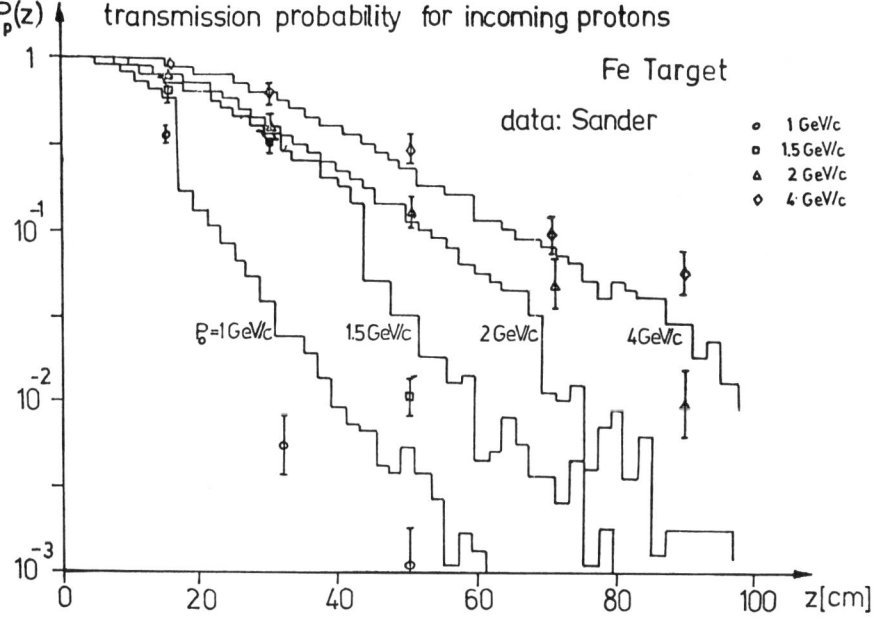

Fig. 6 Transmission probability for incoming protons as a function of the depth in an iron absorber, for primary momenta between 1 and 4 GeV/c, and compared with data[35].

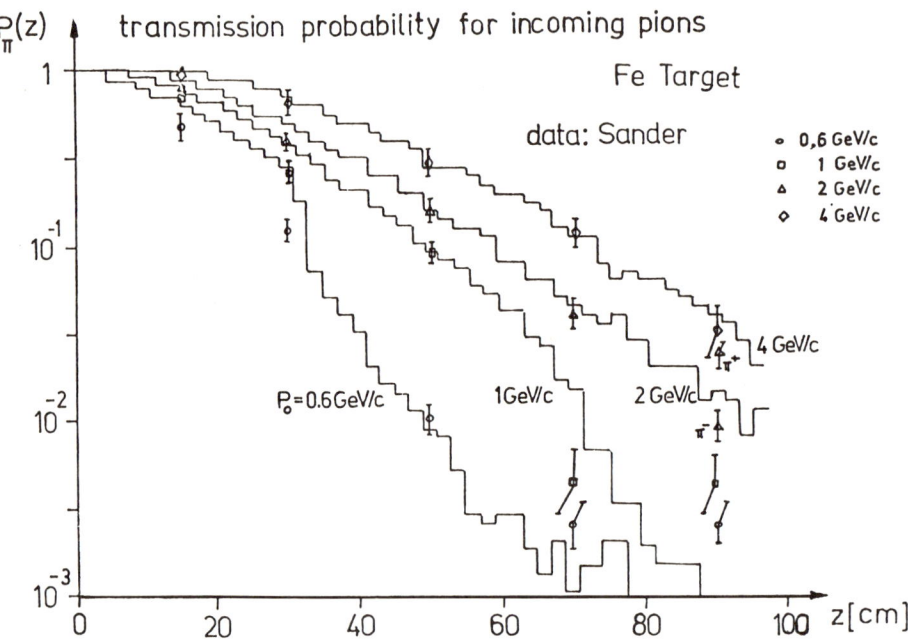

Fig. 7 Transmission probability for incoming pions as a function of the depth in an iron absorber, for primary momenta between 0.6 and 4 GeV/c, and compared with data[35]. The data from π^+ and π^- induced transmission are marked separately only if their difference exceeds the error of the measurement.

Prediction of Radiation Effects Around Accelerators Using Hadron Cascade Calculations

Many applications of hadron cascade calculations for estimating radiation problems around high-energy accelerators are discussed in Refs. 1 and 36. Such applications include estimates of dose to components, induced radioactivity in ejection regions and external target areas of proton accelerators, prediction of target heating and heat deposition in beam dumps, and the calculation of longitudinal and transverse hadron shielding requirements. We discuss here some recent applications, mostly for problems around the 400 GeV CERN Super Proton Synchrotron (SPS).

The thermal effects which occur in external targets when irradiated by fast and slow extracted beams of 400 GeV/c and 10^{13} protons per pulse have been studied by Kalbreier et al.[37] using the program FLUKA. Temperature and thermal stress distributions in targets of different materials were derived from the energy deposition density. The incident beam was assumed to have a Gaussian distribution with 92% of the protons within a diameter of 2 mm. In Table 2 we give calculated temperatures in °C for one nuclear interaction length targets with a 2 mm diameter. In this table T_0 is the maximum temperature rise in the centre of the target for the case of a fast extracted beam, $\bar{T}$ is the average temperature rise at the end of a slow extracted pulse in the absence of cooling, and T_R is the steady-state temperature of the target at the end of the thermal cycle evaluated for cooling by radiation only.

The maximum target temperatures are $T_R + \bar{T}$ for slow extraction and $T_R + T_0$ for fast extraction. The numbers in Table 2 exclude the use of heavy target materials, and forced cooling by convection seems to be advisable. The dynamic stresses created in the targets owing to the rapid heating in fast extracted beams, have to be reduced by subdividing the total target length into several parts. Also, the quasi-static stress produced in the target, as a result of the radial temperature gradient, exceeds the elastic limit in such beams. Thermal problems arising in beam dumps have also been investigated using FLUKA snd MAGKA[37].

Dose to components and remanent dose rates from induced radioactivity have been estimated at the CERN SPS for the ejection region of the main ring and for the target areas in the West, North, and Neutrino zones[38-41]. The results indicate very high dose and radiation levels in critical positions. Therefore it seemed advisable to develop special radiation-hard components for these regions. The experience gained since then indicates that most of the estimates were rather well predicted.

Table 2

Temperature data (in °C) by Kalbreier et al.[37]
(see text for explanation)

Material	T_0	$\overline{T}$	T_R	T_{melt}
Be	490	190	730	1280
BeO	950	420	830	2570
B_4C	1190	550	750	2430
C	1250	510	600	3320
SiC	840	310	910	2700
Al	670	400	660	660
Al_2O_3	1060	450	850	2040
Ti	1670	1070	880	1670
Cu	2900	1910		1080
Mo	9300	4000		2610
W	39000	20000		3380

The programs MAGKA and MAGKO (as well as CYLKAZ and CYLKOZ) are well suited for the study of the effect of heating due to beam losses in superconducting synchrotrons and storage rings. A calculation to study the problems expected with superconducting magnets in the CERN Intersecting Storage Rings (ISR) has been performed by Restat et al.[42]. The effect of radiation heating due to beam losses is not important for the refrigeration requirements, but the sudden temperature rise due to the hadron cascade in the windings can lead to a sudden loss of superconductivity. The tolerable beam loss depends on the arrangements of the superconducting magnets, but it seems that a loss of 10^{12} protons in or near to such a magnet could affect its operation. For these reasons the operation of superconducting synchrotrons and storage rings requires strict control of beam losses. In future accelerators using superconducting magnets, such as ISABELLE, the Fermilab energy doubler, or the Serpukhov UNK, these problems become very important. The energy deposition in the superconducting ring downstream of the ISABELLE beam scrapers has been studied by Blumberg et al.[43] using CYLKAZ and KAPRYZ. The requirement was to limit the energy deposition to values below 3 mJ/cm^3 in the superconducting matrix. This is an extremely small value compared to the total stored energy of the beam (20 MJ), or to the maximum energy deposition encountered when (typically) 0.1% of a 1 mm^2, 200 GeV beam of 7 × 10^{14} particles penetrates the copper matrix (namely, 1 kJ/cm^3). Therefore, special design aspects must be incorporated to exclude energetic particles from striking the vacuum wall. Any such design should be checked with hadronic cascade calculations. The CYLKAZ program has also been used by Stevens and Thorndike[44] to estimate the hadron shielding requirements for ISABELLE.

Optimization of Particle Production from Large Targets

There are many situations around high-energy accelerators where one wants to optimize the secondary particle production. Examples of this kind are the target for neutrino beams, where the maximum intensity of secondary pions and kaons is wanted, or the antiproton target for $\bar{p}p$ colliding beam facilities. The secondary particle production from large targets differs in two respects from the thin target case: i) secondary particles are absorbed inside the target, especially in the case of high-energy secondaries produced at small angles; ii) more secondaries are produced in secondary collisions, but this only helps at low energies and at large production angles. For the CERN SPS neutrino beam the production of pions has been studied using KASPRO[25], where two Be targets (0.3 cm diameter) were considered: length a) 100 cm, and b) 10 pieces (each 10 cm) spread out over a total length of 300 cm (this second target was represented in the calculation by a 300 cm Be target of reduced density $\rho = 0.61$ g/cm^3). The primary proton beam was about half the target diameter. For secondary momenta (20-40 GeV/c) the production from both targets is about equal; for lower momenta the shorter target gives higher production owing to secondary interactions in the cascade. For higher momenta the 300 cm target gives larger production because of the higher probability of secondary absorption in the shorter target.

The optimization of the target to produce antiprotons for the $\bar{p}p$ storage ring facility planned at Novosibirsk was studied[14] using a modified version of KASPRO. The antiproton converter had an intense magnetic field induced by a current pulse along the target. The main point of the investigation was to study whether the focusing action of this magnetic field would increase the phase-space density of the antiprotons by a significant amount. A modified version of KASPRO, taking the action of this magnetic field into account, was used, and the resulting antiproton yield was directly sampled in the form of the phase-space density per given $\Delta p/p$ (see Fig. 8).

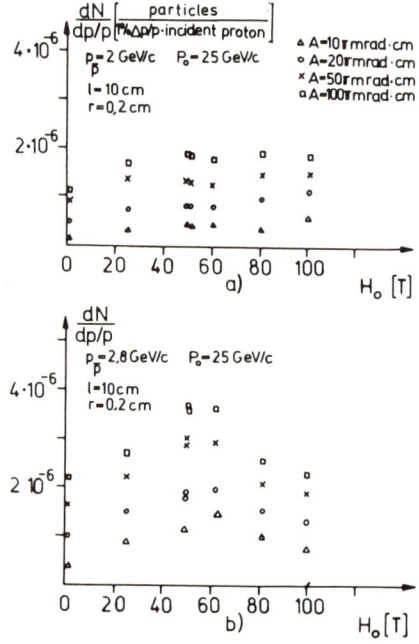

Fig. 8 Number of antiprotons captured in four different phase volumes A, as a function of the magnetic field H_0, for a fixed target.

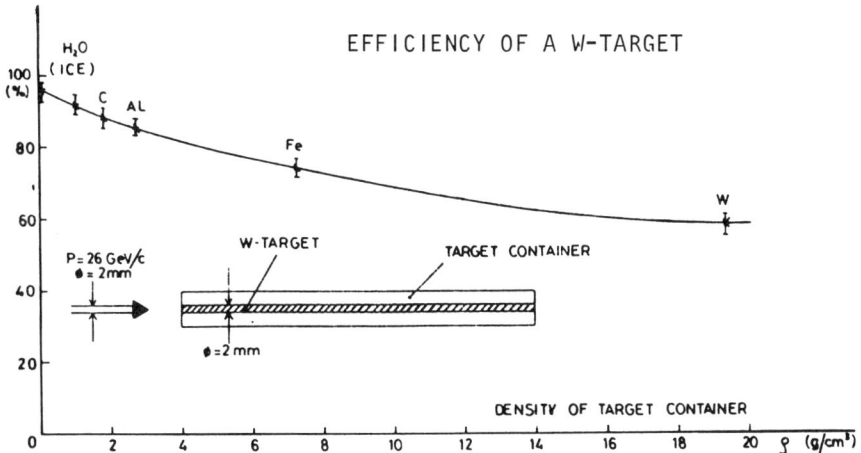

Fig. 9   Target efficiency (escaping $\bar{p}$'s over created $\bar{p}$'s) of a 2 mm diameter tungsten target embedded in various materials. Antiprotons of 3.7 GeV/c produced within a cone of 50 mrad are considered[4,5].

The optimum design of the target to produce antiprotons for the CERN SPS $\bar{p}p$ facility has also been considered by Sievers[4,5]. In Fig. 9 we reproduce the target efficiencies calculated with the program KAPRYZ for thin tungsten targets embedded in materials of different densities.

REFERENCES

1. J. Ranft, Particle Accel. 3, 129 (1972).
2. See Lecture 19.
3. J. Ranft, Nucl. Instrum. Methods 48, 133 (1967); 48, 261 (1967); 81, 29 (1970).
4. J. Ranft and J.T. Routti, Comput. Phys. Commun. 7, 327 (1974).
5. J. Ranft, The computer program FLUKOO, CERN Report SPS-RA/Note/76-1 (1976).
6. V. Schirrmeister and J. Ranft, Nucl. Instrum. Methods 141, 425 (1977).

7. J. Ranft and J.T. Routti, FLUKA and MAGKA, Monte Carlo programs for calculating nucleon-meson cascades in cylindrical geometries, CERN Report LAB II-RA/71-4 (1971).

    J. Ranft, Results of radiation dose calculations with the programs FLUKA and MAGKA and experimental results at 7 and 24 GeV/c, CERN Report LAB II-RA/71-3 (1971).

8. J. Ranft, The computer programme CYLKAZ, Hadron cascade calculation in an array of cylinders out of different materials, CERN Report LAB II-RA/Note/75-12 (1975).

9. J. Ranft, Hadron cascade calculations at large angles to the interaction point, CERN Report LAB II-RA/73-2 (1973).

10. J. Ranft, The computer programme CYLKOZ, CERN Report LAB II-RA/Note/75-12, Add-2 (1977).

11. V.S. Barashenkov, K.K. Gudima, and V.D. Toneev, Energy dependence of the nuclear cross sections for nucleons at energy above 50 MeV, Dubna Report PZ-4183 (1968); V.S. Barashenkov, K.K. Gudima, A.S. Iljinov, and V.D. Toneev, Energy dependence of the interaction cross section of π-mesons with the atomic nuclei at energies more than 50 MeV, Dubna Report P2-4520 (1969).

12. H. Schönbacher, Short write-up of standard RA-Group Monte Carlo Programs available on permanent file in the 7600 computer library, CERN Report LAB II-RA/TM/74-5 (1974).

13. J. Ranft, KASPRO, a Monte Carlo programme to calculate: a) particle fluxes as modified by the hadron cascade in targets, and b) Energy deposition and star densities in the material, CERN Report LAB II-RA/75-1 (1975).

14. B.V. Chirikov, V.A. Tayurski, H.-J. Möhring, J. Ranft, and V. Schirrmeister, Nucl. Instrum. Methods $\underline{144}$, 129 (1977).

15. J. Ranft, The computer program KAPRYZ, CERN Report LAB II-RA/Note/76-2 (1976).

16. J. Ranft, The computer programme KAPRYM, CERN Report HS-RP/IR/77-16 (1977).

17. J. Ranft and J.T. Routti, Particle Accel. 4, 101 (1972).
18. H.E. Bertini, Calculation of nuclear reactions for incident nucleons and π-mesons in the energy range 30 to 2700 MeV, Proc. II Internat. Conf. on Accelerator Dosimetry and Experience, Stanford, 1969, p. 42.
19. V.S. Barashenkov, N.M. Sobolevsky, and V.D. Toneev, Interaction of high-energy radiation with substance, Dubna Report P2-5719 (1971).
20. W.A. Coleman and T.W. Armstrong, The nucleon-meson transport code NMTC, Oak Ridge National Laboratory Report ORNL-4606 (1970).
21. R.L. Ford and W.R. Nelson, The EGS code system: Computer Programs for the Monte Carlo simulation of electromagnetic cascade showers (version 3), Stanford Linear Accelerator Center Report SLAC-210 (1978).
22. J. Ranft and W.R. Nelson, Monte Carlo calculation of photon, electron and positron production from primary proton beams, CERN Report HS-RP/031/PP (1979).
23. V.I. Balbekov et al., The IHEP accelerator and storage complex, Proc. 10th Internat. Conf. on High-Energy Accelerators, Protvino, July 1977; Vol. 1, p. 127.
24. J. Ellis, B.H. Wiik, and K. Hübner, CHEEP, an e-p facility in the SPS, CERN Report CERN 78-02 (1978).
25. H. Hahn, M. Month, R.R. Rau, Rev. Mod. Phys. 49, 625 (1977).
26. Same proceedings as Ref. 23, Vol. 1, p. 177.
27. L. Henny and J.P. Potier, Mesure de l'échauffement de cibles extrêmes par un faisceau de protons à haute énergie, CERN Report MPS/CO/73-2 (1973).
28. M. Awschalom, P.J. Gollon, C. Moore, and A. Van Ginneken, Energy deposition in thick targets by high energy protons, measurements and calculations, Fermilab Report FN-278 (1975).
29. J. Ranft, Comparison of the energy deposition in targets as obtained by hadron cascade calculations with a recent FNAL experiment, CERN Report LAB II-RA/Note/75-13 (1975).

30. P. Sievers, Measurements of the energy deposition of 200 and 400 GeV/c protons in aluminium and copper, CERN Report SPS/ABT/77-1 (1977).
31. J. Engler, W. Flauger, B. Gibbard, F. Mönnig, K. Runge, and H. Schopper, Nucl. Instrum. Methods 106, 189 (1973).
32. G.W. Bennett, G.S. Levine and W.H. Moore, Particle Accel. 2, 251 (1971).
33. J. Ranft, Hadron cascade calculations at large angles to the interaction point, CERN Report LAB II-RA/73-2 (1973).
34. S.P. Kruglov, I.V. Lopatin, K.F. Mus, and V.D. Savelier, Leningrad A.F. Joffe Institute Report (1971).
35. H.G. Sander, Myon-Identifikation mit Hilfe eines Reichweitedetektors, Diplomarbeit, TH Aachen (1975).
36. K. Goebel (editor), Radiation problems encountered in the design of multi-GeV research facilities, CERN 71-21 (1971).
37. W. Kalbreier, W.C. Middelkoop, and P. Sievers, External targets at the SPS, CERN Report LAB II-BT/74-1 (1974).
38. M. Ellefsplass, K. Goebel, J. Ranft, and H. Schönbacher, Estimations of dose to components and dose rates from remanent activity in the West target area and in the neutrino cave, CERN Report LAB II-RA/74-1 (1974).
39. H. Schönbacher, Calculation of energy deposition and star densities with the computer program FLUKA for beryllium, carbon, titanium, copper, and wolfram at 10 and 400 GeV/c, CERN Report LAB II-RA/Note/73-21 (1973).
40. K. Goebel, J. Ranft, J.T. Routti, and G.R. Stevenson, Estimation of remanent dose rates from induced radioactivity in the SPS ring and target area, CERN Report LAB II-RA/73-5 (1973).
41. M. Ellefsplass, K. Goebel, J. Ranft, H. Schönbacher, and G.R. Stevenson, Dose to accelerator components and remanent γ dose rates in the SPS primary beam areas, CERN Report LAB II-RA/75-4 (1975).

42. C. Restat, H. Schönbacher, and M. Van de Voorde, The effect of radiation heating on superconducting coils, CERN Report ISR-MA/75-20 (1975).
43. L. Blumberg, J. Ranft, A. Stevens, and A. Van Steenbergen, Caloric deposition downstream of the ISA beam scrapers, Brookhaven National Laboratory Report BNL-20550 (1975); p.452.
44. A.J. Stevens and A.M. Thorndike, Estimating ISABELLE shielding requirements, Brookhaven National Laboratory Report BNL-50540 (1976).
45. P. Sievers, Some remarks concerning the $\bar{p}$-production target, CERN Report SPS/ABT/TN/78-11 (1978).

LECTURE 23: THE HETC HADRONIC CASCADE CODE

Tony W. Armstrong

Science Applications, Inc.
La Jolla, CA 92038
USA

INTRODUCTION

There are basically four computer codes in the world today that treat the three-dimensional development of hadronic cascades in thick targets, and these codes can be classified into two categories according to their treatment of nonelastic nuclear interactions. All four of these codes use Monte Carlo methods.

There are two transport codes which use the intranuclear-cascade-evaporation model (Lecture 20) to determine the products (energy and angular distributions and multiplicities) from nonelastic collisions: the HETC code (Ref. 1), developed in the U.S., and the code written by Barashenkov, et al., (Ref. 2) at Dubna, U.S.S.R. The capabilities of these two codes are quite similar.

There are two distinguishing characteristics of these codes: first, they treat nuclear interactions and the particle transport in considerable detail - e.g., at each nuclear interaction a Monte Carlo calculation is performed to "follow" the incident particle inside the nucleus to determine the intranuclear cascade and the collision products. These products include low energy particles, so, for example, the effects of low energy (down to thermal) neutrons can be determined. Secondly, these codes are presently not applicable for beam energies greater than a few tens of GeV.

The other hadronic cascade codes are those which use semi-empirical formula, with parameters chosen to represent what accelerator data are available at high energies to determine the description of hadron-nucleus collision products. There are two computer codes in this category: 1) those of J. Ranft of Karl-Marx-University (Ref. 3 and Lecture 22), which have, and continue to be, applied extensively at CERN, and 2) CASIM, which has been developed at Fermilab, by Van Ginneken et al., over the past few years (Ref. 4 and Lecture 21). The important feature of these codes is that the production models used provide the high-energy (> few GeV) secondary particle spectra but not the production at low energies. Also, they calculate the collision products using simple analytical expressions and table look-ups for parameters, so these codes require much less computer time, and are more amenable to incorporating importance sampling techniques to improve computation efficiency than the HETC and Dubna codes.

In this paper, the basic features of the HETC code are discussed. Like many large computer codes HETC has been developed over many years and represents the contributions of numerous people. HETC evolved from the earlier codes NTC and NMTC (Ref. 5). There are three versions of HETC today: the one available from RSIC (Lecture 9), and two versions maintained at Oak Ridge National Laboratory and at Science Applications, Inc. Some of the updated features of the ORNL and SAI versions are indicated in Table 1.

Some of the basic features of the HETC code are summarized in Table 2.

PHYSICS ASPECTS

Atomic Interactions

Rather standard formulas and procedures are incorporated in HETC

TABLE 1. EVOLUTION OF HETC CODE

1. NTC (~early 1960's) — $E_0 \gtrsim 400$ MeV; now obsolete

2. NMTC (~1971) — $E_0 \gtrsim 3$ GeV; now obsolete

3. HETC/RSIC Version (~1973)
   - E_0 above 3 GeV
   - Incorporates MECC-7, EVAP IV

Post 1973 Developments:

4. HETC/ORNL Version
   - Update of nuclear data for evaporation
   - Incorporation of range straggling
   - Update of multiple coulomb scattering treatment

5. HETC/SAI Version
   - All of above physics updates, plus:
     - Time dependence
     - Extension to transport of "Light Heavy Ion" beams (deuterons through alpha particles)
   - Several programming updates e.g.,
     - IBM & CDC versions
     - Combinatorial geometry interface
     - Additional options for speed/storage

TABLE 2. OVERVIEW OF HETC CODE

- Code Name: HETC (High Energy Radiation Transport)

- Particles Transported: Neutrons ($\gtrsim$15 MeV), Protons, $\pi^\pm$ $\mu^\pm$

- Particle Energies Allowed: $\lesssim$100 GeV

- Mechanisms Included:
  - Ionization and Excitation
  - Multiple Coulomb Scattering
  - Range Straggling
  - Pion and Muon Decay
  - Nuclear Interactions

- Calculational Method:
  - Monte Carlo

- Nuclear Collision Model:
  - Intranuclear-Cascade-Evaporation

- Geometry:
  - Three Dimensional

- Materials Allowed:
  - Essentially Arbitrary

- Restrictions:
  - π°, γ, and Heavy Particle (H^2, H^3, He^3, He^4, Residual Nuclei) Products) Computed But Not Transported.

for treating stopping powers (including density effect corrections), mean ranges and range straggling, and multiple coulomb scattering (See Ref. 6).

Nuclear Interactions

The method used in treating nuclear collisions depends upon: 1) whether the collision is elastic or nonelastic, 2) whether the struck nucleus is hydrogen or not, and 3) whether the energy of particle is above or below energy $E_c \sim 3$ GeV (See Table 3). For nonelastic collisions, A>1, and $E<E_c$, the intranuclear-cascade-evaporation model codes MECC-7 and EVAP-IV are used as subroutines (See Lecture 20). For nonelastic collisions with hydrogen and $E<E_c$, the same proton differential cross sections as contained in MECC-7 are used.

For nonelastic collisions, A>1, and $E>E_c$, an extrapolation method (Refs. 7,8) is used in which scaling relations are applied to intranuclear-cascade-evaporation results at $E=E_c$.

Decay

Charged pion and muon decay in-flight are taken into account. Muons and π^+ that reach the end of their range are assumed to decay, and the π^- that stop are allowed to either decay or undergo nuclear capture, depending upon an input option. The π^- capture products are computed and transported.

MISCELLANEOUS ALGORITHMS

At each nonelastic interaction that takes place during the Monte Carlo transport, the intranuclear-cascade-evaporation routines are entered to obtain the collision products for subsequent transport. This avoids the need for handling a large data base for particle-nucleus differential cross sections, and allows very general material compositions to be easily accomodated, at the expense (usually) of some additional computation time.

TABLE 3. HETC TREATMENT OF NUCLEAR COLLISIONS[a,b,c]

NONELASTIC COLLISIONS				ELASTIC COLLISIONS			
A = 1		A > 1		A = 1		A > 1	
$E < E_c$	$E > E_c$	$E < E_c$	$E > E_c$	$E < E_c$	$E > E_c$	$E < E_c$	$E > E_c$
Same n-p, p-p, and π-p cross sections as used in INC	Ranft distributions	INC model	INC plus extrapolation model	Same p cross sections as used in INC	Parametric fits to experimental data	Neglected except for neutrons (d)	Neglected

a. INC ≡ intranuclear-cascade-evaporation model (MECC-7 and EVAP-IV, see Lecture 20).
b. A=1 denotes collisions with hydrogen nuclei; A>1 denotes all nonhydrogeneous target nuclei.
c. E_c=3.5 GeV for nucleons, 2.5 GeV for charged pions.
d. Proton-nucleus and pion-nucleus elastic collisions neglected; neutron-nucleus elastic collisions included optionally below 100 MeV.

HETC is basically an analog Monte Carlo code and contains minimal importance sampling. However, statistical weights are incorporated, so that some simple biasing techniques which are often very effective (source distributions, spatial distribution of initial collisions, etc.) can be easily applied by the user.

CAPABILITIES AND CONSTRAINTS

In terms of allowable target configurations and compositions, HETC is quite general. Up to 15 different material compositions and up to 11 different isotopes per medium are allowed in describing the target. HETC has been interfaced with several general, three-dimensional geometry modules.

The upper particle energy limit for HETC is not well defined. It is expected to be most accurate below ~ 3 GeV where only the intranuclear-cascade model is used. Extension to higher energies using the extrapolation model is expected to provide reasonably accurate results for most problems up to ~ 10 GeV. (For example, good agreement with experimental data has been obtained for 30-GeV protons, Ref. 6). In some cases (e.g., interest in high-energy particle fluxes or star densities), HETC has given results in good agreement with other calculational methods for proton energies up to several hundred GeV (e.g., Ref. 6). However, it is known that the extrapolation model used underestimates the $\pi°$ production (and, therefore, their subsequent effects, such as energy deposition) in this high energy region.

OUTPUT PROVIDED

HETC provides the following information at each "event" which occurs during the transport calculation:

- o   The "state" of the particle being transported at the present and previous event (energy, position, direction cosines, and statistical weight).

- o   Information on the particles produced in nuclear collisions (type, energy, direction, etc.).

o   The charge and mass of the residual nucleus left after the collision.

The above information is made available to the user for analysis for various type of events which occur: nuclear interactions, material boundary crossings, particle escapes from the system, etc.

CODE STRUCTURE

Figure 1 illustrates the interfacing of HETC with other codes.

HETC computes the production of low energy ($\lesssim 15$ MeV) neutrons at each collision site during the transport calculation, but these neutrons are not transported by HETC because the intranuclear cascade model is not applicable at such low energies. However, for problems where these neutrons are important, the source description generated by HETC can be used as input to available Monte Carlo (e.g., MORSE, Lecture 6) or discrete ordinates (e.g., DOT, Lecture 5) low-energy neutron transport codes.

HETC does not transport the photons created in the hadronic cascade. However, the energy and direction of the neutral pions produced at each collision site are provided, so the photon source is defined. Also, HETC provides information so that the electron-positron source from muon decay can be determined. Thus, HETC can be easily interfaced with electron-photon shower codes (e.g., EGS, Lecture 12) to determine the effects of the electromagnetic cascade where important.

The detailed transport results provided by HETC for each "event" that occurs during the Monte Carlo calculation are analyzed by user-written analysis programs to determine the results of interest for a particular problem set-up. A user package of subroutines for some of the more common quantities of interest (particle fluxes, dose, etc.) is available (Ref. 9).

HETC APPLICATIONS

To illustrate the general applicability of the HETC code, Table 4 indicates some of the applications in various technical areas that have been made. Most of these applications have involved comparisons with experimental data.

TABLE 4. SAMPLE HETC APPLICATIONS

1. Accelerator Shielding
    - Prompt Radiations (e.g., Ref. 10)
    - Activation (e.g., Refs. 11-14)

2. Space Radiations
    - Spacecraft Shielding (e.g., Ref. 15)
    - Induced Radioactivity in Meteorites (Ref. 16)
    - Moon
        - Radionuclide Production (Ref. 17)
        - γ-ray Leakage (Ref. 18)
    - Jupiter
        - Effects of radiation belts on induced radiations interior to Jupiter's moons (Ref. 19)
    - Earth's Atmosphere
        - Neutron Cosmic Ray Background (Ref. 20)
        - Dose produced at aircraft altitudes by solar flares (Ref. 21)

3. Radiobiology
    - Comparisons of dose, LET distributions, cell survival, RBE, etc. for various particles (n, p, π^-, etc.) (Refs. 22-24)

4. Dosimetry
    - Calorimeters (Refs. 25-31)
    - Microdosimetry - LET Chambers
    - Silicon detectors (Refs. 32, 33)
    - Tissue Equivalence (Ref. 34)

5. Miscellaneous
    - Flux-to-Dose Factors (MeV-TeV Energies) (Refs. 35-38)
    - Accelerator Breeding of Fissile Materials
    - Transmutation of Nuclear Waste Materials
    - Deuteron and Alpha Particle Transport in Thick Targets (Ref. 39)

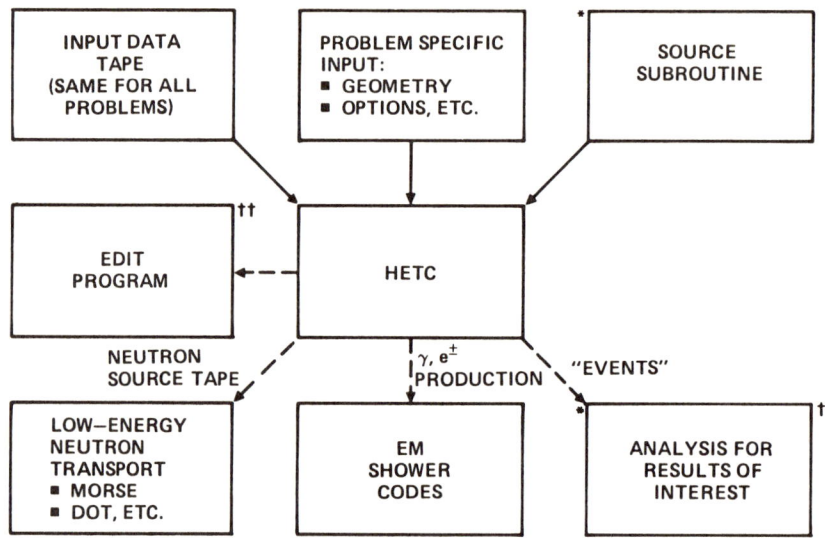

* USER WRITTEN
† SOME "STANDARD" SUBROUTINES AVAILABLE
†† PROVIDES PRINTED OUTPUT OF VARIABLES AT EACH EVENT FOR DEBUGGING.

FIGURE 1. HETC CODE INTERFACES

REFERENCES

1. T. W. Armstrong and K. C. Chandler, "HETC - A High Energy Transport Code", Nucl. Sci. and Engr. 49, 110 (1972).

2. V. S. Barashenkov, N. M. Sobolevskii, and V. D. Toneev, "The Penetration of Beams of High-Energy Particles Through Thick Layers of Material", Atomnaya Energiya 32, 217 (1972).

3. J. Ranft, "Estimation of Radiation Problems Around High Energy Accelerators Using Calculations of the Hadronic Cascade in Matter", Particle Accelerators 3, 129 (1972).

4. A. Van Ginneken, "CASIM, Program to Simulate Hadronic Cascades in Bulk Matter", Fermilab Report FN-272 (1975).

5. W. A. Coleman and T. W. Armstrong, "NMTC - A Nucleon - Meson Transport Code", Nucl. Sci. and Engr. 43, 353 (1971).

6. T. W. Armstrong, R. G. Alsmiller, Jr., K. C. Chandler, and B. L. Bishop, "Monte Carlo Calculations of High-Energy Nucleon-Meson Cascades and Comparison with Experiment," Nucl. Sci. and Engr. 49, 82 (1972).

7. T. A. Gabriel, R. G. Alsmiller, Jr., and M. P. Guthrie, "An Extrapolation Method for Predicting Nucleon and Pion Differential Production Cross Sections from High-Energy (>3 GeV) Nucleon-Nucleus Collisions," Oak Ridge National Laboratory Report ORNL-4542 (1970).

8. T. A. Gabriel, R. T. Santoro and J. Barish, "A Calculational Method for Predicting Particle Spectra From High Energy Nucleon-Pion Collisions (>3 GeV) with Protons", Oak Ridge National Laboratory Report ORNL-TM-3615 (1971).

9. T. W. Armstrong and K. C. Chandler, "Analysis Subroutines for the Nucleon-Meson Transport Code NMTC", Oak Ridge National Laboratory Report ORNL-4736 (1972).

10. T. W. Armstrong and R. G. Alsmiller, Jr., "Calculation of the Residual Proton Dose Rate Around High-Energy Proton Accelerators," Nuclear Science and Engineering 38, 53 (1969).

11. R. G. Alsmiller, Jr., et al, "Photon Dose Rate From Induced Activity in the Beam Stop of a 400 GeV Proton Accelerator", Nucl. Instr. Meth. 155, 399 (1978).

12. T. W. Armstrong and J. Barish, "Calculation of the Residual-Photon Dose Rate Due to the Activation of Concrete by Neutrons from a 3-GeV Proton Beam in Iron," Nucl. Sci. and Engr. 38, 265 (1969).

13. T. W. Armstrong and J. Barish, "Calculated Activation of Copper and Iron by 3-GeV Protons, "Nucl. Sci. and Engr. 41, 443 (1970).

14. T. W. Armstrong and J. Barish, "Reduction of the Residual Photon Dose Rate Around High-Energy Proton Accelerators," Nucl. Sci. and Engr. 40, 128 (1970).

15. R. G. Alsmiller, Jr., R. T. Santoro, J. Barish, and H. C. Clairborne, "Shielding of Manned Space Vehicles Against Protons and Alpha Particles", Oak Ridge Radiation Shielding Information Center Report ORNL-RSIC-35 (1972).

16. T. W. Armstrong, "Monte Carlo Calculations of Residual Nuclei Production in Thick Iron Targets Bombarded by 1- and 3-GeV Protons and Comparison with Experiment," J. of Geophys. Res. 74, 1361 (1969).

17. T. W. Armstrong and R. G. Alsmiller, Jr., "Calculation of Cosomogenic Radionuclides in the Moon and Comparison with Apollo Measurements," Proc. 2nd Lunar Sci. Conf., Geochim Cosmochim. Acta Suppl. 2, 2, 1729 (1971).

18. T. W. Armstrong, "Calculation of Lunar Photon Albedo from Galactic and Solar Proton Bombardment,". Geophy. Res. 77, 524 (1972).

19. T. W. Armstrong, B. L. Colborn and P. Read, "Calculations of the Radiation Background Produced in Jovian Satellites by Trapped Proton Bombardment," SAI-76-658-LJ (1976).

20. T. W. Armstrong and K. C. Chandler, "Calculations of Neutron Flux Spectra Induced in the Earth's Atmosphere by Galactic Cosmic Rays," Geophys. Res. 78, 2715 (1973).

21. T. W. Armstrong, R. G. Alsmiller, Jr., and J. Barish, "Calculation of the Radiation Hazard at Supersonic Aircraft Altitudes Produced by an Energetic Solar Flare," Nucl. Sci. and Engr. 37, 337 (1969).

22. T. W. Armstrong and K. C. Chandler, "Calculations Related to the Application of Negatively Charged Pions in Radiotherapy: Absorbed Dose, LET Spectra, and Cell Survival," Rad. Res. 58, 293 (1974).

23. T. W. Armstrong and K. C. Chandler, "Monte Carlo Calculations of the Dose Induced by Charged Pions and Comparison with Experiment," Rad. Res. 52, 247 (1972).

24. R. G. Alsmiller, et al., "Calculations Related to the Possible Use of Photons, Neutrons, Negatively Charged Pions, Protons, and Alpha Particles in Radiotherapy," Rad. Res. 60, 369 (1974).

25. T. A. Gabriel and B. L. Bishop, "Calculated Hadronic Transmission Through Iron Absorbers," Nucl. Instr. Meth. 155, 81 (1978).

26. T. A. Gabriel, et al., "Calculated Performance of a Mineral-Oil-Iron Ionization Spectrometer," Nucl. Instr. Meth. 129, 409 (1975).

27. T. A. Gabriel, et al., "Preliminary Design Calculations for an Ionization Spectrometer for Use in Colliding Beam Experiments," Nucl. Instr. Meth. 130, 375 (1975).

28. J. D. Amburgey and T. A. Gabriel, "Calculated Performance of a Segmented Pyramid-Shaped Calorimeter of Iron and Plastic," Nucl. Instr. Meth. 133, 75 (1976).

29. R. T. Santoro, et al., "The Calculated Response of a Liquid Scintillator Total-Absorption Hadron Calorimeter," Nucl. Instr. Meth. 134, 87 (1976).

30. T. A. Gabriel and W. Schmidt, "Calculated Performance of Iron-Plastic Calorimeters for Incident Hadrons with Energies of 5 to 75 GeV," Nucl. Instr. Meth. 134, 271 (1976).

31. T. A. Gabriel, "Uranium Liquid-Argon Calorimeters: A Calculational Investigation," Nucl. Instr. Meth. 150, 145 (1978).

32. T. W. Armstrong and K. C. Chandler, "Calculations Related to the Application of Silicon Detectors in Pion Radiobiology," Nucl. Instr. and Meth. 120, 93 (1974).

33. T. W. Armstrong and K. C. Chandler, "Calculations of the Response of Silicon Detectors of Finite Radius to π^- Mesons Stopping in Tissue," Nucl. Instr. and Meth. 123, 363 (1975).

34. T. W. Armstrong and K. C. Chandler, "Calculations on Tissue Equivalence for Stopping π^- Mesons," Oak Ridge National Laboratory Report ORNL-TM-4499.

35. R. G. Alsmiller, Jr., T. W. Armstrong, and B. L. Bishop, "The Absorbed Dose and Dose Equivalent from Negatively and Positively Charged Pions, in the Energy Range 10 to 2000 MeV," Nucl. Sci. Engr. 43, 257 (1971).

36. R. G. Alsmiller, Jr., T. W. Armstrong, and W. A. Coleman, "The Absorbed Dose and Dose Equivalent from Neutrons in the Energy Range 60 to 3000 MeV and Protons in the Energy Range 400 to 3000 MeV," Nucl. Sci. and Engr. 42, 367 (1970).

37. T. W. Armstrong and B. L. Bishop, "Calculation of the Absorbed Dose and Dose Equivalent Induced by Medium-Energy Neutrons and Protons and Comparison with Experiment," Rad. Res. 47, 581 (1971).

38. T. W. Armstrong and K. C. Chandler, "Calculations of the Absorbed Dose and Dose Equivalent from Neutrons and Protons in the Energy Range from 3.5 GeV to 1.0 TeV," Health Physics, 24, 277 (1973).

39. T. W. Armstrong and B. L. Colborn, "LHI Code Development Study", Science Applications, Inc. Report SAI-79-1040-LJ (1979).

UNFOLDING METHODS AND SPECTRUM ANALYSIS

LECTURE 24:   UNFOLDING TECHNIQUES FOR

ACTIVATION DETECTOR ANALYSIS

J.T. Routti and J.V. Sandberg

Helsinki University of Technology
Department of Technical Physics
SF-02150 Espoo 15, Finland

INTRODUCTION

Activation detectors are widely used in measurements of radiation fields. In particular, the energy dependent radiation spectrum or integral quantities, such as the total flux, dose rate or reaction rates, are of interest in these measurements.

Activation detectors do not directly yield the quantities of interest. Solving the radiation spectrum from measured activities requires unfolding or deconvolution. No unique solutions typically exist and prior knowledge is also applied to obtain physically meaningful solutions.

Prior knowledge available depends on the application. In reactor neutron spectroscopy theoretical estimates for the shape of the spectrum are available and the unfolding problem may even be reduced to modifying a trial spectrum to match the measured activities. For accelerators or cosmic radiation spectra good trial spectra are often not available, and more general unfolding methods must be used.

Dose rate and reaction rates can be calculated from unfolded spectra or, in some cases, directly from measurements.

FORMULATION OF THE UNFOLDING PROBLEM

Activation Equation

The general unfolding problem can be expressed as a Fredholm integral equation of the first kind

$$A(E') = \int_{E_{min}}^{E_{max}} K(E',E)\phi(E)dE + \varepsilon(E') \quad , \tag{1}$$

where $\phi(E)$ is the unknown distribution,
$K(E',E)$ is the known kernel or resolution function,
$A(E')$ is the measured response, and
$\varepsilon(E')$ is the uncertainty.

With discrete activation detectors eq. (1) reduces to

$$A_i = \int_{E_{min}}^{E_{max}} \sigma_i(E)\phi(E)dE + \varepsilon_i \quad , \quad i = 1,\ldots,m \quad , \tag{2}$$

where E is the energy variable,
$\phi(E)$ is the differential flux sought after,
$\sigma_i(E)$ is the energy dependent cross sections,
A_i is the saturation activity of detector i,
ε_i is the uncertainty of activity A_i,
m is the number of detectors.

For numerical solution the continuous energy variable is made discrete. Numerical quadrature transforms eq. (2) to a group of linear equations

$$A_i = \sum_{j=1}^{n} K_{ij} \phi_j + \varepsilon_i \quad , \quad i = 1,\ldots m \quad , \tag{3a}$$

or in matrix form,

$$A = K\phi + \varepsilon \quad , \tag{3b}$$

where n is the number of discrete energy points,
$K_{ij} = w_j\sigma_i(E_j)$ is the cross section of reaction i at energy point E_j multiplied by quadrature weight w_j.

Nonuniqueness of the Solution

Even without uncertainties, $\varepsilon(E')$, eq. (1) would have a unique solution only if the kernel $K(E',E)$ had no zero eigenvalues. The experimental errors make the problem of nonuniqueness even worse. Suppose that $\phi(E)$ were a unique solution of eq. (1) with $\varepsilon(E') = 0$. There might then exist a number of functions $\psi(E)$ for which

$$\left| \int_{E_{min}}^{E_{max}} K(E',E)\psi(E)dE \right| < |\varepsilon'(E')| \qquad (4)$$

and each of functions $\phi(E) + \psi(E)$ satisfies eq. (1) within experimental errors /1/. The solution obtained by unfolding therefore often contains fluctuations, which are not seen by the measuring device.

The discrete form (eq. (3)) can be solved properly if $m = n$. If $m > n$ the equation is overdetermined and it can be solved in the least-squares sense. With activation detector measurements the number of energy points is generally larger than the number of responses (i.e., $n > m$). Even in this case a least-squares solution can, in principle, be obtained using linear estimation theory /2/, as discussed later.

Especially when $n > m$, the unfolding problem is ill-conditioned in the sense that small changes in the measured responses may cause large relative errors in the solution. Hence, special methods have been developed for unfolding activation measurements. Generally they make use of prior knowledge of the spectrum, such as a trial spectrum to which the solution is tied in some way, or smoothness and nonnegativity conditions.

Different Kernels

The choice of the best unfolding method depends on the detector kernel. Problems can be classified as few channel and many channel unfolding problems according to the number of measured responses. Activation detectors typically lead to few channel problems with up to about 10-30 measured responses and 20-100 energy points. In some cases a fine mesh of energy points is used for the pointwise representation of resonance cross sections /3/.

In many channel problems the number of measured responses is about the same or greater than that of energy points in the solution, for example, in measuring neutron spectra with proton recoil emulsions or liquid scintillators /4/.

Measurements with threshold activation reactions and with Bonner spheres lead to few channel problems. If resonance detectors are used, the cross sections can be represented with resonance integrals and the corresponding flux values can be determined, in some cases, directly without unfolding. Figure 1 shows a threshold detector kernel for high energy neutron spectroscopy.

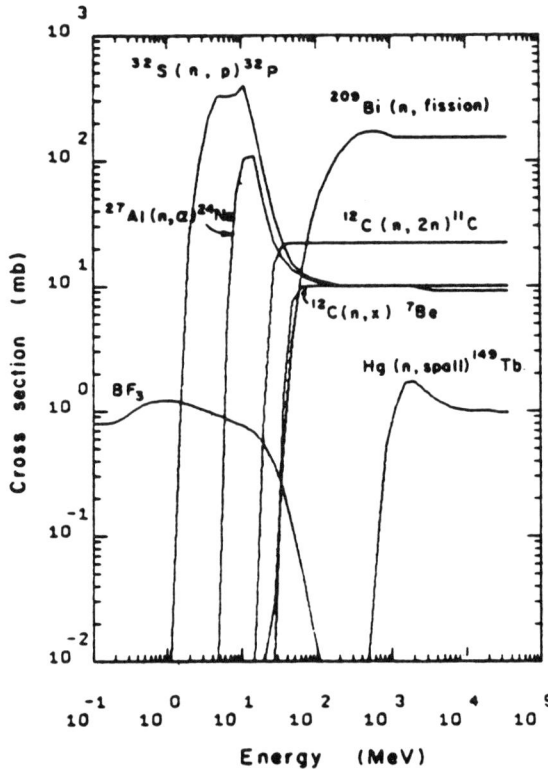

Fig. 1. Cross section of high energy neutron activation detectors.

Figure 2 of lecture 27 shows another threshold kernel, composed of spallation cross sections for producing different reaction products in a copper target. The spallation detectors can be used in measuring high energy neutron fields /5,6/.

The Bonner sphere kernel is presented in Fig. 2. These moderating sphere's have thermal neutron detectors covered with polyethylene of varying thickness. Response functions are calculated for moderators of different sizes.

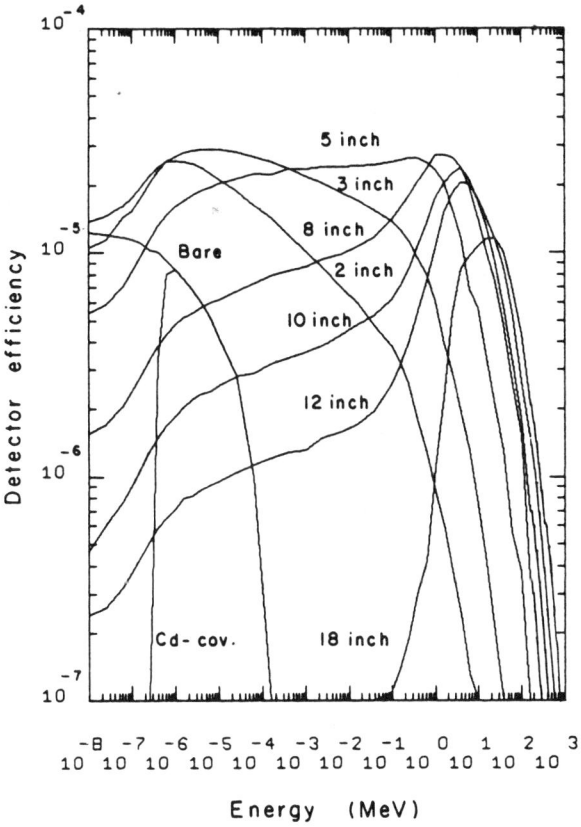

Fig. 2. Response functions of Bonner spheres.

SOLUTION METHODS AND THEIR COMPUTER IMPLEMENTATIONS

Standard methods often employed for unfolding problems include functional representations, orthonormal expansions and least-squares expansion methods. In many applications they have serious shortcomings, however, and special techniques are required. A great number of unfolding codes have been reported for different applications. Some codes are oriented to specific applications while others can be flexibly used for various problems.

Functional Representations

In some cases analytical formulae with a few unknown parameters are available for the solution spectrum. In the simplest case only the normalization of the spectrum is needed and it can be determined with a single detector. Several unknown parameters can be determined by matching the measured and calculated activities in the least-squares sense; in general, with nonlinear minimization programs.

When an appropriate functional representation is available, it is easy to use and the solution remains quite insensitive to measurement errors. Physically acceptable shape of the spectrum can be easily assured. On the other hand, fine structure of the spectrum or deviations from the theoretical model remain unobserved.

The method has been widely used in reactor neutron spectroscopy /7/, with theoretical fission spectrum in the MeV region /8/ and the Maxwellian spectrum in the thermal region /9/. In the intermediate region the 1/E-type model can be used but no fine structure due to resonance absorption can then be observed.

The functional representations have been used also in high energy neutron spectroscopy. Commonly used representations are one- or many-segmented $E^{-\beta}$ shaped spectra.

Series Expansions

In many numerical techniques the spectrum is expressed as a sum of linearly independent orthonormal basis functions

$$\phi(E) = W(E) \sum_{k=1}^{n} \beta_k \psi_k(E) , \qquad (5)$$

where $W(E)$ is a weighting function. Substitution in eq. (2) yields

$$A_i = \sum_{k=1}^{n} \beta_k \int_{E_{min}}^{E_{max}} W(E) \psi_k(E) \sigma_j(E) dE + \varepsilon_i, \quad i = 1,\ldots,m. \qquad (6)$$

Taking the number of basis functions equal to the number of measured activities, $n = m$, and $\varepsilon_i = 0$, yields a well-determined group of linear equations.

In the least squares expansion method with $1 \leq n \leq m$, the coefficients β_k are determined by minimizing the square sum

$$Q = \sum_{i=1}^{m} \left[\frac{A_i - \int_{E_{min}}^{E_{max}} \sigma_i(E) \phi(E) dE}{A_i} \right]^2 . \qquad (7)$$

The optimal value of n corresponds to the smallest value of Q with a physically acceptable solution.

Generally applicable basis functions are difficult to find. The resolution is not very good and experimental errors can cause difficulties. The convergence of the expansion is often slow and negative solution values emerge. Prior information is difficult to apply.

Despite the difficulties, the method has been successfully used in some applications. Ringle /10/, Gold /11/ and di Cola and Rota /12/ have used expansions in determining fast neutron spectra with threshold detectors. The method has been implemented, for example, in the code RDMM /13/.

Nakamura et al. have developed the code LYRA using Laguerre, Hermite or Chebysev polynomials /14,15/. Good results in unfolding bremsstrahlung spectra measured with activation detectors are found.

Linear Estimation Methods

The application of linear estimation to unfolding problems has been discussed by Rust and Burrus /2/. An estimate $\hat{\phi}$ of the solution vector ϕ, which satisfies the activation equation (3), is sought after. Generally the error vector ε is assumed to have normally distributed components with zero mean and covariance matrix S. In point estimation the most probable spectrum is determined. A minimum variance estimate is obtained by minimizing the weighted square sum of errors

$$Q = \varepsilon^T S^{-1} \varepsilon = \sum_{i=1}^{m} \frac{1}{\sigma_i^2} \left[A_i - \sum_{j=1}^{n} K_{ij} \phi_j \right]^2, \qquad (8)$$

where the latter form corresponds to a diagonal S matrix with variances σ_i^2. The solution is then

$$\hat{\phi} = (K^T S^{-1} K)^{-1} K^T S^{-1} A \qquad (9)$$

with the covariance matrix

$$\Sigma = (K^T S^{-1} A)^{-1}. \qquad (10)$$

In the constrained linear estimation the problem is formulated as seeking the minimum and maximum values which a linear function of the spectrum can get, when given error limits are set on the measured responses. The problem can be written as

$$\begin{array}{c} \max \\ \min \end{array} p^k = \sum_{j=1}^{n} u_j^k \phi_j \qquad (11)$$

with one of the following constraints:

a) $\sum_{i=1}^{m} \frac{|\varepsilon_i|}{\sigma_i} < C$, b) $\sum_{i=1}^{m} (\frac{\varepsilon_i}{\sigma_i})^2 < C$, c) $\max_{1 \le i \le m} \frac{|\varepsilon_i|}{\sigma_i} < C$,

where C is a constant determining the confidence level. If the errors are normally distributed, the value C = 1 corresponds to 68% and C = 2 to 95% confidence level.

The conditions a)-c) determine different confidence norms. The ellipsoid condition b) would be physically most realistic but it leads to a time-consuming quadratic programming. Conditions a) and c) lead to linear programming. The preferred condition a) gives the most conservative estimate and leads to a simpler linear programming problem /2/.

Linear estimation theory is well suited for overdetermined cases, such as recoil proton spectrum unfolding. Well known codes are FERDOR, COOLC and FORIST developed by Rust et al. /16,17/ and designed primarily for neutron spectrum measurements with liquid scintillators.

Iterative Methods

In iterative methods a trial spectrum is first chosen using theoretical considerations or previous measurements. It is then modified to match the measured activities.

A widely used iterative unfolding code is SAND II, developed by McElroy and Simons /18/. The solution spectrum $\phi_j^{(k)}$, j = 1,....,n, of iteration step k, is obtained from the trial spectrum $\phi_j^{(k-1)}$ as follows. The ratio R_i of measured and calculated activity is determined for each detector

$$R_i^{(k)} = A_i / \sum_{j=1}^{n} K_{ij} \phi_j^{(k-1)}, \quad i = 1,\ldots,m . \qquad (12)$$

The spectrum is then modified by setting

$$\phi_j^{(k)} = M_j^{(k)} \phi_j^{(k-1)}, \quad j = 1,\ldots,n \qquad (13)$$

where the modifying factor is defined by the equation

$$\ln M_j^{(k)} = \frac{\sum_{i=1}^{m} W_{ij}^{(k)} \ln R_i^{(k)}}{\sum_{i=1}^{m} W_{ij}^{(k)}} \quad . \tag{14}$$

The weighting factors W_{ij} are determined by the relative response of detector i in the energy bin j,

$$W_{ij}^{(k)} = \sigma_{ij} \phi_j^{(k)} / \sum_{i=1}^{n} \sigma_{ij} \phi_j^{(k)} \quad . \tag{15}$$

SAND II gives good results when the trial spectrum is fairly good and the response functions sufficiently cover the whole energy range. In regions not covered by the response functions the solution is shaped as the trial spectrum. With poor trial spectra the number of iterations is large and the solution may develop spurious peaks. The modifying procedure is not necessarily optimal and alternative schemes with possibly faster convergence have been proposed /19/.

The SAND II program package includes cross sections for 40 reactions and a library of 60 trial spectra, both with 620 energy bins. A large number of energy groups can be used with reasonable computing times, which is particularly useful with resonance reactions. The code also includes options for cadmium covers on the detectors. Monte Carlo error simulations have also been carried out /20/.

SAND II has been used mainly in reactor spectroscopy. Hargreaves and Stevenson /21/ have applied a simpler iterative method to high energy neutron spectroscopy, and Sanna /22/ has developed the code BON for unfolding Bonner sphere measurements.

Regularization Methods

A wide class of unfolding methods employ so called regularization. They are least-squares methods with prior conditions expressed as constraints on the solution. Typical conditions require a resemblance of the solution to a given trial spectrum, and smoothness and nonnegativity. Such techniques have been proposed by Phillips /23/, Twomey /24/ and Tikhonov /25/.

The regularization problem can be written as

$$\min Q = \sum_{i=1}^{m} r_i^{\varepsilon} \varepsilon_i^2 + \gamma L(\phi) , \qquad (16)$$

where ε_i is the difference between measured and calculated response,

r_i^{ε} is the weight of response i,

$L(\phi)$ is the regularization functional containing the prior information,

γ is the regularization parameter.

Smoothness cna be required by minimizing the square integral of the second derivative. For the discrete solution, the functional is

$$L_d(\phi) = \sum_{j=2}^{n-1} r_j^d (\phi_{j-1} - 2\phi_j + \phi_{j+1})^2 . \qquad (17)$$

The resemblance to a trial solution P can be put in the form

$$L_p(\phi) = \sum_{j=1}^{n} r_j^P (\phi_j - P_j)^2 . \qquad (18)$$

The coefficients r_j^d and r_j^P are weights at different energy points. The parameter γ, the total weight of the prior conditions, should be chosen so that the activities are matched with reasonable experimental errors.

The solution can be defined in different ways. Instead of minimizing the sum in eq. (16), an upper limit may be set on the square sum of matching errors, while the regularization functional is minimized, or vice versa.

In the following, the unfolding codes SPECTRA, RFSP, CRYSTAL BALL, LOUHI and STAY'SL are discussed briefly.

In SPECTRA, by Greer, Halbleib and Walker /26/, the solution is tied to a trial spectrum. This can also be done iteratively by using the solution $\phi^{(k-1)}$ as a trial spectrum for the next step. The problem is formulated as

$$\min Q = \sum_{i=1}^{m} W_i \left[\frac{A_i - \sum_{i=1}^{n} \sigma_{ij} \phi_j^{(k)}}{A_i} \right]^2 + \sum_{j=1}^{n} \left[\frac{\phi_j^{(k)} - \phi_j^{(k-1)}}{\phi_j^{(k-1)}} \right]^2 . \quad (19)$$

By setting $\partial Q/\partial \phi^{(k)} = 0$ a group of linear equations is obtained. In principle, the result of the complete iteration could be obtained in closed form, but in practice numerical difficulties make stepwise solution easier. SPECTRA uses 30-100 energy points with fewer measured activities.

RFSP, deveoped by Fischer and Turi /27/, resembles SPECTRA, but $E\phi(E)$ is considered as unknown instead of $\phi(E)$. This reduces numerical difficulties, since $E\phi(E)$ generally varies less than $\phi(E)$.

CRYSTAL BALL is a more recent unfolding code developed by Kam and Stallmann /28/. The solution is tied to a trial spectrum and uses methods of direct approximation of Dirac delta-function with integral operators. In addition to the matching errors, the solution minimizes the functional

$$S = \int_0^{\infty} \frac{d}{dE} \left(\frac{\phi(E)}{P(E)}\right)^2 W(E) dE , \quad (20)$$

with the weighting function usually $W(E) = 1/E$.

The method has been developed to avoid numerical instabilities and oscillations. The same energy group structure as in SAND II is used.

LOUHI uses the regularization method /5,29/ with very flexible prior conditions combined in the latest version LOUHI78 /30/.

The linear LOUHI solves the least squares problem

UNFOLDING TECHNIQUES FOR ACTIVATION DETECTOR ANALYSIS

$$\min Q = Q_0 + \sum_{k=1}^{5} W_k Q_k , \qquad (21)$$

where

$$Q_0 = \sum_{i=1}^{m} r_i^{\varepsilon} \varepsilon_i^2 = \sum_{i=1}^{m} r_i^{\varepsilon} [A_i - \sum_{j=1}^{n} K_{ij} \phi_j]^2 ,$$

$$Q_1 = \sum_{j=1}^{n} r_j^P (\phi_j - P_j)^2 , \quad Q_2 = \sum_{j=1}^{n} r_j^f \phi_j^2 ,$$

$$Q_3 = \sum_{j=1}^{n-1} r_j^t (\phi_j - \phi_{j-1})^2 , \quad Q_4 = \sum_{j=2}^{n-1} r_j^d (\phi_{j-1} - 2\phi_j + \phi_{j+1})^2 ,$$

$$Q_5 = \sum_{j=1}^{n-1} r_j^s (\phi_j - z_j \phi_{j+1})^2 .$$

Here r_j are weights defined in each energy point, P is a trial solution and z_j' s are coefficients defining the assumed shape of the solution by the relation $F_j / F_{j+1} = z_j$.

Q is a positive definite function and reaches its minimum at the zero of its gradient. Solution is obtained through matrix inversion. Up to 40 energy points and responses can be used.

In nonlinear LOUHI the nonnegativity of the solution is guaranteed by setting $\phi_j = X_j^2$. The minimization is solved with an iterative gradient type algorithm with variable metric /31/. The prior conditions can be specified also on logarithmic or relative scales; for instance,

$$Q_1^{\log} = \sum_{j=1}^{n} r_i^P (\log X_j^2 - \log P_j)^2 . \qquad (22)$$

It is our intention to extend LOUHI to include more unfolding algorithms to make it a truly general purpose code which would also facilitate easy intercomparisons of different methods.

The linear method can be used if a good trial spectrum is

available or if the number of measured responses is large as in proton recoil spectroscopy. In all other cases the nonlinear method with nonnegativity condition is preferred.

Since the solutions are nonunique and a number of weighting parameters are chosen, LOUHI includes options for internal testing and error simulation, as presented in Fig. 3. The user can specify a test solution and the corresponding responses are computed. The responses can be perturbed to simulate errors. The solution obtained is then compared with the test solution, as are the total flux, average energy, dose rate and other test quantities. The minimum requirement set on any unfolding method is the close match of the responses when no perturbations are applied. The solution should not differ too much from the test function when physically realistic perturbations are made. The perturbations can also be applied to measured activities to simulate the effect of experimental errors to yield a confidence band of the solution.

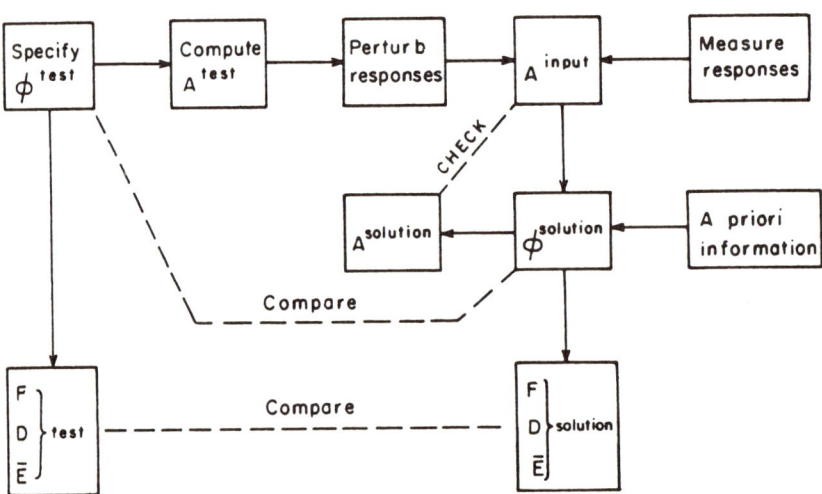

Fig. 3. Block diagram of the procedures used in testing the solution methods and analysing measured data with LOUHI.

Figure 4 presents the neutron spectrum at the CERN Proton Synchrotron as measured with copper spallation reactions and unfolded with LOUHI /6/.

One of the latest few-channel unfolding programs is STAY'SL by Perey /32/. Matching errors and deviation from a trial spectrum are minimized. The uncertainties of the cross sections can be considered and the cross sections can be modified. The unceratinties are represented with covariance matrices obtained from the ENDF/B nuclear data file.

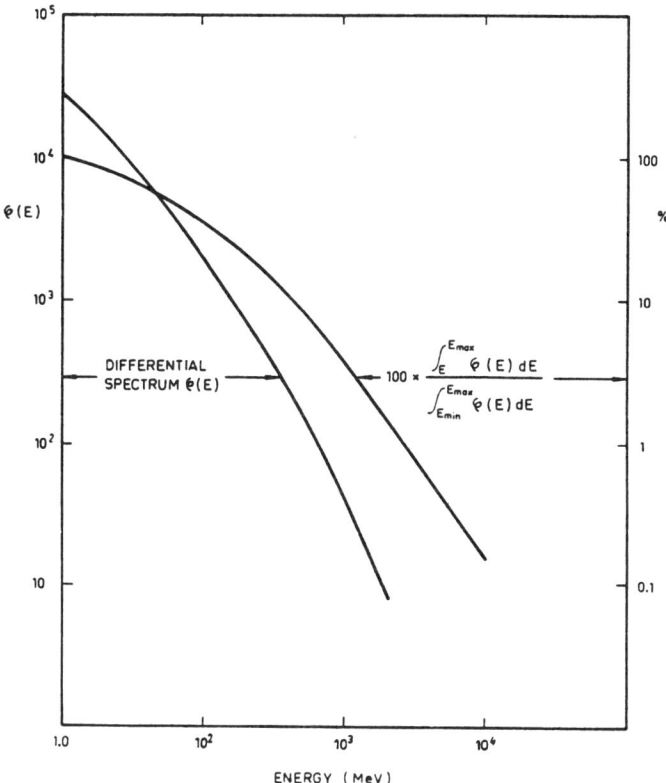

Fig. 4. Differential and integral spectra of particles in a stray radiation field near an internal target at the 24 GeV/c CERN Proton Synchrotron. The spectrum is unfolded with LOUHI using activities of 10 spallation products of Cu.

CONCLUSIONS

Choice of unfolding methods depends largely on the kernel, the number of responses and the prior information available. Other reviews of these topics can be found in references /33/ and /34/. A general comparison of various methods in different applications still remains to be carried out. However, a few comparisons of the existing codes for reactor neutron spectroscopy have been performed /33-36/. Amongst the codes tested, SAND II, SPECTRA and CRYSTAL BALL are the best ones in the usual reactor applications.

In cases where the number of measured activities is equal to or greater than the number of energy bins, codes like FERDOR can be used. If the number of unknowns is about 40 or less LOUHI is well suited. The regularization type codes can also be used in more general unfolding cases with a small number of measured responses and little prior knowledge of the spectrum.

Sometimes it is difficult to decide whether the structure of the solution spectrum is real or only due to the solution method. This problem arises especialy when the response functions do not properly cover the whole energy range. Therefore, an unfolding code should never be used as a black box. The user should always keep in mind the inherent assumption and restrictions of the code and examine the effect of changes in trial spectra and other prior conditions.

REFERENCES

1. W.R. Burrus, Utilization of a Priori Information by Means of Mathematical Programming in the Statistical Interpretation of Measured Distributions (Ph.D thesis), Report ORNL-3743, Oak Ridge National Laboratory (1965).

2. B.W. Rust and W.R. Burrus, Mathematical Programming and the Numerical Solution of Linear Equations, American Elsevier, New York (1972).

3. W.N. McElroy and S. Berg, A Computer-Automated Iterative Method for Neutron Flux Spectra Determination by Foil Activation, Volume 1, 2, 4, Report AFWL-TR-67-41, Air Force Weapons Laboratory (1967).
4. V.V. Verbinski, W.R. Burrus, T.A. Love, W. Zobel, N.W. Hill and R. Textor, Nucl. Instr. Meth. 65, 8 (1968).
5. J.T. Routti, Neutron Spectroscopy with Activation Detectors, Incorporating Methods for Analysis of Ge(Li) γ-ray Spectra and Solution of Fredholm Integral Equations, Report UCRL-18514, Lawrence Berkeley Laboratory (1969).
6. J.T. Routti, Physica Scripta 10, 107 (1974).
7. Neutron Fluence Measurements, Technical Reports Series No. 107, IAEA, Vienna (1970).
8. B.E. Watt, Phys. Rev. 87, 1037 (1952).
9. H. Ibarra and R. Sher, Nucl. Sci. Eng. 29, 15 (1967).
10. John C. Ringle, Measuring Neutron Spectra in Range 2.5-30 MeV, Report UCRL-10732, Lawrence Berkeley Laboratory (1963).
11. Raymond Gold, Nucl. Sci. Eng. 20, 493 (1964).
12. G. Di Cola and A. Rota, Nucl. Sci. Eng. 23, 344 (1965).
13. G. Di Cola and A. Rota, RDMM, Code for Fast Neutron Spectra Determination by Activation Analysis, Report EUR-2985.e, European Atomic Energy Community (Ispra).
14. H. Hirayama and T. Nakamura, Nucl. Sci. Eng. 50, 248 (1972).
15. T. Nakamura, T. Nishimoto and H. Hirayama, J. Nucl. Sci. Tech. 14, 31 (1977).
16. COOLC and FERDOR Spectra Unfolding Codes, RSIC Computer Code Collection PSR-17, Oak Ridge National Laboratory.
17. FORIST Spectra Unfolding Code, RSIC Computer Code Collection PSR-92, Oak Ridge National Laboratory.
18. SAND Neutron Flux Spectra Determination by Multiple Foil Activation - Iterative Method, RSIC Computer Code Collection CCC-112, Oak Ridge National Laboratory.
19. J.K. Schmotzer and S.H. Levine, Trans. Am. Nucl. Soc. 18, 370 (1974).

20. C.A. Oster, W.N. McElroy, R.L. Simons, E.P. Lippincot and G.R. Odette, Modified Monte Carlo Program for SAND-II with Solution Weighting and Error Analysis, Report HEDL-TME-76-60, Hanford Engineering Development Laboratory (1976).

21. D.M. Hargreaves and G.R. Stevenson, Unfolding Neutron Energy Spectra Using the "ALFIE" Routine, Report RHEL/M-147, Rutherford High Energy Laboratory (1968).

22. R.S. Sanna, Modification of an Iterative Code for Unfolding Neutron Spectra from Multisphere Data, Report HASL-311, ERDA Health and Safety Laboratory (1976).

23. David L. Phillips, J. ACM $\underline{9}$, 84 (1962).

24. S. Twomey, J. ACM $\underline{10}$, 97 (1963).

25. A.N. Tikhonov, Soviet Mathematics Doklady, Academy of Sciences of the U.S.S.R. $\underline{151}$, 510 (1963)

26. C.R. Greer, J.A. Halbleib and J.V. Walker, A Technique for Unfolding Neutron Spectra from Activation Detector Measurements, Report SC-RR-67-746, Sandia Laboratory (1967).

27. A. Fischer and L. Turi, International Nuclear Data Committee Report INDC(HUN)-8/U, IAEA, Vienna (1972).

28. F.B.K. Kam and F.W. Stallman, CRYSTAL BALL: A Computer Program for Determining Neutron Spectra from Activation Measurements, Report ORNL-TM-4601, Oak Ridge National Laboratory (1974).

29. J.T. Routti, Comp. Phys. Comm. $\underline{4}$, 33 (1972).

30. J.T. Routti and J.V. Sandberg, General Purpose Unfolding Program LOUHI 78 with Linear and Nonlinear Regularizations, Report TKK-F-A359, Helsinki University of Technology, Department of Technical Physics (1978).

31. W.C. Davidon, Variable Metric Method for Minimization, Report ANL-5990, Argonne National Laboratory (1959).

32. F.G. Perey, Least-Squares Dosimetry Unfolding; the Program STAY'SL, Report ORNL/TM-6062, Oak Ridge National Laboratory (1977).

33. A Review of Radiation Energy Spectra Unfolding, Proceedings of a Seminar-Workshop April 12-13, 1976, Report ORNL/RSIC-40, Oak Ridge National Laboratory (1976).

34. Proceedings of the First ASTM-EURATOM Symposium on Reactor Dosimetry, Petten (Holland) September 22-26, 1975, Report EUR 5667 e/f, Part I (1977).

35. R. Dierckx, International Nuclear Data Committee Report INDC(HUN)-8/U, IAEA, Vienna (1973).

36. A. Fischer, International Intercomparison of Neutron Spectra Evaluation Methods Using Activation Detectors, Report JUL-1196, Kernforschungsanlage Jülich (1975).

LECTURE 25: BREMSSTRAHLUNG SPECTRUM ANALYSIS

BY ACTIVATION METHOD (LYRA, DIBRE, REFUM)[1-6]

Takashi Nakamura

Institute for Nuclear Study, University of Tokyo
Midori-cho 3-2-1, Tanashi, Tokyo
Japan

INTRODUCTION

In the spectrum measurement of bremsstrahlung and photoneutrons generated from the electron accelerator, it is very difficult to detect them with pulse-type detectors, since a large number of gamma-ray bursts are produced. The bremsstrahlung spectra have been measured by using various large, complicated, counter systems for thin and thick targets; nevertheless, in such a high gamma-ray flux field, activation detectors are an effective means of spectrum measurement of gamma-rays and neutrons.

The use of threshold detectors for the determination of fast-neutron spectra has been a common practice for many years, and its one advantage is that the threshold detectors are not sensitive to the high gamma-ray fluxes found in reactors. In the spectrum measurement of photoneutrons generated from targets in the linear accelerator, the fluxes of bremsstrahlung bursts are dominant in comparison with the photoneutron fluxes, so the effect of photonuclear reactions in the threshold detectors cannot be neglected. For example, the neutron reaction $^{24}Mg(n,p)^{24}Na$ and the photonuclear reaction $^{25}Mg(\gamma,p)^{24}Na$ produce the same radioisotope in the magnesium detector. Accordingly, for the determination of a neutron spectrum it is necessary to obtain the photon energy spectrum at the same time.

We have developed a new method of determining photon energy spectra from the activities induced by various photonuclear reactions, applying the detecting method of fast neutron spectra with threshold detectors[3,4]. This bremsstrahlung spectrum evaluation method mathematically reverses the usual method in which the bremsstrahlung cross-section is assumed and the photonuclear reaction is calculated. This method has two great advantages: first, the energy spectrum of bremsstrahlung bursts from a linear accelerator is very easily measurable, whereas its measurement is difficult with a scintillation spectrometer because of the accumulation of the output pulses. Secondly, the photon energy spectrum in the bulk medium is measurable owing to the small volume of activation detectors.

PHOTONUCLEAR AND NEUTRON REACTIONS AVAILABLE AS ACTIVATION DETECTORS

A target bombarded by electrons generates bremsstrahlung radiation and successively photoneutrons, and, as a result, the activities induced in activation detectors by photonuclear reactions include the contribution of these by neutron reactions in some cases. The information is summarized on photonuclear reactions which can be used as gamma-ray detectors in a mixed neutron gamma-ray field, and on neutron reactions used as neutron detectors in a high gamma-ray flux field.

Photonuclear Reactions Available as Threshold Detectors

The photonuclear reactions which can be used in threshold detectors to obtain the photon spectrum have to satisfy the following conditions:

1) the isotope has a relatively large natural abundance;

2) the solid-state material is obtainable at room temperature;

3) the half-life of the daughter nucleus produced is above about 10 minutes and below about 100 days;

4) the daughter nucleus emits gamma-rays having a large and known branching ratio.

Furthermore, another condition has to be satisfied when the gamma-rays are detected in a mixed neutron gamma-ray field. For a nucleus A having one or more stable isotopes, the competing photonuclear and neutron reactions as shown in Table 1 produce the same radio-isotope in the nucleus A.

We now consider the bremsstrahlung spectrum measurement with activation detectors. When charged particles of energy E strike the target, the bremsstrahlung of continuous energy from 0 to E is generated and the photoneutrons are successively produced by the photonuclear reaction. The photoneutron energy is equal to the bremsstrahlung energy minus the threshold energy of the photonuclear reaction of the target nucleus (the neutron binding energy), about 10 MeV. The result is that the activation detector placed near the target has the effective threshold energy of its neutron reaction, which is the sum of the threshold energy of the neutron reaction in the detector and that of the photonuclear one in the target.
The threshold energies of the (γ,n), (γ,p), and (γ,α) reactions are several to 10 MeV, while those of the (γ,np), (γ,t), $(\gamma,2p)$, and $(\gamma,2n)$ reactions are 10 to 20 MeV. Moreover, the threshold energies of the (n,n'), (n,p), (n,α), (n,t), and $(n,^3He)$ reactions are several MeV, while those of the $(n,2n)$, (n,np), $(n,n\alpha)$, and (n,nt) reactions are about 10 MeV. Then, taking into consideration only the difference of the effective threshold energies between photonuclear and neutron reactions without regarding the difference of their cross-sections, the following photonuclear reactions can be used as gamma-ray detectors in a neutron-mixed field:

Table 1 Competing photonuclear and neutron reactions producing the same isotopes

Produced nucleus	Target nucleus				
	$_Z^{m-2}A$	$_Z^{m-1}A$	$_Z^{m}A$	$_Z^{m+1}A$	$_Z^{m+2}A$
$_Z^{m-1}A$	(n,γ)	(γ,γ') (n,n')	(γ,n) (n,2n)	(γ,2n) (n,3n)	(γ,3n)
$_{Z-1}^{m-1}A$			(γ,p) (n,p)	(γ,np) (n,np)	(γ,t) (n,nd)
$_{Z-2}^{m-4}A$	(γ,2p) (n,³He)	(γ,³He) (n,α)	(γ,α) (n,nα)		

m: mass number; Z: atomic number

5) i) for one kind of stable isotope
  (γ,n), (γ,p), (γ,t), (γ,2n), (γ,α);

  ii) for more than two kinds of stable isotopes
   a) the isotope of the smallest mass number
     (γ,n), (γ,p), (γ,t), (γ,2n), (γ,α);

   b) the isotope of the largest mass number m, adjacent to which the isotope has the mass number of m-2
     (γ,p), (γ,α).

The list of photonuclear reactions that satisfy these conditions (1) to (5) is shown in Refs. 2 and 5. The energy range which can be measured with these photonuclear reactions is from about 5 MeV (near the lowest threshold energy) to about 30 MeV, above which the influence of the neutron reaction may not be negligible. However, when the bremsstrahlung radiation is dominant compared with the neutrons, the gamma-ray spectrum measurement by this method may be possible at high energies.

Neutron Reactions Available as Threshold Detectors

In the photoneutron spectrum measurement with activation detectors as described here, it is necessary to choose a neutron reaction that is completely insensitive to the influence of photons, since the detector is irradiated by strong bremsstrahlung bursts. As shown in Table 1, there is only one neutron reaction that does not suffer from the effect of a photonuclear reaction; namely,

the (n,p) reaction for one kind of stable isotope and for the isotope of the largest mass number among more than two kinds of stable isotopes.

The list of the (n,p) reactions satisfying the above condition and, moreover, the conditions (1) to (4) described above, is also shown in Refs. 2 and 5.

METHOD OF SPECTRUM EVALUATION

Principle of the Unfolding Method

The activities induced by the photonuclear reactions in the detector were inverted into the photon energy spectra with a computer code LYRA, which was improved from the spectrum evaluation code by the orthonormal expansion method LUNA 5 [7] with respect to the following: the form of the weighting function $W(E)$, the introduction of a variable maximum energy E_0, and the estimation of the errors.

The conventional orthornormal expansion method for the evaluation of a neutron energy spectrum was applied to determine the bremsstrahlung spectrum from the induced activities. In this method, the bremsstrahlung flux $\phi(E)$ is expanded in the following series of orthonormal functions $\psi_k(E)$,

$$\phi(E) = W(E) \sum_{k=1}^{n} a_k \psi_k(E) , \qquad (1)$$

where n is the number of detectors, $W(E)$ the weighting function, and a_k the coefficients of the expansion. The function $W(E)$ was selected as Wyard's approximation formula[8] to the bremsstrahlung spectrum

$$W(E) = \frac{1}{E} - \frac{1}{E_0} + \frac{3}{4E_0} \ln \frac{E}{E_0} , \qquad (2)$$

where E_0 is the maximum energy of the bremsstrahlung corresponding to the incident electron energy. The activation rate A_i is given as

$$A_i = \int_0^{E_0} \sigma_i(E)\phi(E) \, dE . \qquad (3)$$

The quantity A_i indicates the reaction probability per second for one nucleus of the ith isotope, having $\sigma_i(E)$ as the differential cross-section for the reaction under investigation. By introducing Eq. (1) into Eq. (3) we obtain

$$A_i = \sum_{k=1}^{n} a_k S_{ik} , \qquad i = 1, \ldots, n , \qquad (4)$$

where

$$S_{ik} = \int_0^{E_0} W(E)\psi_k(E)\sigma_i(E) \, dE . \qquad (5)$$

The solution of the linear equation system, Eq. (4), uniquely gives the coefficients of the expansion in Eq. (1).

The above method for the evaluation of the photon spectrum with photonuclear reactions is the same as that of the neutron spectrum with neutron reactions. But the physical characteristics of photonuclear reactions are different from those of neutron reactions in the following points:

i) The photonuclear reaction cross-sections have a giant resonance peak near about 20 MeV.

ii) The photonuclear reactions, such as (γ,n) and (γ,p), which are practical and available as activation detectors, have threshold energies above about 6 MeV.

iii) Many of the daughter nuclei produced by the photonuclear reactions have either short or very long half-lives, thereby limiting the photonuclear reactions which can be used for activation analysis.

As an example, the photonuclear cross-section is shown in Fig. 1 together with the neutron reaction cross-section. The difference in the shapes of the two cross-sections has no significant influence on the evaluated spectrum, but in the low-energy region the reliability of the photon spectrum decreases because of high threshold energies of photonuclear reactions.

Next, we have evaluated the statistical errors of the experimental bremsstrahlung spectra obtained by this evaluation method. The difference between the true photon spectrum and the spectrum obtained from the measured activation rate by the orthonormal expansion method may be due to the following reasons[10]:

i) inaccuracy of the hypothesis expressed by Eq. (1);

ii) experimental errors in the activation rates;

iii) inaccuracy of the cross-section data.

The first error accounts for the difference between the experimental activation rates A_i^{exp} and the activation rates A_i' calculated with the evaluated spectrum $\phi_t(E)$ through the following equation:

$$A_i' = \int_0^{E_0} \sigma_i(E) \phi_t(E) \, dE \,, \tag{6}$$

where

$$\phi_t(E) = W(E) \sum_{k=1}^{t} a_k \psi_k(E) \,, \qquad t = 1, \ldots, n \,. \tag{7}$$

Accordingly, in the LYRA code the ratio of A_i' to A_i^{exp},

$$R_i = \frac{A_i'}{A_i^{exp}} \,, \tag{8}$$

is calculated for each reaction. The best approximation to the

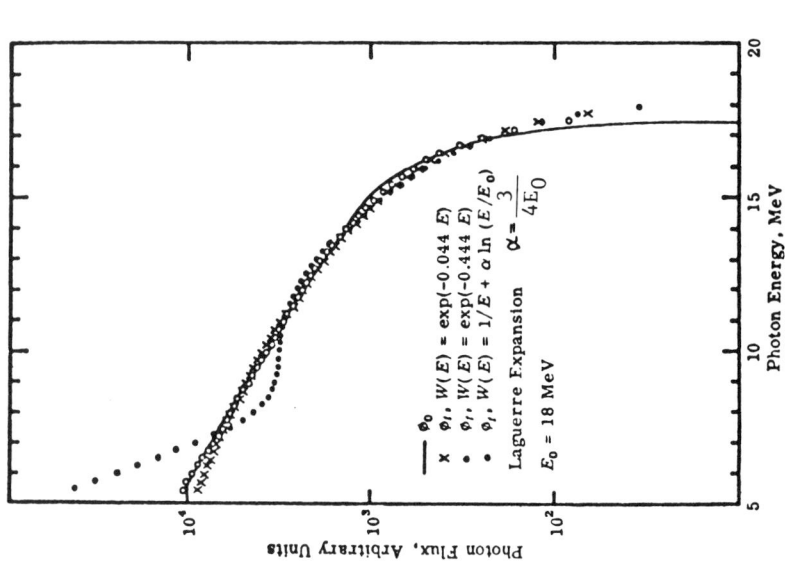

Fig. 2. Comparison of a given bremsstrahlung-like spectrum with that evaluated for several weighting functions W(E) with the LYRA code.

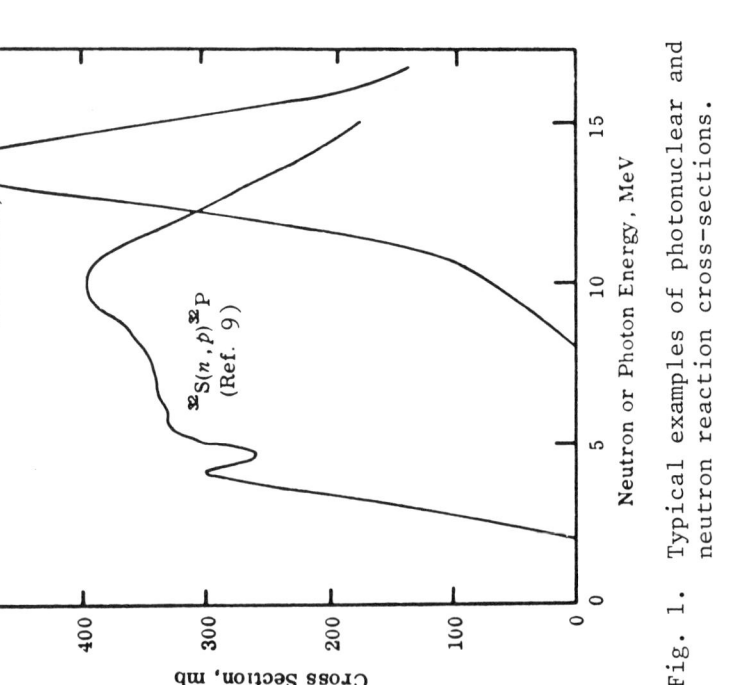

Fig. 1. Typical examples of photonuclear and neutron reaction cross-sections.

true spectrum is the value of $\phi_t(E)$ for which R_i is nearly equal to unity for all detectors.

With regard to the last two items, in the LYRA code the errors were obtained by using the Monte Carlo method which was reported by Di Cola and Rota[10]. In their method, the error distribution in the input data of A_i and σ_i is supposed to be normal. The 2g randomly varied data sets of A_i and σ_i generate the g-different functions $\phi_t^k(E)$. The standard deviation of the average of $\phi_t^k(E)$, $\bar{\phi}(E)$, is obtained by random sampling from the normal probability distribution function.

Test Calculations of the Orthonormal Expansion Method

Test calculations, in which the activation rates and the cross-section data are given as error-free, have been performed to examine the accuracy of this spectrum evaluation method. In these calculations the weighting function, the maximum energy, and the orthonormal function have been varied. They have been carried out as follows:

i) calculation of the activation rate A_i for a given photon energy spectrum $\phi_0(E)$ by Eq. (3);

ii) calculation of the photon energy spectrum $\phi_t(E)$ by Eqs. (4) and (7) using a set of reaction rates A_i;

iii) comparison of $\phi_0(E)$ with $\phi_t(E)$.

The photonuclear reactions used in these test calculations are shown in Table 2.

The bremsstrahlung-like energy spectrum $\phi_0(E)$ can be compared to the LYRA-unfolded spectrum $\phi_t(E)$. The dependence of $\phi_t(E)$ on the weighting function is shown in Fig. 2. For this calculation the Laguerre polynomials and a maximum energy of 18 MeV were used. Another weighting function

$$W(E) = e^{-\alpha E}, \tag{9}$$

Table 2 Physical characteristics of activation detectors

Detector	Reaction	Abundance (%)	Threshold energy (MeV)	Half-life	Gamma-ray energy (MeV)	Branching ratio	Reference
Au	^{197}Au(γ,n)^{196}Au	100.0	8.1	6.18 d	0.356	0.94	62FU 70VE
Fe	^{54}Fe(γ,n)^{53}Fe	5.84	13.6	8.51 m	0.511	1.96	53MO 57CA
Ni	^{58}Ni(γ,n)^{57}Ni	67.76	12.2	36.0 h	1.37	0.86	68MI 73FU
In	^{115}In(γ,γ')^{115m}In	95.77	1.0	4.5 h	0.335	0.50	57BO 65KR
	^{115}In(γ,n)^{114m}In	95.77	9.0	50.0 d	0.192	0.17	53GO 62BO 69FU
Th	^{232}Th(γ,n)^{231}Th	100.0	6.4	25.5 h	complex		57KA
Cr	^{50}Cr(γ,n)^{49}Cr	4.31	12.9	41.9 m	0.511	1.86	54GO
Mn	^{55}Mn(γ,n)^{54}Mn	100.0	10.2	300 d	0.835	1.0	59PA 73AL1
Co	^{59}Co(γ,n)^{58}Co	100.0	10.5	71 d	0.810	0.99	73AL2
C	^{12}C(γ,n)^{11}C	98.89	18.7	20.4 m	0.511	2.0	57CA 66BA 66FU 71IS
Ag	^{107}Ag(γ,2n)^{105}Ag	51.35	17.8	40 d	0.280	0.32	69BE

was selected, where α is a constant. Figure 2 shows that for the determination of the bremsstrahlung energy spectrum the weighting function of the type of Eq. (2) more nearly expresses the approximate bremsstrahlung spectrum, although above 10 MeV there is no great difference in the photon spectra based on the different weighting functions. The dependence of the evaluated photon spectrum on the maximum energy E_0 is shown in Fig. 3. The spectra $\phi_t(E)$ obtained for maximum energies of 18 and 20 MeV were compared with $\phi_0(E)$ having a maximum energy of 17.5 MeV. This figure shows that the orthonormal expansion method is very sensitive to the selection of the maximum energy E_0. By comparison with the results obtained from Chebyshev polynomials, the spectra were shown to be insensitive to variations in the orthonormal function.

When the weighting function of Eq. (2)-type, a maximum energy of 20 MeV, and Laguerre polynomials were used in the calculation, comparison between $\phi_0(E)$ and $\phi_t(E)$ is shown in Fig. 4, for two other types of photon energy spectra; one is a spectrum uniformly distributed from 0 to 17.5 MeV, and the other is an exponential function truncated at 17.5 MeV. This figure shows that the agreement between $\phi_0(E)$ and $\phi_t(E)$ is good, especially above 8 MeV. This orthonormal expansion method may also evaluate the spectrum when it deviates from the bremsstrahlung energy spectrum, provided that the maximum photon energy is known.

The LYRA unfolding code is, of course, applicable to the neutron spectrum evaluation with an appropriate weighting function, and it was tested on the international intercomparison of neutron spectra evaluating methods using activation detectors[11].

ANALYTICAL CALCULATION OF THICK-TARGET BREMSSTRAHLUNG SPECTRUM

Method of Calculation

For a monoenergetic and uni-directional beam of electrons incident on the thick target, the various analyses of thick-target

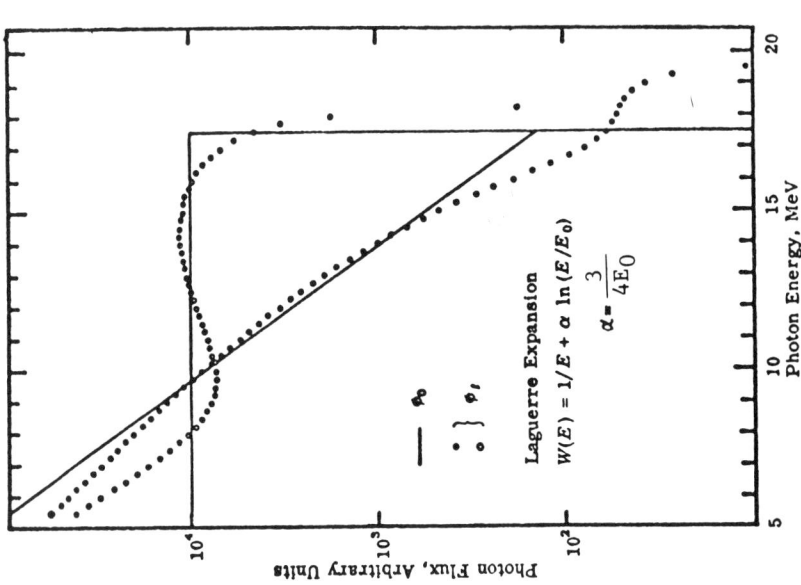

Fig. 4. Comparison between two types of given spectra and the spectra evaluated with the LYRA code.

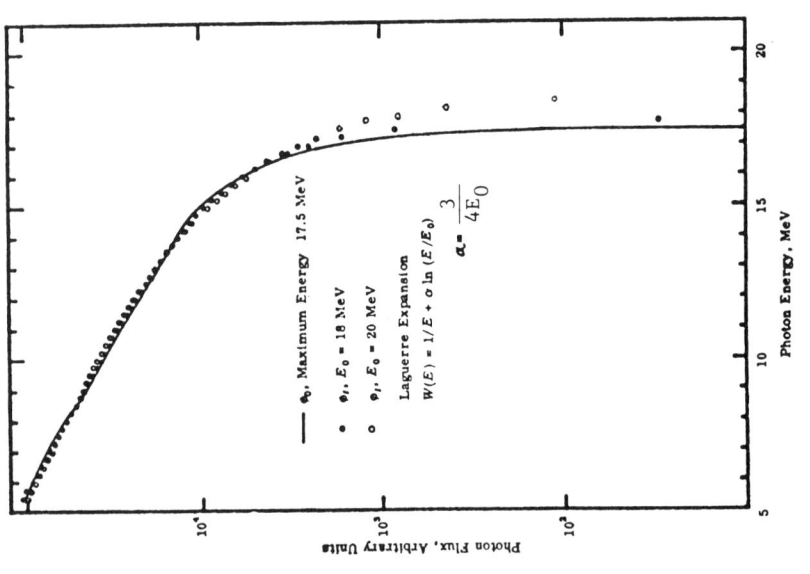

Fig. 3. Comparison of a given bremsstrahlung-like spectrum with that evaluated for two values of maximum energy E_0 with the LYRA code.

bremsstrahlung have been summarized by Zerby and Keller[12]. The
DIBRE code has been developed[1] to calculate the thick-target
bremsstrahlung spectrum by an analytical method, based on the works
by Scott[13] and Ferdinande et al.[14]. The calculation is based on
the continuous slowing down model of the electron energy, which
neglects the lateral displacement of the electron and the electron-
photon cascade. This means that the DIBRE code has high accuracy
for an incident electron energy below about 50 MeV and at any emer-
gent angle of bremsstrahlung smaller than about 30° to the electron
beam axis, as seen later.

Consider an electron with kinetic energy E_0, incident normally
on a target with total thickness T. The target has been subdivided
into a number of thin slabs n, each of thickness Δt, as shown in
Fig. 5. The ith slab lies at an average depth $t_i = (i - 0.5)\Delta t$.
Within each thin target the electron will travel in some direction
defined by the polar angle θ and azimuthal angle ψ after multiple
scattering, under the assumption that the lateral displacement of
the electron within the target is negligibly small compared with
the distance between the target and the detector. A schematic
representation of a radiative scattering in any slab i is also shown
in Fig. 5. In the figure, θ_i is the angle between the electron
velocity vector in the ith slab and the incident electron direction.
The angle ω_i is the angle between the electron velocity vector and
the ray of the emitted photon. Note that only photons travelling
at an angle α with respect to the incident electron direction will
reach the detector. These angles are related by

$$\cos \omega_i = \cos \alpha \cos \theta_i + \sin \alpha \sin \theta_i \cos \psi_i . \qquad (10)$$

For all angles of electron direction θ and ψ, the total probability
that this electron produces a photon of energy k in the direction
ω_i, and that this photon will reach the detector within the solid
angle $d\Omega$ at angle α, is given by the summation over all thin slabs:

$$\frac{d^2N}{dkd\Omega} = \sum_{i=1}^{n} \int_0^{2\pi} d\psi \int_0^{\pi} \sin\theta \, d\theta \, \tau(E_0,t_i)\nu_i F(E_i,\theta,t_i) \times \quad (11)$$

$$\times \frac{d^2\sigma}{dkd\Omega}(E_i,k,\omega_i) Y_i(k,\rho_i) , \qquad (E_i \geq k) ,$$

under the above approximation.

As the electrons penetrate the scatterer, they lose their energy according to the continuous slowing-down model, and their average kinetic energy in the ith slab E_i is given by

$$E_1 = E_0 - \frac{\Delta t}{2}\left(\frac{dE}{dt}\right)_{E_0} , \qquad (12a)$$

$$E_i = E_{i-1} - \frac{\Delta t}{\cos(\langle\theta^2\rangle_{i-1})^{\frac{1}{2}}}\left(\frac{dE}{dt}\right)_{i-1} , \qquad (i = 2, \ldots, n) , \quad (12b)$$

where $(\langle\theta^2\rangle_{i-1})^{\frac{1}{2}}$ is the r.m.s. angle for electron scattering up to the ith slab[15]. According to Ref. 14, a path-length correction is included in Eq. (12b) instead of neglecting the electron lateral displacement. The quantity ν_i stands for the number of atoms per unit cross-section in the ith slab and is given by

$$\nu_i = N_A \rho \Delta t / [A \cos(\langle\theta^2\rangle_i)^{\frac{1}{2}}] , \qquad (13)$$

where N_A is Avogadro's number, A is the mass number of the scatterer, and ρ is the density of the scatterer.

The transmission factor $\tau(E_0,t_i)$ accounts for the fraction of the incident electron beam intensity which reaches the centre of the ith slab t_i without any collision. In this calculation, Mar's empirical transmission formula[16], which is fitted to his Monte Carlo calculation, was used for electron energies below 4.0 MeV, and a more accurate formula by Ebert et al.[17] was used above 4.0 MeV. The electron angular distribution $F(E_i,\theta,t_i)$ is the probability that an electron of energy E_i is scattered toward the direction θ in the ith slab, and is divided into two regions. On the incident-

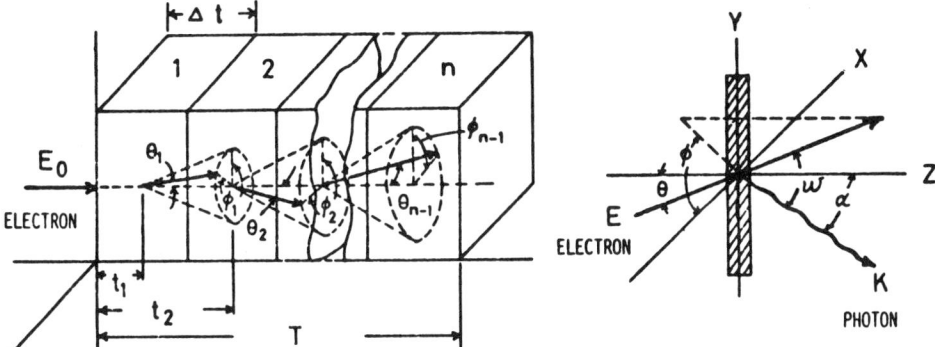

Fig. 5. Schematic diagram of calculation procedure.

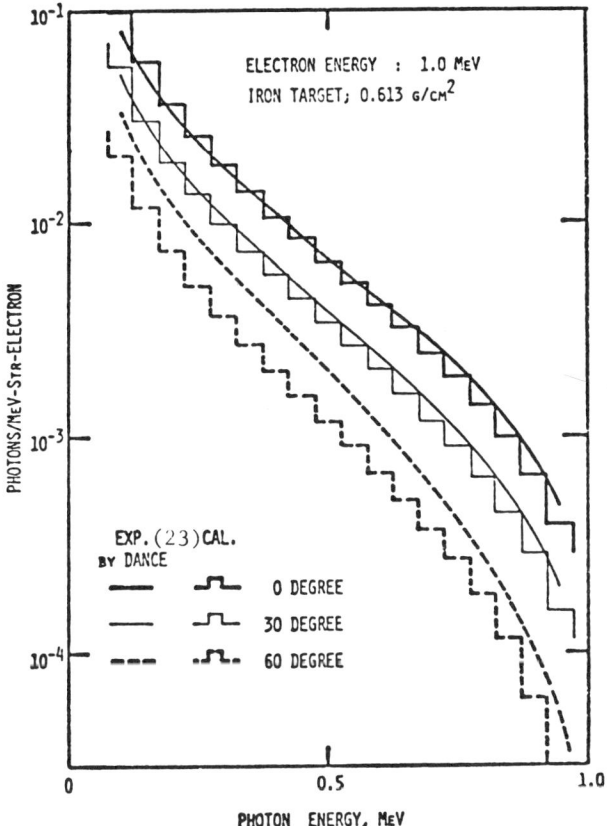

Fig. 6. Comparison of calculated and experimental bremsstrahlung spectra from 0.613 g/cm² thick iron target in 0°, 30°, and 60° directions by 1 MeV electrons.

beam side in which the r.m.s. angle $(\langle\theta^2\rangle_i)^{\frac{1}{2}}$ is less than 30°, Molière's theory[18] of multiple scattering modified to be applicable to a large angle by Bethe[19], was used for each slab. On the outgoing-beam side, in which $(\langle\theta^2\rangle_i)^{\frac{1}{2}}$ is larger than 30°, it was approximated that $F(E_i,\theta,t_i)$ obeys the $\cos^2\theta$ distribution having $(\langle\theta^2\rangle_i)^{\frac{1}{2}}$ of 45° according to Frank[20].

The differential bremsstrahlung cross-section, $(d^2\sigma/dkd\Omega)$ (E_i,k,ω_i), is given[21] for various ranges of the electron energy E_i as

$$
\begin{array}{lll}
E_i \leq 4\ mc^2 & C_R f_E \sigma(2BN) & \\
4\ mc^2 < E_i \leq 30\ mc^2 & C_R f_E \sigma(2BN) & \text{if } \gamma > 15 \\
 & C_R f_E \sigma(2BS) & \text{if } \gamma < 15 \qquad (14)\\
E_i > 30\ mc^2 & \sigma(2BN) & \text{if } \gamma > 15 \\
 & \sigma(2CS) & \text{if } \gamma < 15,
\end{array}
$$

where $\gamma = 100\ mc^2\ k\left[(E_i + mc^2)(E_i + mc^2 - k)Z^{1/3}\right]^{-1}$, mc^2 is the electron rest mass energy, Z the atomic number of target material, C_R an empirical correction factor, f_E the Elwert factor, $\sigma(2BN)$ the non-screened Born approximation cross-section, $\sigma(2BS)$ the screened Born approximation cross-section, and $\sigma(2CS)$ the screened Coulomb correction cross-section.

The quantity $Y_i(k,\rho_i)$ indicates the attenuation of bremsstrahlung transmitted through the scatterer; that is,

$$Y_i(k,\rho_i) = B(k,\rho_i)\ \exp\left[-\mu(k)\rho_i\right], \qquad (15)$$

where ρ_i is the path length of bremsstrahlung as it traverses the rest of the scatterer, $\mu(k)$ the linear attenuation coefficient, and $B(k,\rho_i)$ the build-up factor of the Berger form, which is summarized by Trubey[22].

The energy spectrum of bremsstrahlung, dN/dk, at a specified detector position behind the target, is expressed from Eq. (11) as follows:

$$\frac{dN}{dk} = \sum_{i=1}^{n} \int_0^{2\pi} d\psi \int_0^{\pi} \sin\theta \, d\theta \, \tau(E_0, t_i) \nu_i F(E_i, \theta, t_i) \frac{d^2\sigma}{dkd\Omega}(E_i, k, \omega_i) \times$$

$$\times \frac{S}{(T - t_i + z)^2 + r^2} B(k, \rho_i) \exp[-\mu(k)\rho_i], \qquad (E_i \geq k), \qquad (16)$$

where z is the longitudinal distance from the target face of the outgoing-beam side to the detector, r is the lateral distance from the electron beam axis to the detector, S is the detector area, and

$$\rho_i = (T - t_i)[(T - t_i + z)^2 + r^2]^{\frac{1}{2}} / (T - t_i + z). \qquad (17)$$

The practical value of 10^{-3} radiation length units was selected for Δt, according to Ferdinande et al.[14].

Comparison with Other Calculated and Experimental Results

The bremsstrahlung spectra calculated by this method were compared with other experimental and calculated results. Figure 6 shows the comparison between calculated and experimental[23] bremsstrahlung spectra in three emission angles from the iron target bombarded by 1 MeV electron beams. The figure shows that the calculated spectra are in very good agreement with the experimental ones at 0° and indicates a tendency to become smaller than the experimental ones with increasing angle, mainly because of the neglect of the lateral displacement of transmitted electrons in the DIBRE code. Figure 7 shows the comparison of the bremsstrahlung spectra at 0° and 20° emission angles for 10 MeV electrons incident on a 4.67 g/cm² aluminum target. The DIBRE calculation shows good agreement with the Monte Carlo calculation[24] by the ETRAN code[25] but a little lower than the experimental data by Jupiter et al.[24], especially at 20°. Figure 8 shows the comparison of the bremsstrahlung spectra

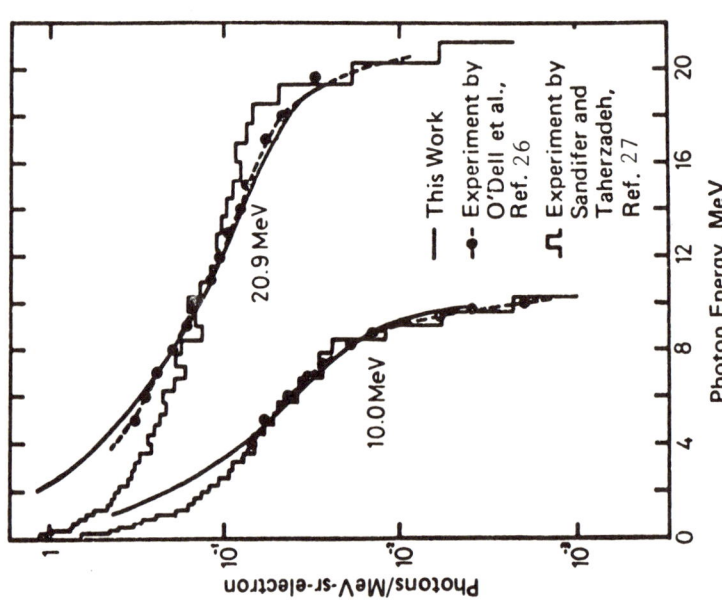

Fig. 8. Comparison of calculated and experimental bremsstrahlung spectra for 10.0 and 20.9 MeV electrons on a 0.735 g/cm² thick gold-tungsten converter target with a 7.72 g/cm² thick aluminum filter.

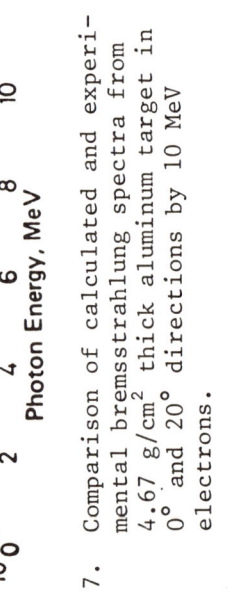

Fig. 7. Comparison of calculated and experimental bremsstrahlung spectra from 4.67 g/cm² thick aluminum target in 0° and 20° directions by 10 MeV electrons.

emitted in the forward direction for 10 and 20.9 MeV electrons incident on a 0.735 g/cm^2 gold-tungsten target backed by a 7.72 g/cm^2 aluminum filter. The DIBRE results also show good agreement with the experimental data[26,27], especially above about 4 MeV. The measurement by Sandifer and Taherzadeh[27] does not give the absolute values, and their results were normalized to an appropriate point in the figure. Figure 9 shows the comparison of the bremsstrahlung spectra at 0°-5° and 25°-30° emission angles for 30 MeV electrons incident on a 3.5 radiation lengths tungsten target. The DIBRE results show good agreement with the two Monte Carlo (ETRAN[25,28] and EGS[29]) results at 0°-5° (the DIBRE calculation was carried out at 2.5°); the DIBRE results, done at 25°, are somewhat below those calculated by ETRAN and EGS.

These figures show that this analytical method is sufficiently accurate for calculation of thick and very thick target bremsstrahlung spectra, but it has a tendency to underestimate the results at bremsstrahlung angles that are large, relative to the beam direction. The DIBRE code was also applied to estimate the bremsstrahlung contribution of the beta-ray spectrometry for fission-product decay heating[30].

MEASUREMENTS OF THICK TARGET BREMSSTRAHLUNG SPECTRA

Experimental Procedure

Two experiments have been performed to prove that the activation method can be used for the measurement of the bremsstrahlung spectra produced in 1.3 cm thick iron and 0.7 cm thick tungsten targets by 15 MeV electrons[3], and in 5.0 cm and 2.5 cm thick lead targets by 15, 20, and 25 MeV electrons[4]. The experimental arrangement is shown in Fig. 10. The monoenergetic electron beam, analysed by a magnet, struck the target. The bremsstrahlung radiation was measured by activation detectors of Mn, Au, Cr, Ni, Co, Fe, C, and In. Their physical properties are listed in Table 2. These detectors,

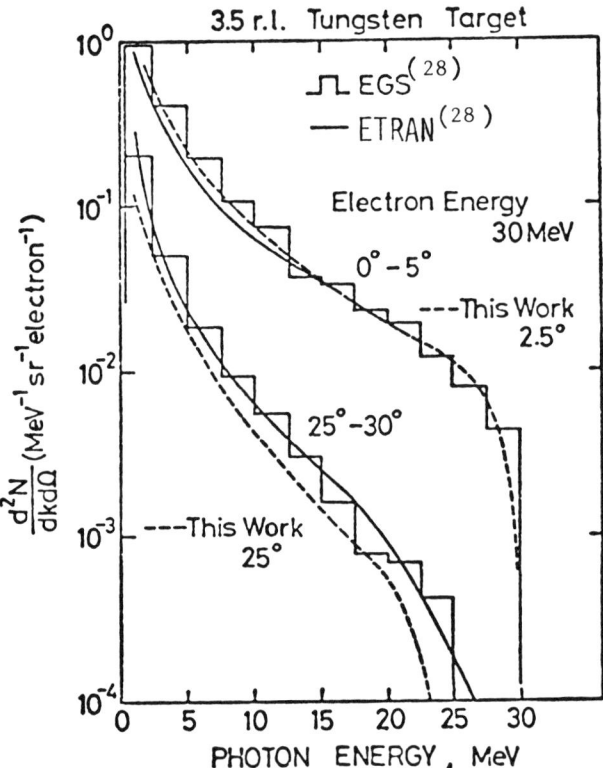

Fig. 9. Comparison of bremsstrahlung spectra calculated by the DIBRE code and two Monte Carlo codes (EGS and ETRAN), for a 3.5 radiation lengths tungsten target irradiated by 30 MeV electrons.

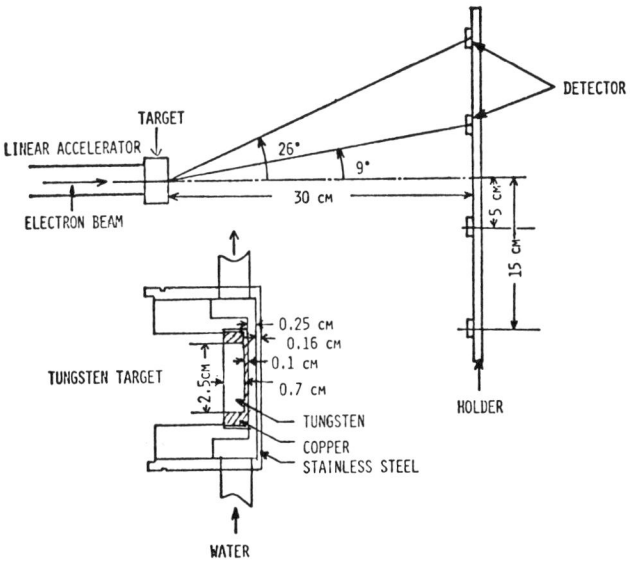

a) Iron and tungsten target

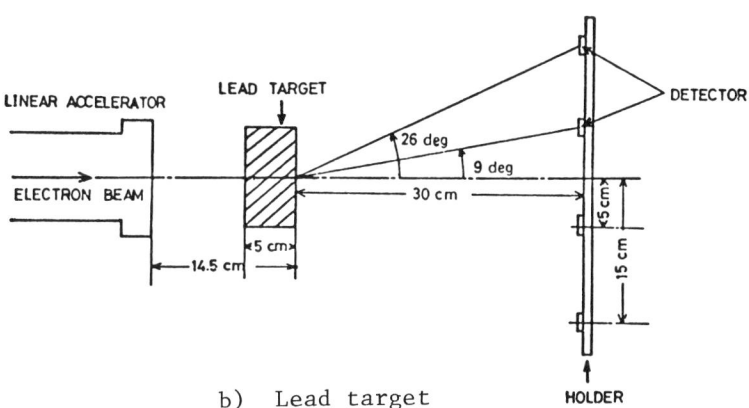

b) Lead target

Fig. 10. Experimental arrangement

except indium, were selected for this experiment from the detectors for gamma-ray detection in a mixed neutron-gamma field listed in Refs. 2 and 5. The $^{115}\text{In}(\gamma,\gamma')^{115}\text{In}$ reaction, of which the threshold energy is 1 MeV, was especially used to get information on the lower part of the bremsstrahlung spectrum, because, in this experimental geometry, the bremsstrahlung radiation is considered to be dominant compared with the photoneutrons and with the negligible contribution from the $^{115}\text{In}(n,n')^{115m}\text{In}$ reaction. The $^{115}\text{In}(\gamma,n)^{114m}\text{In}$ reaction was also used because of the negligible contribution of the $^{113}\text{In}(n,\gamma)^{114m}\text{In}$ reaction. The detectors were attached with plastic tape to the surface of an acrylic plate holder (perpendicular to the electron-beam axis) and were arranged on two coaxial circles with radii of 5 and 15 cm, respectively, and where the centre is the point of intersection between the surface and the beam axis. The polar angles θ between the detectors and the centre of the target, subtended by the 5 and 15 cm radii, were about 9° and 26°, respectively.

The gamma-ray activities of the detectors induced by bremsstrahlung were detected with a 3 in. (diam) × 3 in. NaI(Tl) scintillator and a Ge(Li) detector, coupled to a multichannel pulse-height analyser. The activation rates are obtained by the following equation:

$$A_i^{exp} = \frac{\lambda_i C_p^i}{N_T^i \eta_i \varepsilon_p^i \exp(-\lambda_i T_w)[1 - \exp(-\lambda_i T_{ir})][1 - \exp(-\lambda_i T_c)]} \text{ sec}^{-1},$$

(18)

where λ_i is the decay constant of the ith isotope, η_i the number of photons per decay of the ith isotope, N_T^i the total number of nuclei of the ith isotope in the detector, C_p^i the peak counts, ε_p^i the peak efficiency of the detector, T_w the waiting time, T_{ir} the irradiation time, and T_c the counting time. The values of λ_i and η_i that were used can be obtained from Table 2. The peak efficiencies

ε_p^i of NaI(Tl) and Ge(Li) detectors were calculated with the REFUM code[31,32]. The REFUM code has been developed to calculate peak and total efficiencies and response functions of the detectors for thick, disk-shaped, gamma-ray sources by considering the self-absorption and self-scattering in the sources. The experimental activation rates A_i^{exp} obtained from Eq. (18) are normalized to one incident electron. The average number of electrons incident on the target per second was measured with a Faraday cup at the target.

Evaluation of Bremsstrahlung Spectra

The bremsstrahlung spectrum, obtained by unfolding the induced activation rates, is very much dependent on the photonuclear cross-section data of the detectors. As shown in the next section, the most accurate photonuclear cross-section data currently available were selected in advance of the spectrum unfolding. By use of these photonuclear cross-sections, the evaluation of bremsstrahlung spectra was performed with the LYRA code[3]. For this spectrum unfolding, several appropriate activation rates were selected among the activation detectors used in the experiment. To estimate the error of the evaluated spectrum, it was assumed that all cross-sections are in error by 10%.

Figures 11 and 12 show the experimental bremsstrahlung spectra $\phi_t(E)$ evaluated by the LYRA code for the iron and tungsten targets, respectively, together with the spectrum calculated from Eq. (16), $\phi^{cal}(E)$. In the figures, the experimental errors of the spectra due to unfolding, statistical, and cross-section errors are shown in the error bars. These figures show that for the iron and tungsten targets the shapes of experimental and calculated spectra resemble each other for both angles, 9° and 26°, and the absolute values of the calculated spectra are a little larger than the experimental ones for 9°. Figures 13, 14, and 15 show the experimental bremsstrahlung spectra $\phi_t(E)$ evaluated by the LYRA code for a lead target bombarded by 15, 20 and 25 MeV electrons, respectively,

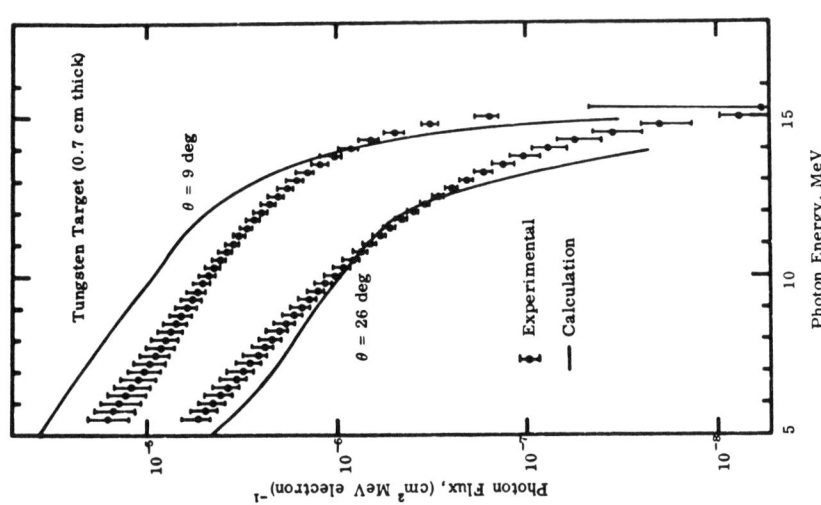

Fig. 12. Comparison between the experimental and the calculated bremsstrahlung spectra in 9° and 26° directions for the tungsten target.

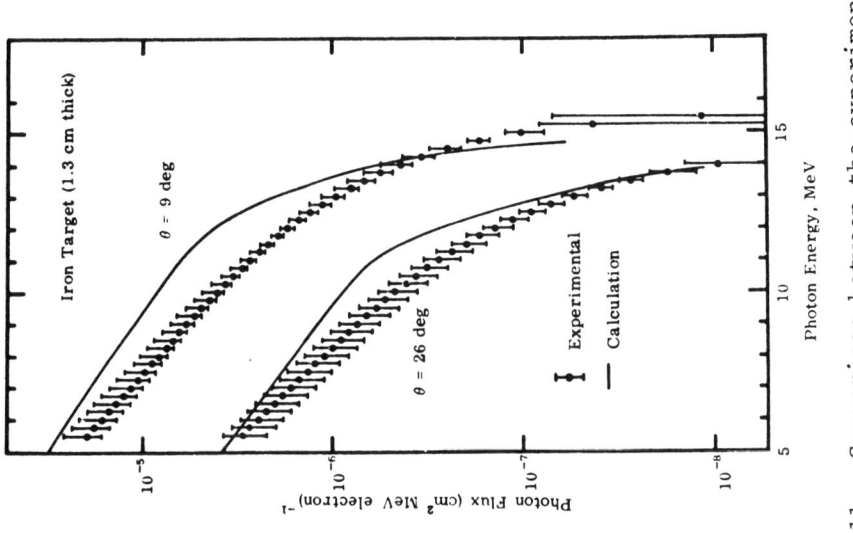

Fig. 11. Comparison between the experimental and the calculated bremsstrahlung spectra in 9° and 26° directions for the iron target.

BREMSSTRAHLUNG SPECTRUM ANALYSIS BY ACTIVATION METHOD

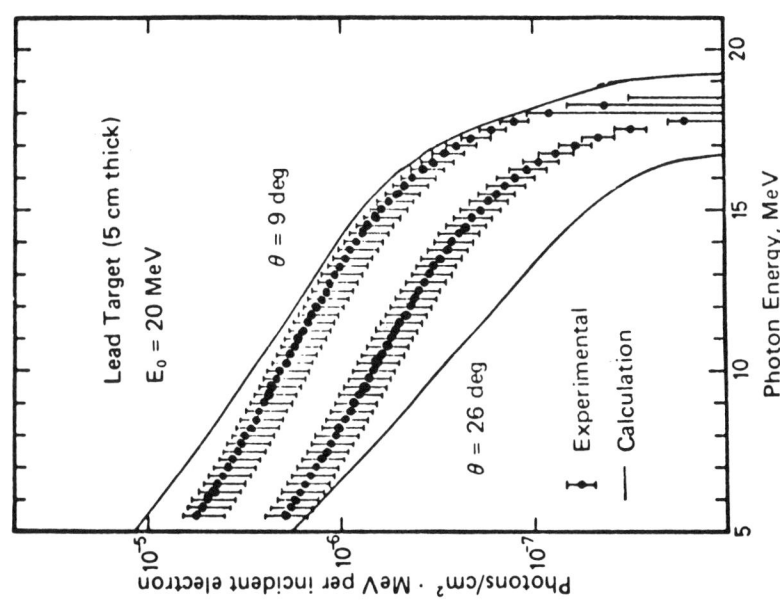

Fig. 14. Comparison of experimental and calculated bremsstrahlung spectra from a lead target in 9° and 26° directions by 20 MeV electrons.

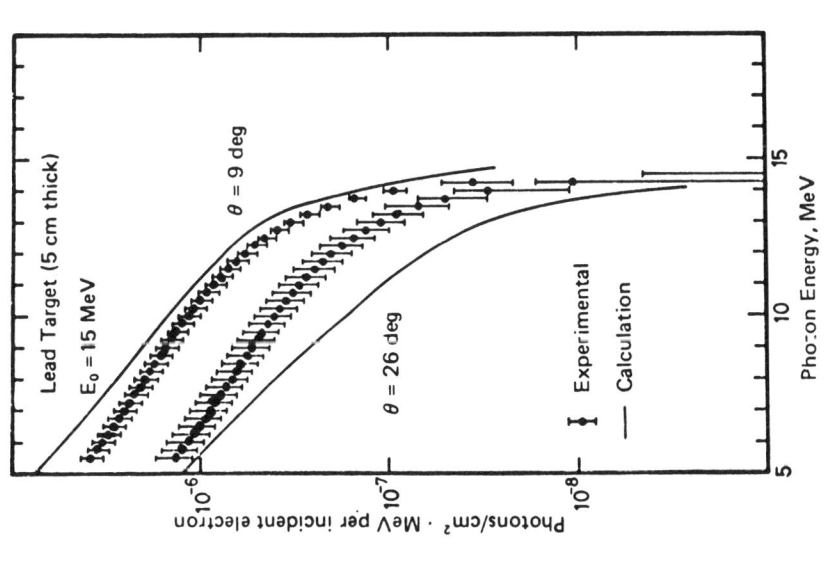

Fig. 13. Comparison of experimental and calculated bremsstrahlung spectra from a lead target in 9° and 26° directions by 15 MeV electrons.

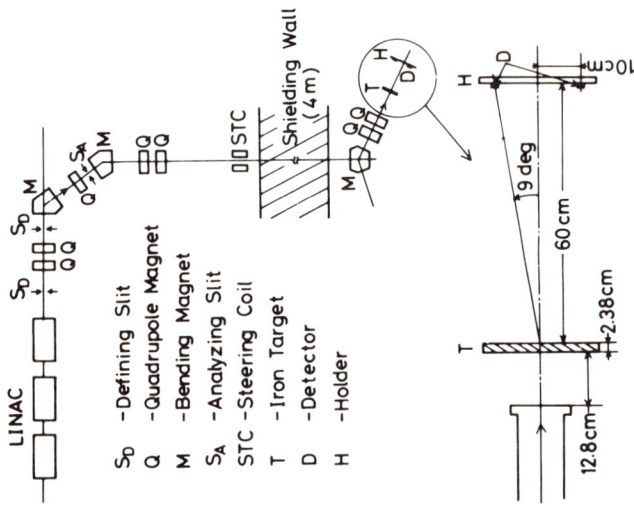

Fig. 16. Experimental arrangement.

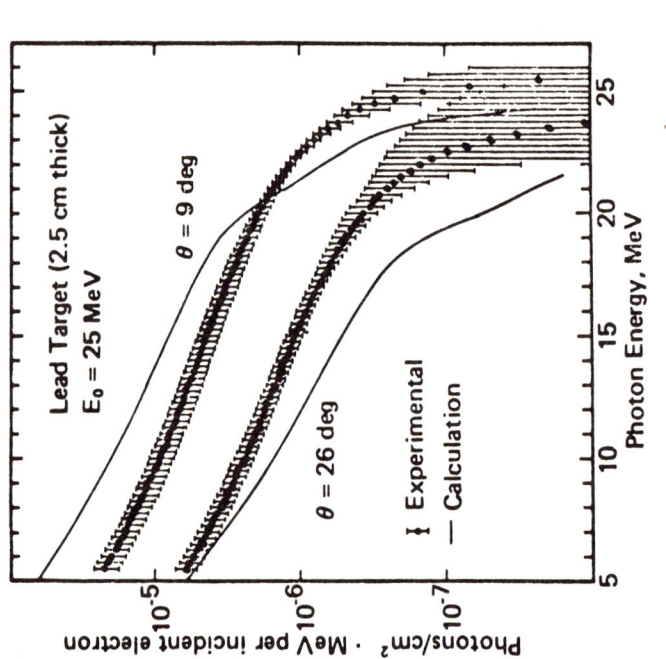

Fig. 15. Comparison of experimental and calculated bremsstrahlung spectra from a lead target in 9° and 26° directions by 25 MeV electrons.

together with the calculated spectrum $\phi^{cal}(E)$. The agreement of the experimental and the calculated spectra is rather good in absolute values for the 9° angle, especially at 15 and 20 MeV.

The noticeable difference between experiment and calculation may be explained partly by the errors in the photonuclear cross-sections and partly by the fact that the electron beam analysed by a magnet was not a monoenergetic spot but had a tail containing lower energy components, as opposed to the calculational model which assumed a narrow monoenergetic beam.

There are no other experimental data for such thick targets as used in this experiment. It has been confirmed from these figures that this new activation method can be used to measure the bremsstrahlung spectrum of very thick targets within about 50% error and, consequently, the spatial distribution of bremsstrahlung in the bulk medium, and that it can well describe variation in a bremsstrahlung spectrum as a function of incident electron energy.

INTEGRAL CHECK OF PHOTONUCLEAR CROSS-SECTIONS

The spectrum obtained by this activation technique is strongly dependent on the photonuclear cross-section data of the activation detectors used. The photonuclear cross-section data are generally not well compiled and are less accurate than neutron cross-section data. Berman compiled[33] some photoneutron cross-sections obtained with monoenergetic photons from positron annihilation, and the Photonuclear Data Center at the National Bureau of Standards is proceeding with a comprehensive compilation and evaluation of selected photonuclear cross-section data[34]. Activation detectors are potentially useful for integral experiments; namely, as sensitivity checks on the photonuclear cross-section data, by comparing the experimental induced activity with the integrated value of the product of the photonuclear cross-section and the calculated bremsstrahlung spectrum.

Integral Experiment

The integral experiment for a sensitivity check of photonuclear cross-section data of C, Mn, Fe, In, and Au was performed by using the bremsstrahlung radiation produced in a 2.38 cm thick iron target by 18, 22, 26, and 30 MeV electrons[6]. The physical properties of activation detectors used are listed in Table 2. The electron beam was extracted in the direction of 90° into the experimental room through a 4 m thick concrete wall. The electron beam, having 1.4% energy spread and a diameter of 3 mm, directly struck the iron target fixed 12.8 cm backward from a thin titanium window. The experimental arrangement is shown in Fig. 16.

The integral check of photonuclear cross-section can be done by using the ratio of the measured activation rate A_i^{exp}, and the calculated one A_i^{cal}:

$$R_i' = A_i^{exp}/A_i^{cal} . \qquad (19)$$

Since the bremsstrahlung spectrum can be obtained from the DIBRE calculation[1] with sufficient accuracy, the calculated activation rate A_i^{cal} is obtained from Eq. (20):

$$A_i^{cal} = \int_{E_{th}^i}^{E_0} \sigma_i(E) \phi^{cal}(E) \, dE , \qquad (20)$$

where $\phi^{cal}(E)$ = bremsstrahlung spectrum calculated from Eq. (16),
E_0 = energy of incident electron,
E_{th}^i = threshold energy of photonuclear reaction,
$\sigma_i(E)$ = photonuclear cross-section to be checked.

The photonuclear cross-sections to be tested by this integral experiment are shown in Fig. 17. The ratio R_i' of the measured activation rate A_i^{exp}, to the calculated one A_i^{cal}, is listed in Table 3. In this table, "Detector" indicates the method of the cross-section measurement, "ACT" means that the cross-section was

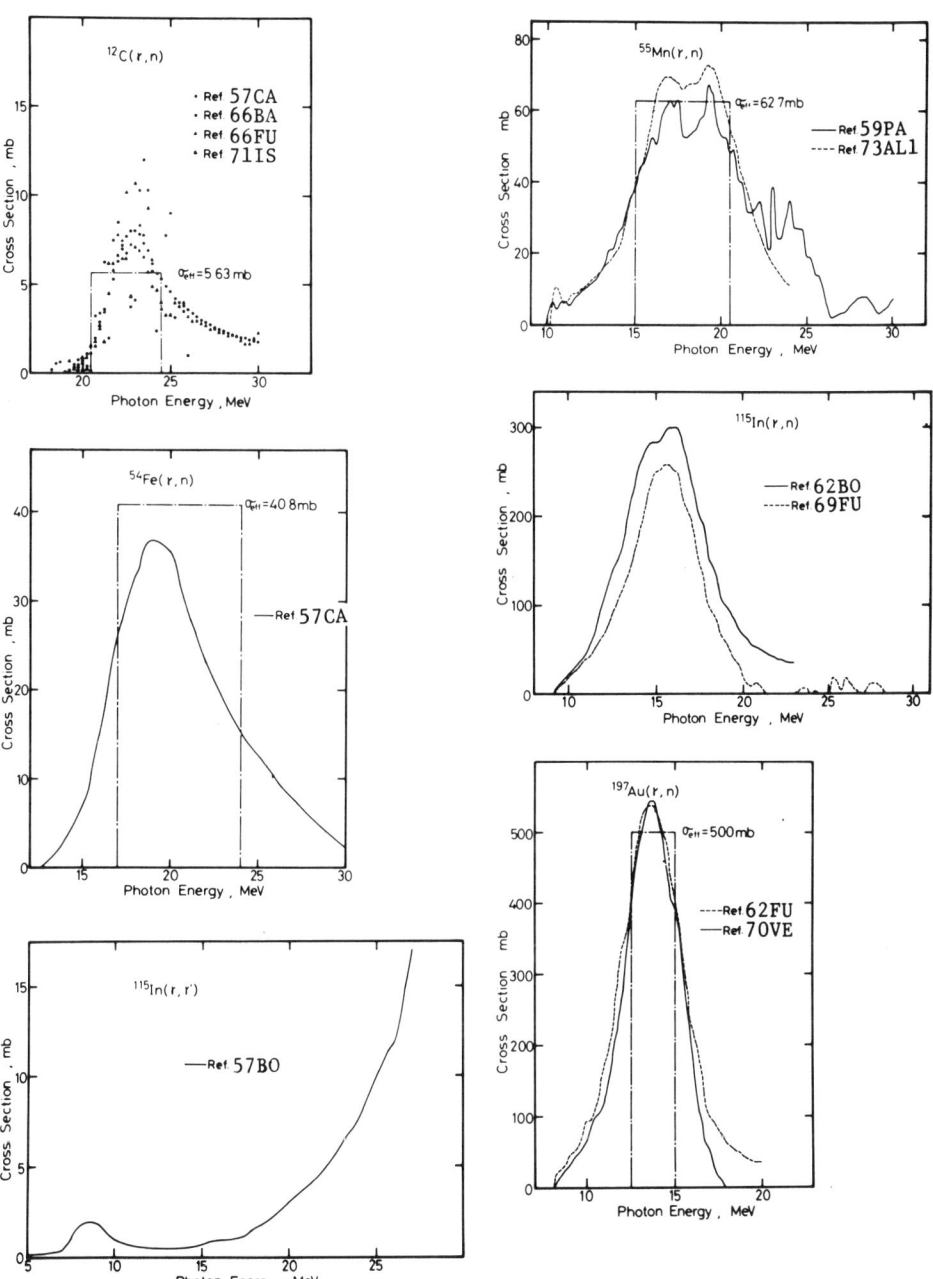

Fig. 17. Photonuclear reaction cross-section data to be tested in this experiment, the effective energy range and the effective cross-section.

Table 3 Ratio of measured and calculated saturated activities at 9°, $R'_i = A_i^{exp}/A_i^{cal}$

Reaction	Reference	Detector	Electron energy				Threshold energy of competing (γ,np) reaction (MeV)
			18 MeV	22 MeV	26 MeV	30 MeV	
$^{12}C(\gamma,n)^{11}C$	57CA	ACT*		1.205	1.09	0.824	27.4
	66BA	BF3**		0.757	0.931	–	
	66FU	BF3		1.492	1.189	0.914	
	71IS	BF3		1.418	1.159	–	
$^{55}Mn(\gamma,n)^{54}Mn$	59PA	BF3	0.575	0.810	0.598	–	17.8
	73AL1		0.595	0.875	0.654	0.609	
$^{54}Fe(\gamma,n)^{53}Fe$	57CA	ACT	0.542	0.783	0.722	1.12	20.9
$^{115}In(\gamma,n)^{114m}In$	62BO	BF3	0.448	0.511	–	–	15.9
	69FU	BF3	0.603	0.672	0.586	1.00	
$^{115}In(\gamma,\gamma')^{115m}In$	57BO	ACT	0.829	1.00	0.852	0.995	
$^{197}Au(\gamma,n)^{196}Au$	62FU	BF3	0.743	0.553	0.529	–	13.7
	70VE	BF3	0.842	0.628	0.608	0.476	

* ACT: measurement of radioactivity of the target

** BF3: BF$_3$ neutron counter with moderator

obtained from the measurement of radioactivity of the target, and "BF3" from the measurement of produced photoneutrons by the moderated BF_3 counter. Table 3 clearly reveals the following facts: 1) the new data of $^{55}Mn(\gamma,n)$, $^{115}In(\gamma,n)$, and $^{197}Au(\gamma,n)$ cross-sections are much better than the old data; 2) the cross-section data measured by the activation method show better results for all incident electron energies than those by the photoneutron method; 3) for the method of photoneutron measurement, the experimental activity A_i^{exp} becomes smaller than the calculated one A_i^{cal}, with increasing incident electron energy, except for the case of the $^{115}In(\gamma,n)$ reaction at 30 MeV electron energy.

These observations may be explained by the fact that all (γ,n) cross-section data measured by the photoneutron method include the competing (γ,np) reaction above its threshold energy, which is shown in the last column of Table 3, and they become larger than the true (γ,n) cross-section. The one exception of the $^{115}In(\gamma,n)$ reaction may be attributed to the contribution of the neutron capture reaction $^{113}In(n,\gamma)^{114m}In$, which produces the same radioisotope as for the $^{115}In(\gamma,n)^{114m}In$ reaction.

It is concluded from this result that the photonuclear cross-sections, to be used for the bremsstrahlung spectrum measurement with activation detectors, must be obtained by the method of the radioactivity measurement of the target.

Effective Energy Range and Effective Cross-Section

The shapes of bremsstrahlung spectra produced by electrons of various energies are close together except at the high-energy ends. It is expected, then, that the ratio of the measured activation rate to the integral bremsstrahlung flux in the giant resonance region is almost constant for various electron energies when the maximum bremsstrahlung energy is higher than the giant resonance region. By defining this ratio as the effective cross-section σ_i^{eff}, the following equation is given:

Table 4 Effective cross-section and effective energy range

Reaction	Effective energy range (MeV)	Effective cross-section (mb)			
		Electron energy			Average
		22 MeV	26 MeV	30 MeV	
$^{12}C(\gamma,n)^{11}C$	20.5 – 24.5	–	5.98	5.28	5.63 ± 0.35
$^{55}Mn(\gamma,n)^{54}Mn$	15.0 – 20.5	–	56.2	52.7	54.5 ± 1.7
$^{54}Fe(\gamma,n)^{53}Fe$	17.0 – 24.0	–	33.6	47.9	40.8 ± 7.2
$^{197}Au(\gamma,n)^{196}Au$	12.5 – 15.0	553.5	532.5	415.0	500 ± 85

$$A_{ij}^{exp} = \int_{E_{th}^i}^{E_{0j}} \sigma_i(E) \phi_j^{cal}(E) \, dE = \sigma_i^{cal} \int_{E_1^{eff}}^{E_2^{eff}} \phi_j^{cal}(E) \, dE, \qquad (21)$$

where A_{ij}^{exp} = activation rate of ith reaction for j^{th} incident electron energy;

$\phi_j^{cal}(E)$ = bremsstrahlung spectrum for j^{th} incident electron energy;

E_1^{eff} = minimum energy of effective energy range;

E_2^{eff} = maximum energy of effective energy range.

The effective energy range from E_1^{eff} to E_2^{eff}, and the effective cross-section, are determined by a simultaneous parametric survey, such that the ratio of the measured activity A_{ij}^{exp} to the DIBRE-calculated bremsstrahlung flux, integrated over several different energy regions including the giant resonance $\int \phi_j^{cal}(E) \, dE$, is almost constant for 22, 26, and 30 MeV electrons (for each photonuclear reaction). The effective energy range and the effective cross-section determined by this method are shown for $^{12}C(\gamma,n)$, $^{55}Mn(\gamma,n)$, $^{54}Fe(\gamma,n)$, and $^{197}Au(\gamma,n)$ reactions in Table 4 and in Fig. 17. The $^{115}In(\gamma,n)$ reaction was excluded owing to the influence of the $^{113}In(n,\gamma)$ reaction, as described before.

It is necessary, for the bremsstrahlung spectrum evaluation by the activation method, to get accurate cross-section data of (γ,n), (γ,np), (γ,p), $(\gamma,2n)$ reactions, etc., separately. Otherwise, the gross structure of the bremsstrahlung spectrum can be obtained by using the effective energy range and the effective cross-sections for those reactions.

REFERENCES

The references are listed in the last part of the next lecture (Lecture 26).

LECTURE 26: APPLICATION OF ACTIVATION-SPECTRUM ANALYSIS METHOD

TO SHIELDING (TAURUS, LYRA, DIBRE, SAND-II)[35-38]

Takashi Nakamura

Institute for Nuclear Study, University of Tokyo
Midori-cho 3-2-1, Tanashi, Tokyo
Japan

INTRODUCTION

Various kinds of high-energy electron accelerators have recently been used in practice for medical, industrial, and scientific research applications. It is a very important problem to protect mankind and instruments from a large amount of bremsstrahlung generated in a target bombarded by electrons. Many works have been performed on the depth-dose and the energy-deposition distributions due to thick-target bremsstrahlung, which are summarized in some books[39-41]. However, there were no theoretical or experimental studies on the space-energy distribution of bremsstrahlung in and through a bulk medium directly bombarded by electrons, in spite of the great importance of such work in the study of radiation shielding, residual activity, and radiation damage to accelerator and target materials.

We have recently published several papers on the spatial distributions of bremsstrahlung in water and water-iron[36], aluminum, and iron[35] by 22 MeV electron bombardment, measured with our new method[3], as described in the preceding lecture. Those experimental results were compared with the analytical calculation by the DIBRE code[1] and the Monte Carlo calculation by the TAURUS code[37].

The Monte Carlo method has been widely studied as the most efficient way to calculate the electron and photon penetration in matter[25,28,42-44]. The ETRAN code[25] has been used as a standard because of its good accuracy and wide applicability, but it requires a lot of core memory and computing time. So we have developed an electron-photon cascade Monte Carlo code, TAURUS, based upon the code by Sugiyama[44].

ELECTRON-PHOTON CASCADE MONTE CARLO CALCULATION

Method of Calculation

The Monte Carlo code, TAURUS, has been developed[37] for the calculation of gamma-ray and bremsstrahlung penetration in matter, for cylindrical and plane geometry, bombarded by gamma-rays and electrons. This calculation includes the assumption that photons interact by one of three well-known processes: photoelectric effect, Compton effect, and pair production; and that electrons trace a zigzag path in the medium owing to multiple scattering accompanying the bremsstrahlung emission. This code neglects the collision energy-loss straggling of the electron and the production of knock-on electrons, thereby significantly influencing the slowing down of the electron, decreasing the computation time, and saving core memory. This simplification causes a slightly rough simulation of the electron trajectory, but is accurate enough for photon simulation.

Photon Interaction Processes Simulated

If a photoelectric event is determined, the entire photon energy is assumed to be transferred to a single electron (neglecting an accompanying characteristic X-ray), and the electron emission angle is obtained by sampling Sauter's distribution formula[45] and by assuming azimuthal isotropy. The energy and direction of secondary photons by Compton scattering are sampled by the Latter-Kahn technique[46] of the Klein-Nishina differential cross-section. In pair production events, it is assumed that the kinetic energy shared by the electron-positron pair, $h\nu - 2m_0c^2$, is partitioned such that the energy given to either member of the pair follows Hough's approximate formula[47], and that both the electron and positron are generated in the same direction as the primary photon. Two annihilation photons of energy m_0c^2 are assumed to be emitted isotropically and in opposite directions after the positron has lost all its initial kinetic energy.

Electron Interaction Processes Simulated

The electron transport is handled with the Monte Carlo simulation of the electron path using a large number of small path-length intervals, which are fixed to be the equidivision of the initial electron range. The electron energy loss is determined from a continuous slowing-down model through the medium. Immediately after passing straight through any segment, the electron changes its direction of motion owing to multiple scattering in the segment. The angular deflection of the multiple-scattered electron is determined by sampling from the Molière formula modified by Bethe[19]. The entire trajectory of the electron is traced by combining the results of the successive path segments until the electron is slowed down below a given cut-off energy value. The positron is traced similarly, except that it annihilates when completely stopped.

In order to decrease the computing time and improve the statistical convergence for the calculation of the thick-target bremsstrahlung spectra, it is assumed that the bremsstrahlung production by the electron, of energy lower than the cut-off energy, is included in the ionization loss. Under this assumption, the collision stopping power is given by

$$\left(-\frac{dE}{dX}\right)_{col,E} = \left(-\frac{dE}{dX}\right)_{ion,E} + \int_0^{E_{cut}} kC(E)\left(\frac{d\sigma}{dk}\right)_E dk , \qquad (1)$$

where $(-dE/dX)_{ion,E}$ = ionization and excitation energy loss,

E_{cut} = cut-off energy,

E = electron energy,

$(d\sigma/dk)_E$ = differential bremsstrahlung cross-section by Schiff's formula[21], $\sigma(3BS)$,

$$C(E) = \frac{\left(-\frac{dE}{dX}\right)_{rad,E}}{\int_0^E k\left(\frac{d\sigma}{dk}\right)_E dk},$$

$(-dE/dX)_{rad,E}$ = radiation energy loss.

The electron range with collision energy loss $R(E)$ is then

$$R(E) = \int_{E_{cut}}^{E} \frac{dE'}{\left(-\frac{dE'}{dX}\right)_{col,E'}} + R_0, \tag{2}$$

where R_0 is the electron range of energy E_{cut}.

The treatment of bremsstrahlung production is separated into first production and successive ones. The first position and probability of bremsstrahlung emission is sampled in the following way. The probability $P(E)$ that the electron of energy E produces at least one bremsstrahlung quantum until it has stopped, is

$$P(E) = 1 - e^{-f(E)}, \tag{3}$$

where

$$f(E) = \int_{E_{cut}}^{E} \sigma(E') \frac{dE'}{\left(-\frac{dE'}{dX}\right)_{col,E'}}$$

$$\sigma(E) = \int_{E_{cut}}^{E} C(E')\left(\frac{d\sigma}{dk}\right)_{E'} dk.$$

It is forced to produce bremsstrahlung with the weight $P(E_0)$, once for each incident electron of energy E_0, and the emission position is given by random sampling; namely,

$$f(E_B) = f(E_0) + \ln\left[1 - P(E_0)r\right], \tag{4}$$

where r is a random number and E_B is the electron energy at the first point generating bremsstrahlung. This procedure is especially effective in the case of small probability of bremsstrahlung emission. The successive bremsstrahlung production is traced by usual random sampling. A random number r is compared with the probability distribution function $P(E_i)$, where E_i is the electron energy in the ith segment. The next position of bremsstrahlung emission is also given by $f(E_i) = f(E_{i-1}) + \ln r$. The energy and direction of bremsstrahlung are sampled according to Schiff's differential cross-section formula[21].

Calculation of Photon Spectrum at Detector Position

The photon spectrum at the detector position consists of the Compton-scattered photon and the bremsstrahlung. It is calculated from summation of the product of the statistical weight of the photon, or the bremsstrahlung, and its probability of reaching the detector from each collision or emission point. The bremsstrahlung spectrum decreases rapidly with energy, so the usual sampling causes poor statistics in the high-energy side of the spectrum. To avoid this, a kind of importance sampling is adopted so that the bremsstrahlung energy k distributes uniformly from the cut-off energy E_{cut}, to the kinetic energy of the electron E_i, at the ith position generating bremsstrahlung. Since both the electron and the photon only change their weights on bremsstrahlung emission as described above, the electron or the photon weight W_i, at the ith emission point, in the same way is given by

$$W_i = \prod_{m=0}^{i-1} \frac{\left(\frac{d\sigma}{dk}\right)_{E_m}}{\sigma(E_m)} (E_m - E_{cut}) C(E_m) P(E_0). \tag{5}$$

The bremsstrahlung is generated with the weight W_{Bi} in the unit solid angle around the direction Ω_i towards the detector from the emission position, as follows:

$$W_{Bi} = C(E_i) \left(\frac{d^2\Omega}{dkd\Omega}\right)_{E_i,k,\Omega_i} (E_i - E_{cut}) W_i / \sigma(E_i) , \qquad (6)$$

where

$$k = E_{cut} + r(E_i - E_{cut}) ,$$
$$r = \text{random number} .$$

The photon energy spectrum at the detector position is expressed as

$$\phi(k_\ell) = \frac{1}{N} \sum_{n=1}^{N} \left[\sum_i W_{Bi} \frac{e^{-\mu_i t_i}}{r_i^2} + \sum_j W_j \frac{1}{r_j^2} \left(\frac{d\mu_s}{d\Omega}\right)_{\Omega_j} \frac{e^{-\mu_j t_j}}{\mu_j} \right] \delta(k-k_\ell) , \qquad (7)$$

where

i, j	= number of emission or collision,
μ_i	= photon linear attenuation coefficient,
t_i	= path length in the medium from the ith collision point to the detector,
r_i	= distance from the ith collision point to the detector,
$(d\mu_s/d\Omega)_{\Omega_j}$	= differential Compton-scattering cross-section,
n	= number of history,
N	= maximum number of history,
$\delta(k - k_\ell)$	= $\begin{cases} 1 \text{ if } k_\ell = k \\ 0 \text{ if } k_\ell \neq k \end{cases}$

The first and the second term of Eq. (7) correspond to the bremsstrahlung radiation from the emission point and the scattered photons from the collision point, respectively.

Description of the TAURUS Computer Program

The trace of a cascade process is performed by using a method of so-called lexicographic processing of trees. The trajectory of a particle is finished and that of another particle starts when the particle energy is below the cut-off energy, or the particle escapes from the medium. Each history is finished when all branches of a cascade tree have been traced completely. In the thick-target bremsstrahlung calculation, the cut-off energy and the segment of the electron path were selected as 1.0-5.0 MeV and about 1/40 of the range, respectively. A history number of 10,000-20,000 is required for convergence. The typical computing time is about 50 histories per sec and the core memories are about 45k words for the FACOM 230-75 computer.

Results and Discussion

In order to examine the accuracy of the TAURUS code, several calculations have been performed to compare with other experimental and calculated results.

Figure 1 shows the absolute bremsstrahlung spectra emitted in 0° and 40° directions from a 4.67 g/cm^2 (0.19 radiation length) aluminum target for 10 MeV incident electrons. The spectra calculated with this code are in good agreement with the experimental values by Jupiter et al.[24] and the calculated values[24] by the ETRAN code[25] for both 0° and 40°. To be more specific, however, our calculated result is a little larger than the ETRAN result at 0°. Figure 2 shows the forward bremsstrahlung spectra from a 0.735 g/cm^2 (0.106 r.l.) tungsten target for 10 MeV and 20.9 MeV electron beams. Our calculation agrees in general with the absolute experiment by O'Dell et al.[26] and two Monte Carlo results[28,29] by ETRAN and EGS codes, but it gives a value a little larger than the other three results, especially in the low-energy region.

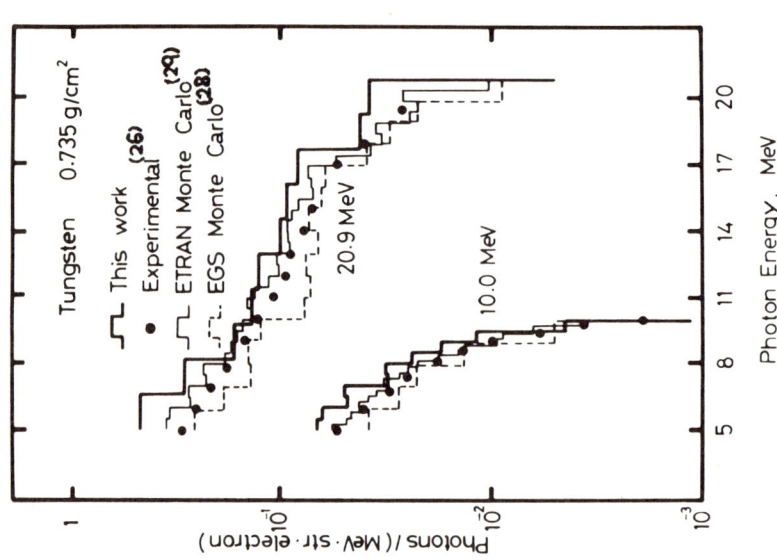

Fig. 2. Comparison of calculated forward bremsstrahlung spectra from a 0.735 g/cm² tungsten target exposed to 10 and 20.9 MeV electrons with other results.

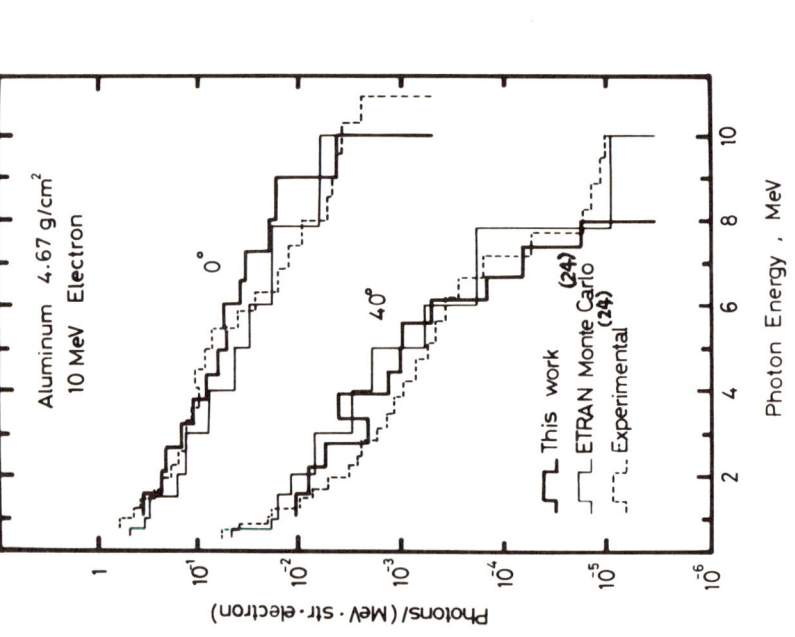

Fig. 1. Comparison of calculated bremsstrahlung spectra emitted at 0° and 40° from a 4.67 g/cm² aluminum target exposed to 10 MeV electrons with other results.

Figure 3 shows the bremsstrahlung spectra for 30 MeV electrons incident on a 24 g/cm^2 (3.5 r.l.) tungsten target. The ETRAN results give the value averaged in angular intervals of 0°-0.5°, 25°-30°, and 55°-60°, while in the TAURUS calculation the medium angle (0.25°, 27.5°, and 57.5°) was selected for each angular interval as shown in Fig. 3. The TAURUS calculation shows good agreement with the ETRAN calculations[29] in three angular regions, excluding some statistical fluctuation of the spectra at larger angle in the TAURUS calculation because of the insufficiency of history number. In Fig. 4, the forward bremsstrahlung spectra from a 35 g/cm^2 (6 r.l.) lead target are shown for 25 MeV electron beam incidence. The EGS result[28] is the average value in a 0° to 5° cone, and the TAURUS result is the value at 2.5°. Our result is in good agreement with the EGS result except at the high-energy end.

SHIELDING EXPERIMENT OF BREMSSTRAHLUNG IN BULK MEDIA WITH ELECTRONS

Experimental Procedure

The spatial distributions of bremsstrahlung in water, water-iron[36], aluminum, and iron[35] assemblies bombarded by 22 MeV electrons have been measured by using the activation-spectrum analysis method. The experimental arrangement is shown in Fig. 5. The dimensions of the water, aluminum, iron, and water-iron systems are 60 × 80 × 100, 60 × 100 × 30, 60 × 80 × 18.5, and 60 × 80 × 66.3 cm^3, respectively. The water was contained in a 0.3 cm thick aluminum tank. The activation detectors of Mn, Ni, Co, In, C, Cr, and Au were used for the iron and aluminum experiments, and Mn, C, Au, and Ag were used for the water and water-iron experiments. Their physical properties are already shown in Table 2 of the preceding lecture. The activation detectors were set at several (r,z) positions in each system, where r is the lateral distance from the electron beam axis and z the longitudinal distance from the front face of the medium. To reduce the contribution of the spurious bremsstrahlung radiation from the accelerating tube, the target end of the tube

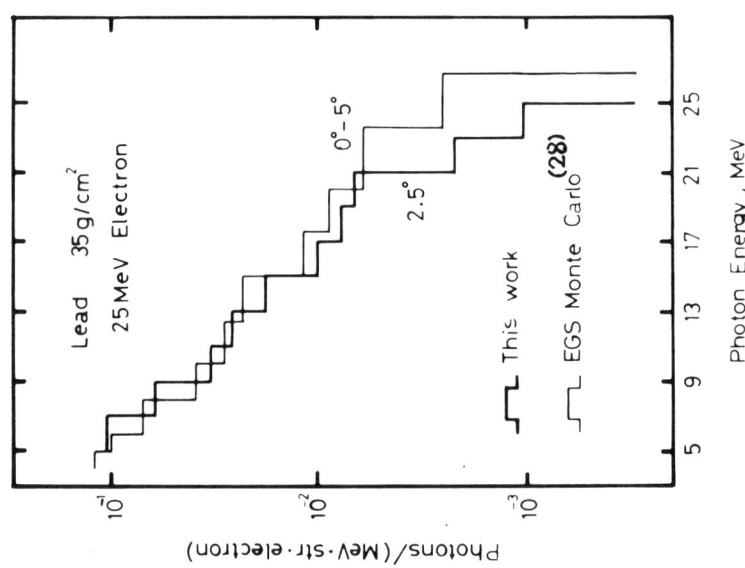

Fig. 4. Comparison of calculated bremsstrahlung spectra at 2.5° from a 35 g/cm² lead target exposed to 25 MeV electrons with other results.

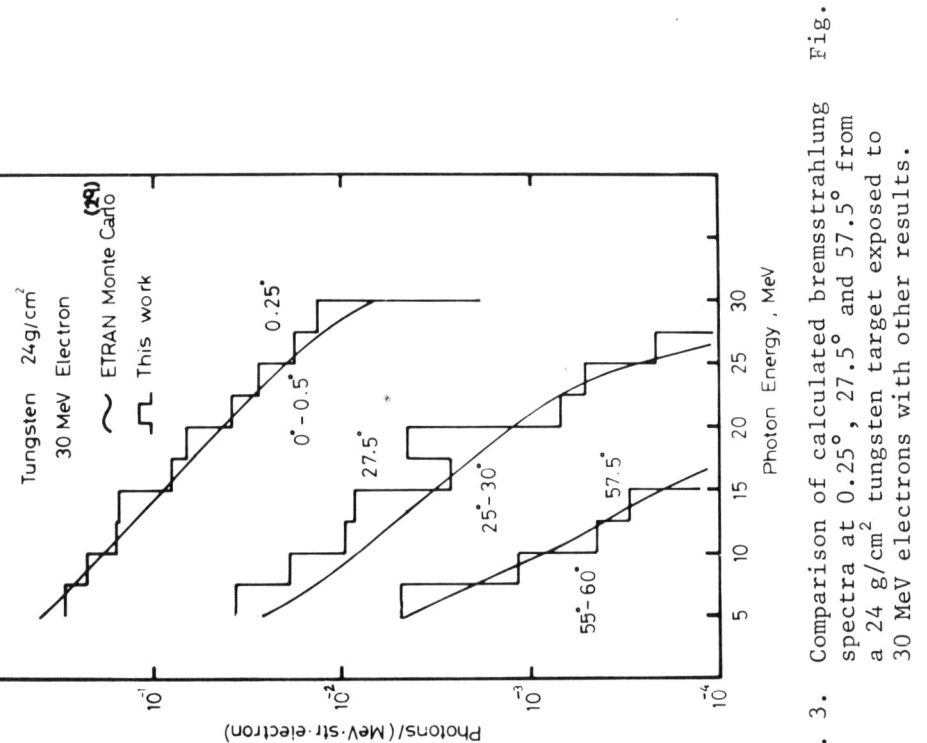

Fig. 3. Comparison of calculated bremsstrahlung spectra at 0.25°, 27.5° and 57.5° from a 24 g/cm² tungsten target exposed to 30 MeV electrons with other results.

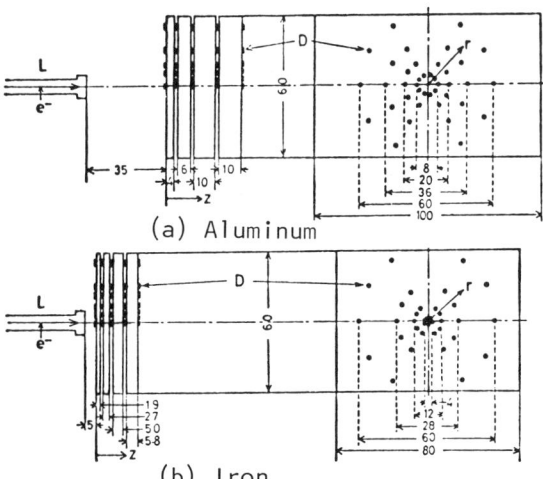

(a) Aluminum

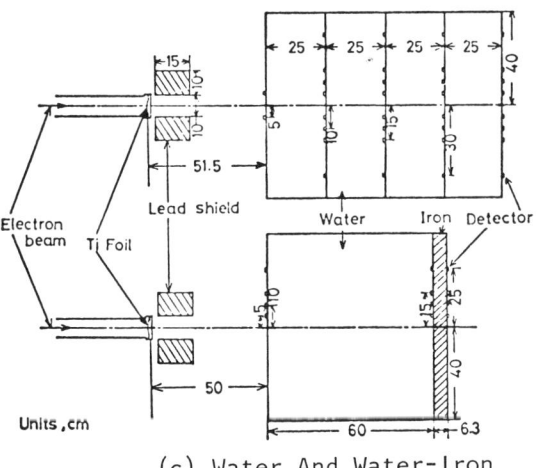

(b) Iron

L: Linear Accelerator, D: Detector

(c) Water And Water-Iron

Fig. 5. Experimental arrangement.

was surrounded with the lead blocks as shown in Fig. 5 (for the water and water-iron systems only). The incident electron beam had a broad energy spectrum having a peak at about 22 MeV, as shown in Fig. 6. The electron energy spectrum was much broader for aluminum and iron experiments than for water and water-iron experiments. The gamma-ray activities of the detectors induced by bremsstrahlung were converted to the bremsstrahlung energy spectrum $\phi_t(E)$ by the LYRA[3] and the SAND-II[48] unfolding codes. The experimental spectra were obtained in the energy ranges above 8 MeV for water, and above 5.5 MeV for aluminum and iron, depending on the different kinds of activation detectors used in these three experiments.

Experimental Results and Discussion

Water System. For water, the experimental results were compared with the analytical calculation by the DIBRE code. The bremsstrahlung flux $\phi^{cal}(E)$, at the (r,z) position in the medium, is given by

$$\phi^{cal}(E) = \sum_{i=1}^{n} \int dE_0 \int_0^{2\pi} d\psi \int_0^{\pi} \sin\theta\, d\theta\, S(E_0)\tau(E_0,t_i)\nu_i F(E_i,\theta,t_i) \times$$

$$\times \frac{d\sigma}{dEd\Omega}(E_i,E,\omega_i) B(E,\rho_i) e^{-\mu(E)\rho_i}\, d\Omega, \qquad (E_i \geq E),$$

(8)

where

$S(E_0)$ = electron energy spectrum,

$$d\Omega = \frac{2\pi \cos\alpha}{\pi\rho_i^2 + \cos\alpha + \rho_i[\pi(\pi\rho_i^2 + \cos\alpha)]^{\frac{1}{2}}},$$

$$\cos\alpha = \frac{z - t_i}{[(z - t_i)^2 + r^2]^{\frac{1}{2}}}.$$

Other functions in Eq. (8) are the same as in the preceding lecture, except that the photon energy is changed from k to E.

APPLICATION OF ACTIVATION-SPECTRUM METHOD TO SHIELDING

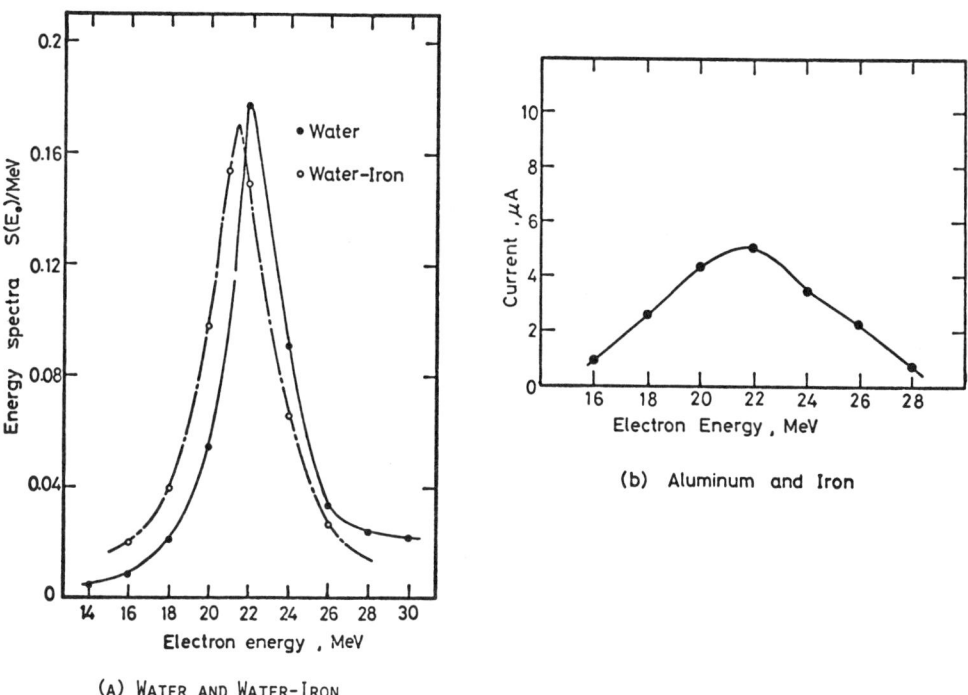

(A) WATER AND WATER-IRON

(b) Aluminum and Iron

Fig. 6. Incident electron energy spectra.

Figures 7(a), (b), and (c) show the experimental bremsstrahlung spectra $\phi_t(E)$ above 8 MeV evaluated by the LYRA and the SAND-II codes at z = 25, 50, and 100 cm in water, respectively, together with the calculated spectra $\phi^{cal}(E)$. The agreement of the experimental spectra by LYRA and SAND-II is quite good for all cases. The spectra unfolded by SAND-II show some irregularity around 17 MeV, corresponding to the dip between the ^{197}Au(γ,n) and ^{55}Mn(γ,n) cross-sections as seen from Fig. 8, because the SAND-II method divides the spectrum into small energy intervals to make it converge for each energy interval. On the other hand, the LYRA spectra give the smooth curves due to the spectrum expansion into the continuous function. These figures show that the bremsstrahlung spectra become softer with distance from the beam axis, but get harder with water depth (at a fixed r position). The agreement of the experimental and the calculated spectra is good in absolute value for all detector positions (except at r = 30 cm and z = 25, 50 cm) over the energy range of 8-17 MeV, where the ^{197}Au(γ,n) reaction has a giant resonance. Beyond 17 MeV, especially beyond 22 MeV, the calculated values are much smaller than the experimental values.

The photon flux $\phi(r,z)$ at each (r,z) position in water was obtained by integrating the energy spectrum above 8 MeV, and is shown in Fig. 9 as a function of lateral distance r at several depths in water. The agreement of the experimental and the calculated flux is very good in absolute value. But the experimental flux at the position of r = 30 cm and z = 25 cm -- that is, the angle to beam direction, $\theta \simeq 50°$ -- is about five times larger than the calculated flux, because the DIBRE calculation gives an underestimation with increasing emission angles as described before. Figure 9 shows that the experimental lateral flux distribution obeys approximately the exponential decrease near the beam axis at every depth z, and its slope varies only slightly at the larger depths.

APPLICATION OF ACTIVATION-SPECTRUM METHOD TO SHIELDING

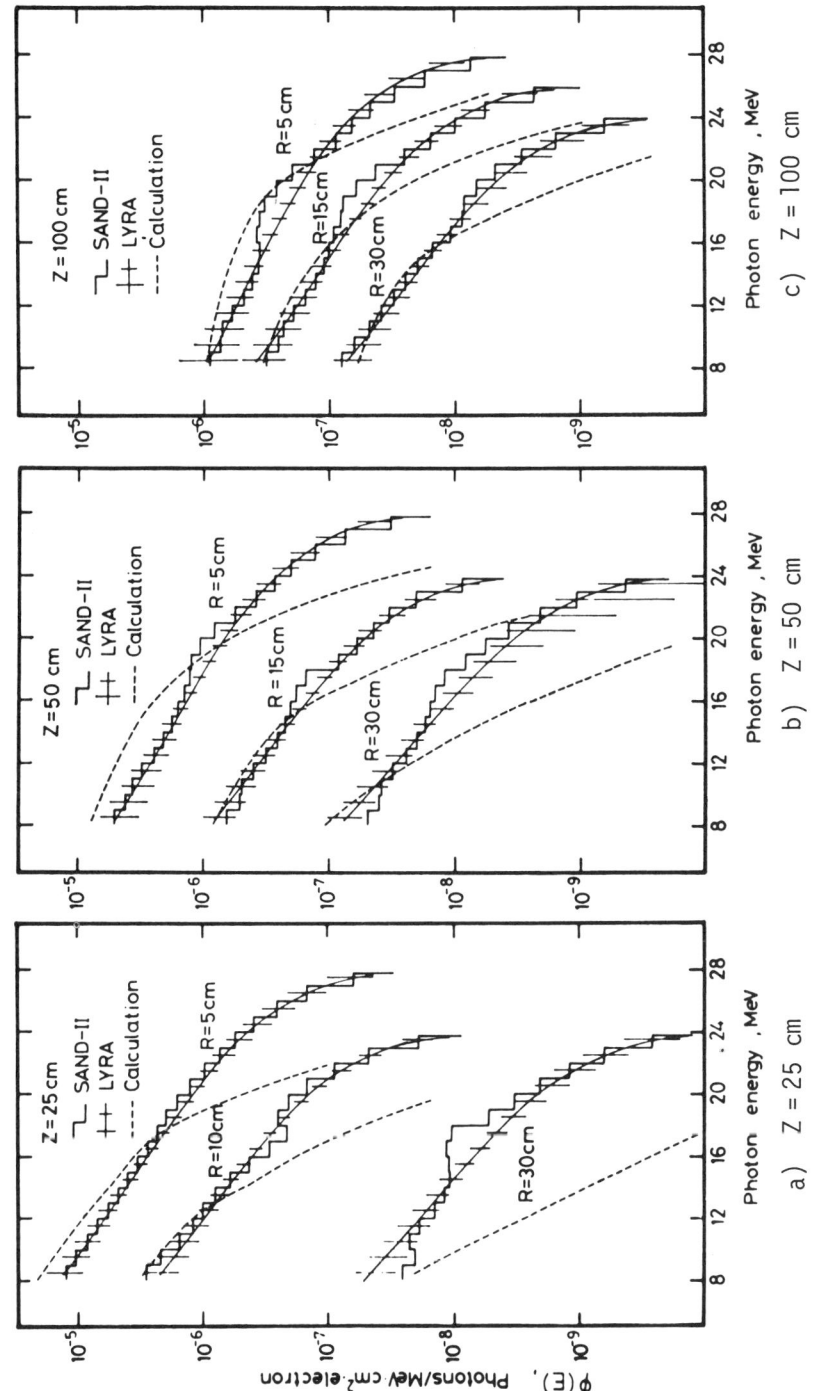

Fig. 7. Comparison of calculated and experimental bremsstrahlung spectra in water.

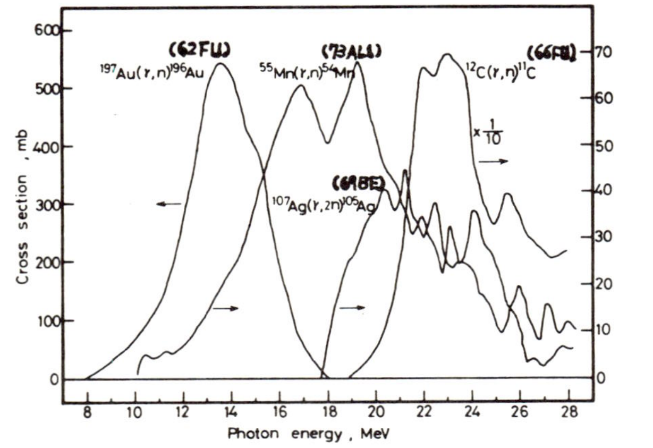

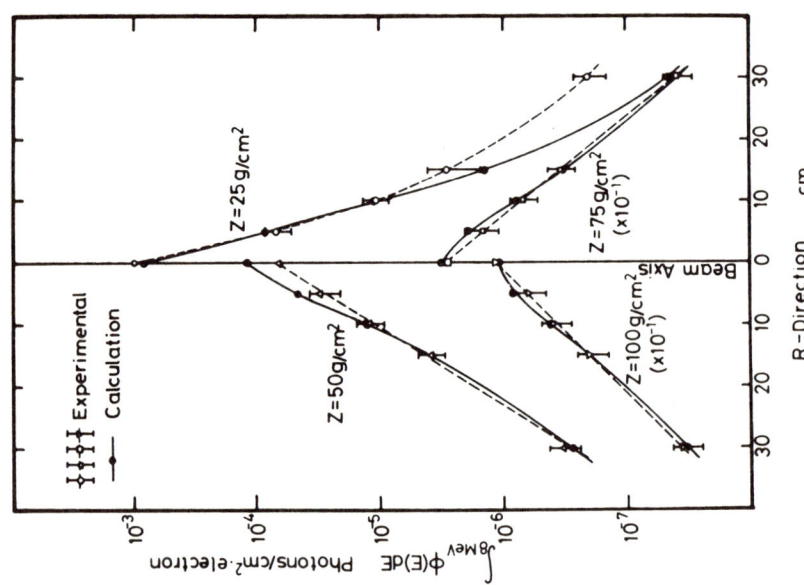

Fig. 9. Lateral distribution of experimental and calculated integrated photon fluxes in water.

Fig. 8. Photonuclear reaction cross-section of ^{197}Au(γ,n), ^{55}Mn(γ,n), ^{107}Ag(γ,2n) and ^{12}C(γ,n).

Figure 10 shows the integrated flux $\phi(r,z)$ as a function of the axial depth z at several radial depths r in water, including the extrapolated value $\phi(0,z)$ (dotted line). The agreement of the experimental and the calculated flux is also good in absolute values, except at $(r,z) = (30,25)$ (see Fig. 9 as well). The figure indicates the build-up and the successive attenuation of photons.

The total number $\Phi(z)$ of photons crossing a plane normal to the beam axis is obtained as

$$\Phi(z) = \int_0^\infty \phi(r,z) 2\pi r \, dr \, . \tag{9}$$

The $\Phi(z)$ distribution along the axial depth z, shown in Fig. 11, clearly gives an exponential attenuation as

$$\Phi(z) \propto e^{-\mu_{eff} z}, \qquad z \gtrsim 25 \text{ g/cm}^2 \, . \tag{10}$$

The effective attenuation coefficient μ_{eff} is obtained from Fig. 11 as

$$\mu_{eff} = 0.0172 \text{ cm}^2/\text{g} \, .$$

<u>Water-Iron System</u>. Only the saturation activities of gold detectors for the water-iron system were obtained with good experimental accuracy. The detectors were placed on the front face of the system, z = 0 cm, on the interface of water and iron, z = 60 cm, and on the three planes in iron, z = 61.8, 63.6, and 66.3 cm. Figure 12 shows the lateral distribution of gold saturation activities on these z-planes. The figure indicates the exponential decrease along the radial depth r more beautifully than in the water system and, moreover, the slopes of the curves in iron are nearly equal to each other, as opposed to those in water (see Fig. 9).

Figure 13 shows the gold saturation activity $A(r,z)$, as a function of z at several radii, in iron following 60 cm water. The figure includes the longitudinal distribution of the gold saturation

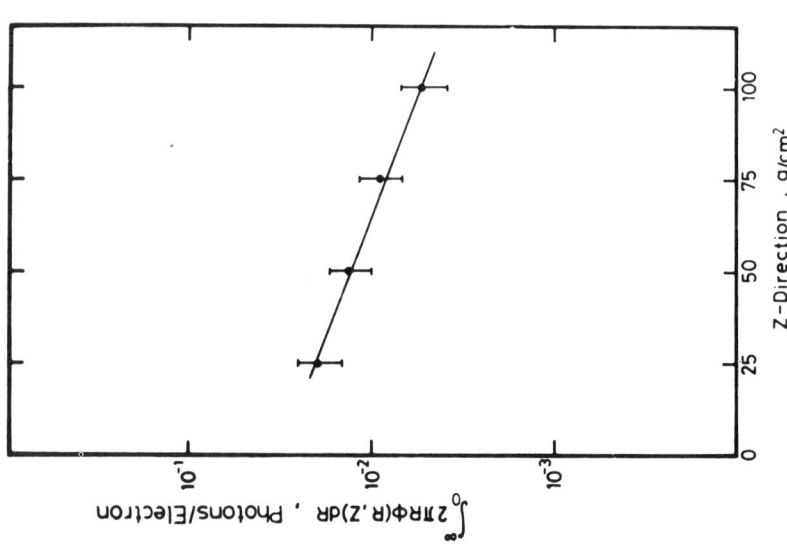

Fig. 11. Attenuation of experimental total photon number crossing a plane normal to the beam axis in water.

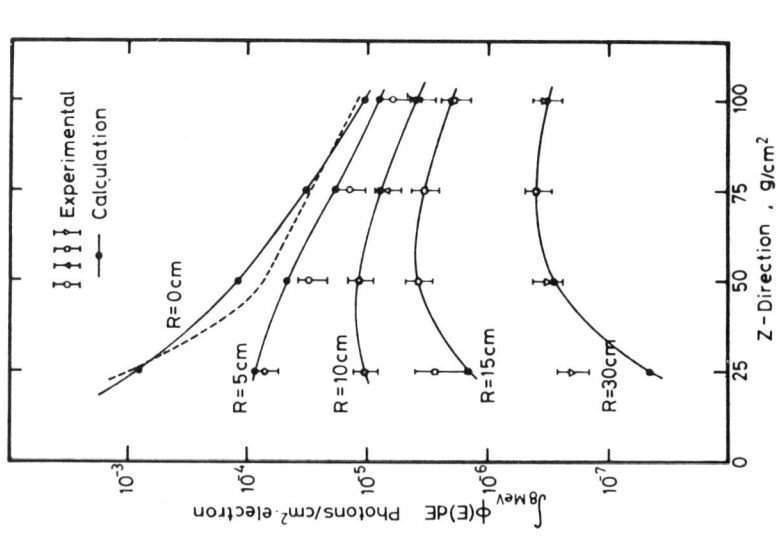

Fig. 10. Longitudinal distribution of experimental and calculated integrated photon fluxes in water.

APPLICATION OF ACTIVATION-SPECTRUM METHOD TO SHIELDING

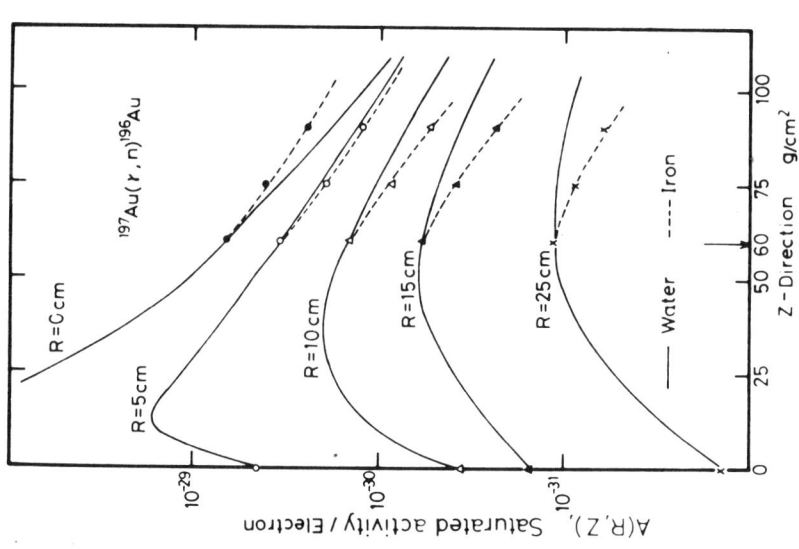

Fig. 13. Comparison of longitudinal distribution of experimental saturated activity of gold detector in water and water-iron.

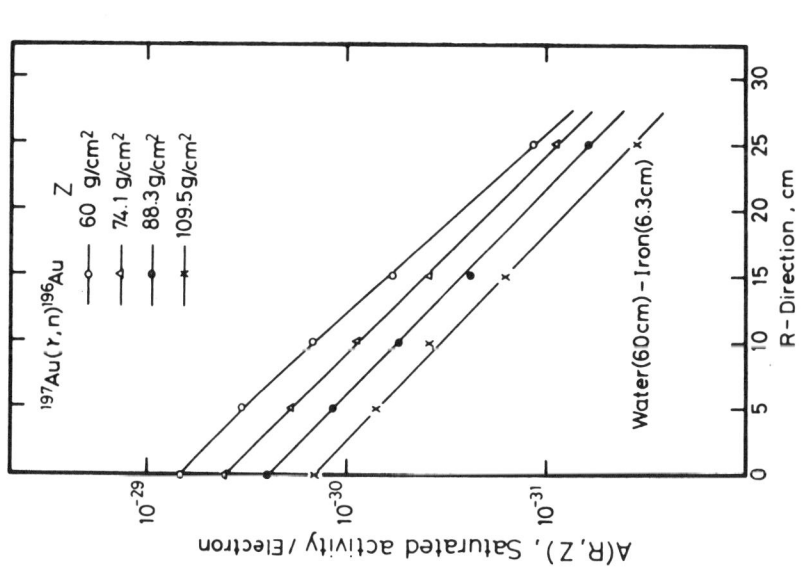

Fig. 12. Lateral distribution of experimental saturated activity of gold detector in water-iron.

activity in water drawn as solid curves. The solid curve at r = 0 cm in water is the extrapolated value and that at r = 25 cm in water is the interpolated value. From the comparison of their longitudinal distributions in water and iron (transmitted through 60 cm water), the attenuation along the axial depth in iron is found to be slower than in water at r = 0 cm owing to the bremsstrahlung emission in iron; and to be faster than in water at r ≠ 0 cm owing to the strong absorption of photons in iron.

The gold saturation activity at the (r,z) position, A(r,z), was integrated over a plane normal to the beam axis similarly to Eq. (9); namely,

$$A(z) = \int_0^\infty A(r,z) 2\pi r \, dr \,. \tag{11}$$

Figure 14 shows the A(z) distributions along the axial depth z in water and in iron. They attenuate exponentially in the same way as Eq. (10), and the effective attenuation coefficients μ_{eff} are obtained from Fig. 14 as

μ_{eff} = 0.0205 cm^2/g for water,

= 0.0328 cm^2/g for iron (following 60 cm water).

The value of μ_{eff} = 0.0205 cm^2/g for the gold saturation activity in water is rather close to μ_{eff} = 0.0172 cm^2/g for the integrated flux in water, since the spatial distribution of the integrated flux is similar to that of the gold saturation activity as observed in the water experiment.

Comparison with the TAURUS Calculation

The bremsstrahlung energy spectrum was calculated from Eq. (7), taking account of the energy spectrum of the incident electron beam (see Fig. 6). Figures 15, 16, and 17 show the calculated bremsstrahlung spectra at representative (r,z) positions in water, aluminum, and iron, respectively. In each of the figures, one example

APPLICATION OF ACTIVATION-SPECTRUM METHOD TO SHIELDING

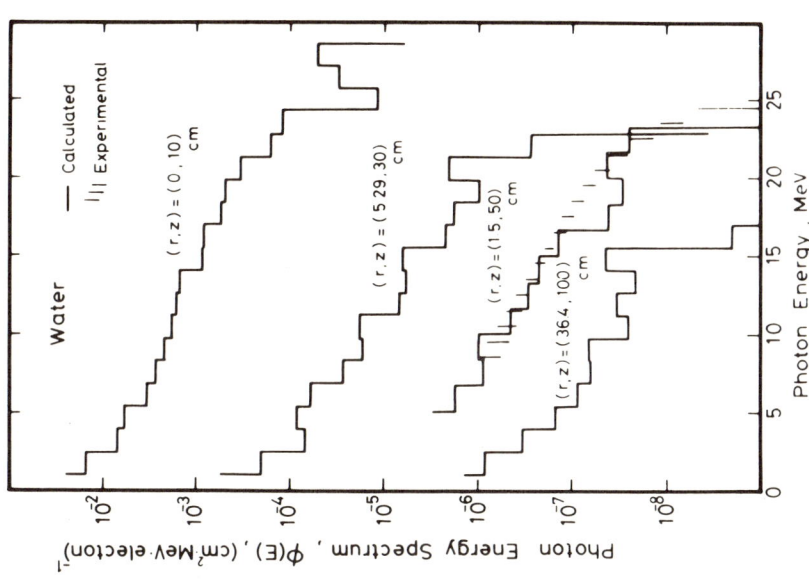

Fig. 15. Comparison of calculated bremsstrahlung spectra at several (r,z) positions in water with our experimental spectrum.

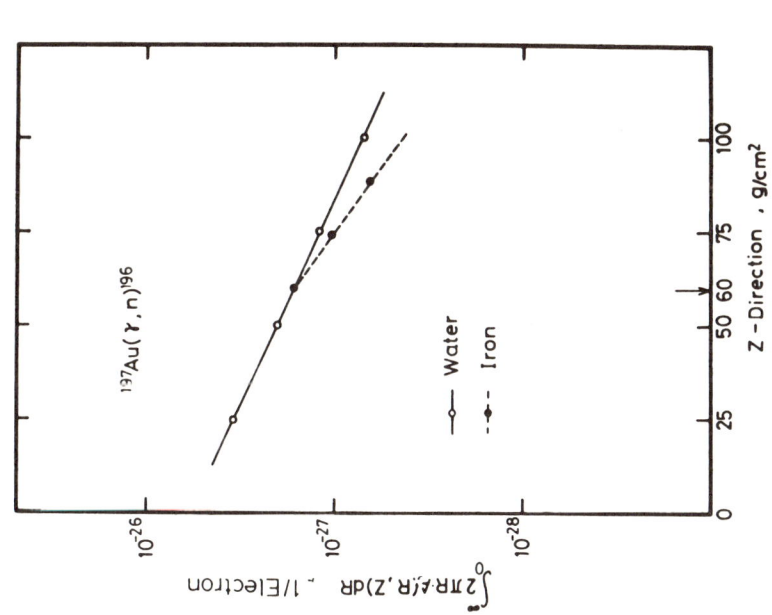

Fig. 14. Attenuation of experimental saturated activity of gold detector integrated over a plane normal to the beam axis in water and water-iron.

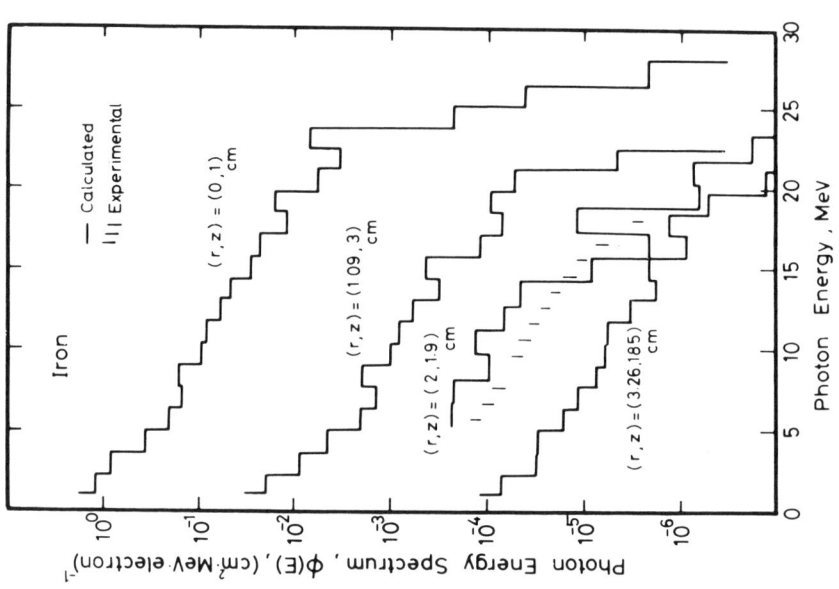

Fig. 17. Comparison of calculated bremsstrahlung spectra at several (r,z) positions in iron with our experimental spectrum.

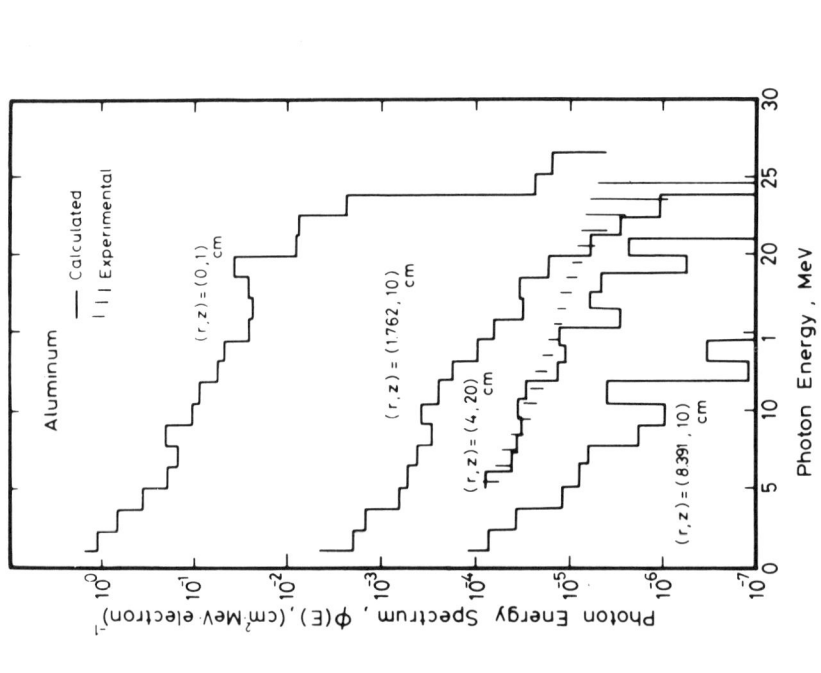

Fig. 16. Comparison of calculated bremsstrahlung spectra at several (r,z) positions in aluminum with our experimental spectrum.

of the experimental spectrum is shown, which is in good agreement with the absolute value of the calculated result. One can clearly see good agreement of the spectrum shapes between experiment and calculation for all three systems.

The photon fluxes $\phi(r,z)$ at each (r,z) position were obtained by integrating the spectra $\phi(E)$ above 8 MeV for water and above 5.5 MeV for aluminum and iron. The lateral distributions of experimental and calculated photon fluxes $\phi(r,z)$ are shown in Figs. 18 to 20 as a function of lateral distance r at several axial depths z in water, aluminum, and iron. Figure 18 clearly reveals that the agreement between experiment and calculation is very good in absolute value for the water system. For aluminum, on the other hand, the experimental photon fluxes are about a factor of 10 larger than the calculated ones at small (r,z) positions ($z = 4$, 10 cm, and $r = 4$, 10 cm), but the agreement between experiment and calculation is rather good (a factor of 2-3) at (r,z) positions away from the electron incidence point ($z = 20$, 30 cm, and $r = 4$, 10, 18 cm), as shown in Fig. 19. The experimental photon fluxes are in good agreement with the calculated ones at small (r,z) positions ($z = 1.9$, 4.6 cm, and $r = 2.6$ cm) in the iron system, but the former is about a factor of 10 smaller than the latter at large (r,z) positions ($z = 9.6$, 15.4 cm, and $r = 2.6$, 14 cm) as shown in Fig. 20, as compared with the aluminum experiment.

The large discrepancy may be explained from the following reasons:

i) The target end of the accelerating tube was not shielded differently from the water experiment, so the spurious bremsstrahlung radiation injected into the system from the tube cannot be neglected.

ii) The calculational model assumed a narrow electron beam, whereas the actual beam spreads in the target prior to reaching the system.

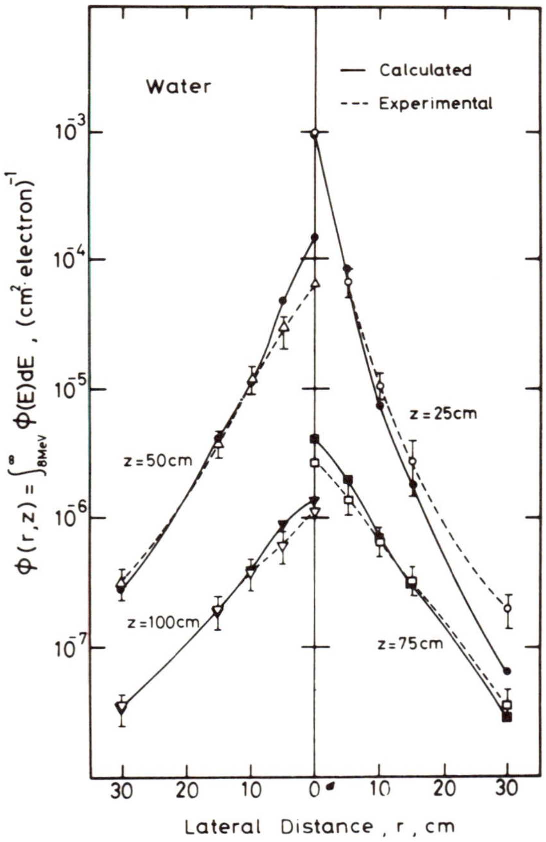

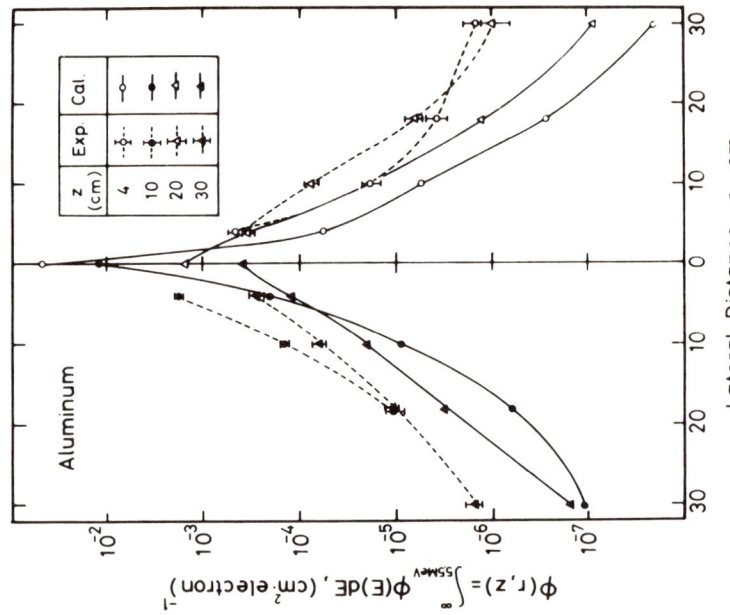

Fig. 18. Comparison of calculated lateral distribution of photon fluxes integrated above 8 MeV in water with our experimental results.

Fig. 19. Comparison of calculated lateral distribution of photon fluxes integrated above 5.5 MeV in aluminum with our experimental results.

APPLICATION OF ACTIVATION-SPECTRUM METHOD TO SHIELDING 467

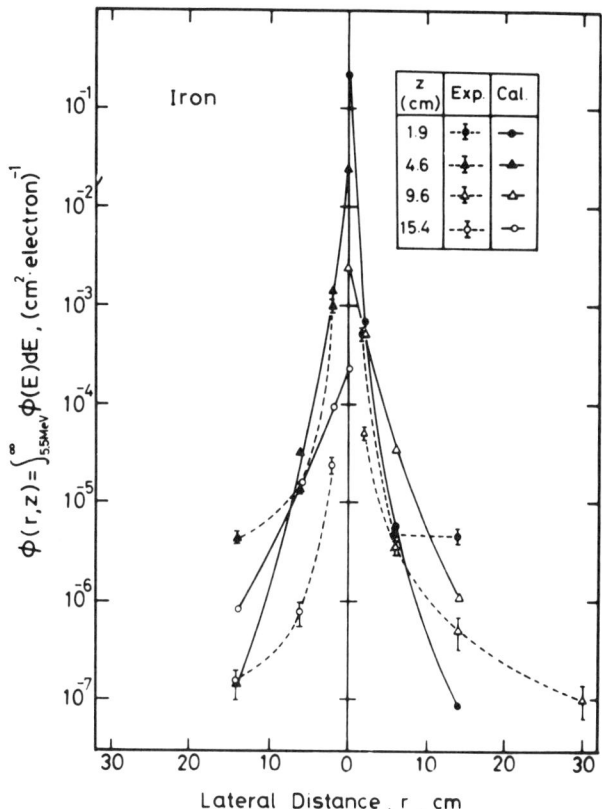

Fig. 20. Comparison of calculated lateral distribution of photon fluxes integrated above 5.5 MeV in iron with our experimental results.

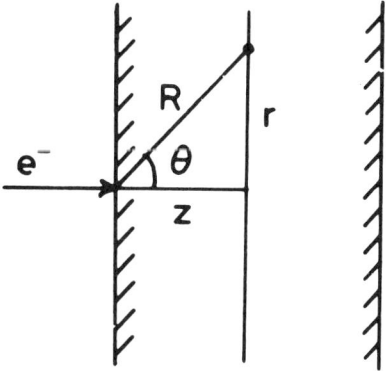

Fig. 21. Schematic diagram of a relation between space parameters.

At large lateral distances r = 20, 30 cm in both aluminum and iron systems, the experimental data have large statistical errors due to small counting rates and include background photons coming from room scattering to some extent. Generally speaking, the estimation of the spatial distribution of bremsstrahlung radiation in bulk media by the TAURUS calculation gives good agreement with our experimental results.

MATERIAL DEPENDENCE OF SPATIAL BREMSSTRAHLUNG DISTRIBUTION

The spatial bremsstrahlung distributions in bulk media of water, aluminum, iron, and lead were calculated in the photon energy range above 1 MeV by the TAURUS code, which was confirmed to give good estimation[38]. In this calculation, it is assumed that an electron beam is narrow and has the energy spectrum shown in Fig. 6a for water and in Fig. 6b for aluminum, iron, and lead. These four materials, commonly used as beam stopper and shield, cover a wide range of atomic numbers from Z = 7.23 (water) to Z = 82 (lead). The dimensions of the water, aluminum, and iron systems in this calculation were set to be equal to those in Fig. 5, and the lead system was selected as 15 cm in thickness and 30 × 30 cm^2 in area.

To get more detailed information on the spatial bremsstrahlung distribution in a bulk medium, the photon flux $\phi(R,\theta)$ was obtained by integrating the calculated spectrum above 1 MeV in the θ direction at the transmission length R, instead of $\phi(r,z)$ integrated above 5.5 MeV and 8 MeV. The relation among R, θ, r, and z is shown in Fig. 21. Figure 22 shows the calculated flux distribution multiplied by the square of the length R^2, as a function of R for water, aluminum, iron, and lead. The quantity $R^2\phi(R,\theta)$ obeys an exponential attenuation law having the same slope for every value of parameter θ for these four media except in the vicinity of the beam incidence point. The slope increases with the atomic number of the system as shown in Table 1, owing to the increase of the photon absorption cross-section. Figure 23 shows the variation of $R^2\phi(R,\theta)$ at a

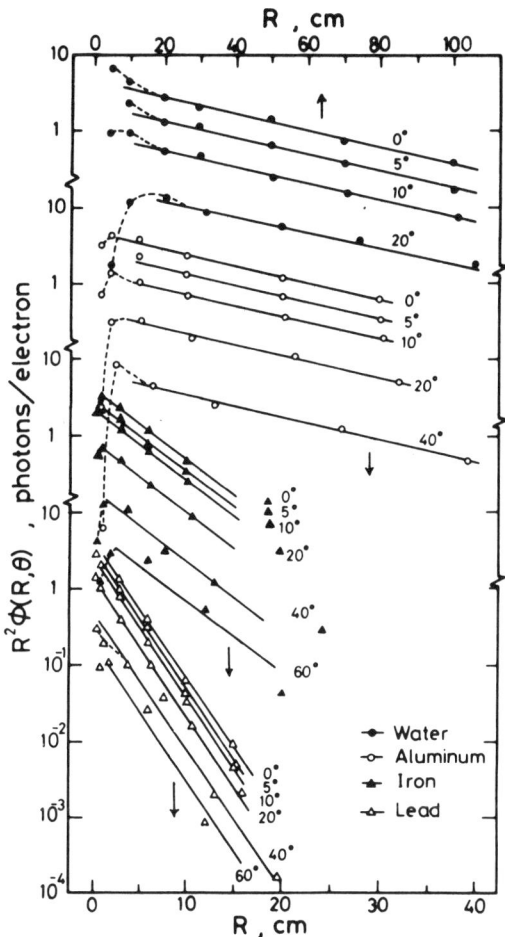

Fig. 22. Calculated spatial distribution of photon fluxes integrated above 1 MeV, $\phi(R,\theta)$ in water, aluminum, iron, and lead as a function of R, where R is the distance from a beam incidence point to a point in a medium. The $R^2\phi(R,\theta)$ distribution is drawn for several values of the parameter θ in the figure.

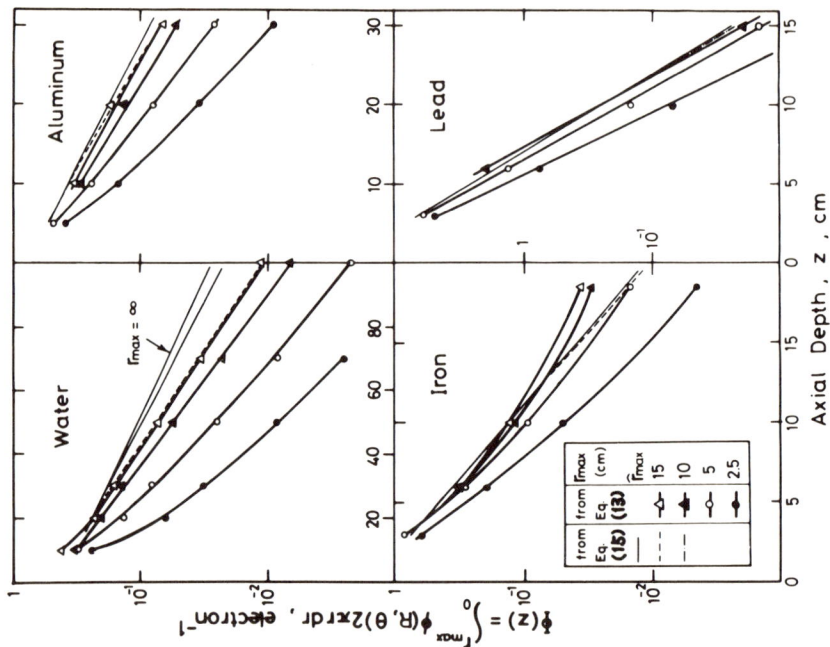

Fig. 24. Longitudinal distribution of photon fluxes, $\Phi(z)$, integrated from Eq. (13) and Eq. (15) for various r_{max} values.

Fig. 23. Calculated spatial photon distribution $R^2\phi(R,\theta)$ at $R = R_p$ in water, aluminum, iron, and lead as a function of θ, where θ is the polar angle from the beam axis. The value of R_p was adopted as 60 cm for water, 15 cm for aluminum, and 6 cm for iron and lead, as described in the text.

certain position P with the angle θ. The value of R_p was fixed to 60 cm for water, 15 cm for Al, and 6 cm for Fe and Pb, at which position $R^2\phi(R,\theta)$ is subject to an exponential attenuation in Fig. 22. The figure clearly indicates also the exponential decrease of $R^2\phi(R,\theta)$ with θ, but the slope of the curve decreases with increasing atomic number Z of the system, as shown in Table 1. From Figs. 22 and 23, the angular photon flux distribution $\phi(R,\theta)$ can be expressed as

$$\phi(R,\theta) = c \frac{e^{-\lambda\theta} e^{-\mu R}}{R^2} \quad \text{for} \quad R \gtrsim R_0 , \qquad (12)$$

where

$R_0 \approx$ 15 cm for water,
$\approx$ 4 cm for Al,
$\approx$ 1 cm for Fe and Pb.

The values of λ and μ are shown in Table 1. It is interesting that the value of μ in cm^2/g stays almost constant from water to iron. Table 1 also includes the calculated values of $\phi(R_e, 0°)$ for these four materials, where R_e is the range of a 22 MeV electron. The quantity $\phi(R_e, 0°)$ is considered (approximately) to be the forward thick-target bremsstrahlung yield which increases with the atomic number Z but deviates from the Z^2 curve.

The longitudinal flux distribution $\Phi(z)$ can be obtained from

$$\Phi(z) = \int_0^{r_{max}} \phi(R,\theta) 2\pi r \, dr , \qquad (13)$$

which corresponds to the photon number above 1 MeV crossing a circle of radius r_{max} normal to the electron beam axis. The $\Phi(z)$ distributions along the axial depth z, integrated from the $\phi(R,\theta)$ values in Fig. 22, are shown in Fig. 24 for several r_{max} values. In Fig. 24, the $\Phi(z)$ distribution approaches an exponential attenuation with increasing r_{max} value, with the exception of the point at z = 18.5 cm

Table 1 Coefficients of analytical representation fitted to spatial bremsstrahlung distribution

System	Atomic number Z	μ a) (cm²/g)	λ a) (rad)⁻¹	α a) (cm²/g)	Initial electron range R_e (cm)	$\phi(R_e, 0°)$ b) (cm² electron)⁻¹
Water	7.23	0.0256	8.66	0.0628 (0.0172) c) (0.0205) c)	10	4.46×10^{-2}
Aluminum	13	0.0254	6.10	0.0275	4.2	2.3×10^{-1}
Iron	26	0.0281	4.16	0.0312	1.4	1.5
Lead	82	0.0377	2.81	0.0410	0.84	3.51

a) The values of μ, λ, and α were obtained by least squares fittings of the values of $R^2\phi(R,\theta)$ in Fig. 22.

b) See text.

c) These figures are the μ_{eff} values in Eq. (10) (see text).

in iron owing to the poor statistics in the Monte Carlo calculation, and at the points near the beam incidence point.

The distribution $\Phi(z)$ can also be obtained from Eq. (12); namely,

$$\Phi(z) = \int_0^{r_{max}} \phi(R,\theta) 2\pi r \, dr = 2\pi c \int_0^{r_{max}} \frac{e^{-\lambda\theta} e^{-\mu R}}{R^2} r \, dr$$

for $R \gtrsim R_0$. (14)

By using the relation $R \cos\theta = z$ from Fig. 21,

$$\Phi(z) = 2\pi c \int_0^{\theta_{max}} e^{-\lambda\theta} e^{-\mu z/\cos\theta} \tan\theta \, d\theta , \quad (15)$$

where

$$\theta_{max} = \cos^{-1}(z/\sqrt{r_{max}^2 + z^2}).$$

In Fig. 24 the $\Phi(z)$ values calculated from Eq. (15) are shown for three r_{max} values for these four systems. The first r_{max} value is fixed to the largest value among those used in the $\Phi(z)$ calculation by Eq. (13) in order to compare both $\Phi(z)$ results, the second r_{max} is fixed to the dimension of the system used in this research $\hat{r}_{max}$, and the third one is infinite. The $\Phi(z)$ values for $r_{max} = \hat{r}_{max}$ and $r_{max} = \infty$ are almost the same in Fig. 24, except for the water system, which means that the sizes of aluminum, iron, and lead used in this study are considered to be infinite. The good agreement between two $\Phi(z)$ values from Eq. (13) and Eq. (15) can be seen for $R \gtrsim R_0$ in Fig. 24, except for $z = 18.5$ cm in iron. When r_{max} is infinite $\Phi(z)$ is the total photon number above 1 MeV crossing a plane normal to the beam axis and is found from Fig. 24 as

$$\Phi(z) = \int_0^\infty \phi(R,\theta) 2\pi r \, dr \propto e^{-\alpha z}, \quad z \gtrsim z_0, \quad (16)$$

where $z_0 \approx R_0$ for water and aluminum and $z_0 \approx 3$ cm for iron and lead. The value of α is shown in Table 1 and increases with

$$\alpha \approx 1.1 \mu, \quad (17)$$

from water to lead.

Table 1 also includes the value of μ_{eff} [in Eq. (10)] for water, together with the value of α [in Eq. (16)]. The quantity μ_{eff} was obtained for photon energies above 8 MeV, while α is for photon energies above 1 MeV. The μ_{eff} value is smaller than the α value corresponding to the decrease of the total attenuation coefficient μ with the photon energy.

REFERENCES

1. T. Nakamura, M. Takemura, H. Hirayama and T. Hyodo, J. Appl. Phys. 43, 5189 (1972).
2. T. Nakamura and H. Hirayama, J. Atomic Energy Soc. Japan 14, 668 (1972) (in Japanese).
3. H. Hirayama and T. Nakamura, Nucl. Sci. Eng. 50, 248 (1973).
4. T. Nakamura and H. Hirayama, Nucl. Sci. Eng. 59, 237 (1976).
5. H. Hirayama, Doctor Thesis, Kyoto Univ. (September 1976).
6. H. Hirayama and T. Nakamura, Nucl. Instrum. Methods 147, 563 (1977).
7. T. Matsui and T. Yamada, Japan Atomic Energy Research Establishment Report JAERI-memo 3000 (1968) (in Japanese).
8. H.H. van Tuyl, Hanford Atomic Products Operation (General Electric Co.) Report HW-83784 (1964).
9. Neutron Cross-Sections, Brookhaven National Laboratory Report BNL-325, 2nd ed., Suppl. No. 2, I (1964); IIA, B (1966); III (1965).
10. G. di Cola and A. Rota, Nucl. Sci. Eng. 23, 344 (1965).
11. A. Fischer, Kernforschungsanlage Jül-1196 (1975).
12. C.D. Zerby and F.L. Keller, Nucl. Sci. Eng. 27, 190 (1967).
13. W.W. Scott, Protection against Space Radiation, National Aeronautics and Space Administration Report NASA-SP 169 (1968), p. 339.
14. H. Ferdinande, G. Knuyt, R. Van de Vijver and R. Jacobs, Nucl. Instrum. Methods 91, 135 (1971).
15. R.D. Birkhoff, Handbuch der Physik (Springer-Verlag, Berlin, 1958), Vol. 34, pp 53-138.
16. B.W. Mar, Nucl. Sci. Eng. 24, 193 (1966).
17. P.J. Ebert, A.F. Lauzon and E.M. Lent, Phys. Rev. 183, 422 (1969).
18. G. Molière, Z. Naturforsch. A3, 78 (1948).
19. H.A. Bethe, Phys. Rev. 89, 1256 (1953).
20. H. Frank, Z. Naturforsch. A14, 247 (1959).

21. H.W. Koch and J.W. Motz, Rev. Mod. Phys. 31, 920 (1959).
22. D.K. Trubey, Oak Ridge National Laboratory Report ORNL-RSIC-10 (1966).
23. W.E. Dance, D.H. Rester, B.J. Farmer, J.H. Johnson and L.L. Baggerly, J. Appl. Phys. 39, 2881 (1968).
24. C.P. Jupiter, J.A. Lonergan and G. Merkel, Protection against Space Radiation, National Aeronautics and Space Administration Report NASA-SP 169 (1968), p. 249.
25. M.J. Berger and S.M. Seltzer, Nat. Bureau of Standards Reports 9836 and 9837 (1968).
26. A.A. O'Dell, Jr., C.W. Sandifer, R.B. Knowlen and W.D. George, Nucl. Instrum. Methods 61, 340 (1968).
27. C.W. Sandifer and M. Taherzadeh, IEEE Trans. Nucl. Sci. NS-15, 336 (1968).
28. M.J. Berger and S.M. Seltzer, Phys. Rev. C2, 621 (1970).
29. R.L. Ford and W.R. Nelson, Stanford Linear Accelerator Center Report SLAC-210 (1978).
30. J.K. Dickens, J.F. Emery, T.A. Love, J.W. McConnell, K.J. Northcutt, R.W. Peelle and H. Weaver, Oak Ridge National Laboratory Report ORNL/NUREG-14 (1977).
31. T. Nakamura, Nucl. Instrum. Methods 105, 77 (1972).
32. T. Nakamura, Nucl. Instrum. Methods 131, 521 (1975).
33. B.L. Berman, Lawrence Berkeley Laboratory Report UCRL-74622 (1973).
34. E.V. Fuller, private communication (1974).
35. H. Hirayama and T. Nakamura, Nucl. Instrum. Methods 133, 355 (1976).
36. T. Nakamura, T. Nishimoto and H. Hirayama, J. Nucl. Sci. Technol. 14, 31 (1977).
37. K. Shin, Y. Hayashida and T. Nakamura, Nucl. Instrum. Methods 151, 271 (1978).
38. T. Nakamura, H. Hirayama and K. Shin, Nucl. Instrum. Methods 151, 277 (1978).

39. F.H. Attix, W.C. Roesch and E. Tochillin, Radiation Dosimetry, 2nd ed., Vol. 3 (Academic Press, New York, 1969).
40. H.W. Patterson and R.H. Thomas, Accelerator Health Physics, (Academic Press, New York, 1973).
41. W.P. Swanson, Radiological Safety Aspects of the Operation of Electron Linear Accelerators (IAEA, Vienna, 1975).
42. C.D. Zerby and H.S. Moran, Oak Ridge National Laboratory Report ORNL-TM-422 (1962).
43. R.G. Alsmiller, Jr. and H.S. Moran, Oak Ridge National Laboratory Report ORNL-TM-1502 (1966).
44. H. Sugiyama, Researches of the Electrotechnical Laboratory, No. 724 (1972) (in Japanese).
45. F. Sauter, Ann. Physik $\underline{11}$, 454 (1931).
46. M.J. Berger, Res. Nat. Bur. Stand. $\underline{55}$ (6), 343 (1955).
47. P.V.C. Hough, Phys. Rev. $\underline{73}$, 266 (1948).
48. W.N. McElroy et al., Air Force Weapons Laboratory Report AFWL-TR 67-41, Vol. I-IV (1967).

Photonuclear Cross-Section Bibliography

53GO J. Goldemberg and L. Katz, Phys. Rev. $\underline{90}$, 308 (1953).
53MO R. Montalbetti, L. Katz and J. Goldemberg, Phys. Rev. $\underline{91}$, 659 (1953).
54GO J. Goldemberg and L. Katz, Canad. J. Phys. $\underline{32}$, 49 (1954).
57CA J.H. Carver and K.H. Lokan, Austral. J. Phys. $\underline{10}$, 312 (1957).
57BO O.V. Bogdankevich, L.E. Lazareva and F.A. Nikolaev, Sov. Phys. JETP $\underline{4}$, 320 (1957).
57KA L. Katz, K.G. McNeill, M. LeBlanc and F. Brown, Canad. J. Phys. $\underline{35}$, 470 (1957).
59PA R.W. Parsons, Canad. J. Phys. $\underline{37}$, 1344 (1959).
62BO O.V. Bogdankevich, B.I. Goryachev and V.A. Zapevalov, Sov. Phys. JETP $\underline{15}$, 1044 (1962).
62FU S.C. Fultz, R.L. Bramblett, J.T. Caldwell, N.E. Hansen and C.P. Jupiter, Phys. Rev. $\underline{127}$, 1273 (1962).

65KR P. Kruger, T.M. Crawford, J. Goldemberg and W.C. Barber, Nucl. Phys. $\underline{62}$, 584 (1965).

66BA E.B. Bazhanov, A.P. Komar, A.V. Kulikov and V.I. Ogurtsov, Sov. J. Nucl. Phys. $\underline{3}$, 522 (1966).

66FU S.C. Fultz, J.T. Caldwell, B.L. Berman, R.L. Bramblett and R.R. Harvey, Phys. Rev. $\underline{143}$, 790 (1966).

68MI K. Min and T.A. White, Phys. Rev. Lett. $\underline{21}$, 1200 (1968).

69BE B.L. Berman, R.L. Bramblett, J.T. Caldwell, H.S. Davis, M.A. Kelly and S.C. Fultz, Phys. Rev. $\underline{177}$, 1745 (1969).

69FU S.C. Fultz, B.L. Berman, J.T. Caldwell, R.L. Bramblett and M.A. Kelly, Phys. Rev. $\underline{186}$, 1255 (1969).

70VE A. Veyssière, H. Beil, R. Bergère, P. Carlos and A. Leprêtre, Nucl. Phys. $\underline{A159}$, 561 (1970).

71IS B.S. Ishkhanov, I.M. Kapitonov, I.M. Piskarev and V.G. Shevchenko, Sov. J. Nucl. Phys. $\underline{14}$, 142 (1972).

73AL1 R.A. Alvarez, B.L. Berman, F.H. Lewis and P. Meyer, Lawrence Berkeley Laboratory Report UCRL-74461, Proc. Internat. Conf. Photonuclear Reactions and Applications (1973).

73AL2 R.A. Alvarez, B.L. Berman, F.H. Lewis and P. Meyer, Lawrence Berkeley Laboratory Report UCRL-74469, Proc. Internat. Conf. Photonuclear Reactions and Applications (1973).

73FU S.C. Fultz, R.A. Alvarez, B.L. Berman and P. Meyer, Lawrence Berkeley Laboratory Report UCRL-74468, Proc. Internat. Conf. Photonuclear Reactions and Applications (1973).

LECTURE 27:   ACTIVATION DETECTORS AND THEIR

GAMMA SPECTRUM ANALYSIS

M.J. Koskelo and J.T. Routti

Department of Technical Physics
Helsinki University of Technology
SF-02150 Espoo 15, Finland

INTRODUCTION

Measurement of radiation fields in or near nuclear reactors and high energy accelerators is of importance in estimating radiation dose rates, radiation damage and shielding problems as well as in verifying results of computational methods. Such measurements must often be made in difficult environments, which excludes the possibility of using elaborate spectrometers, for instance. Activation detectors are well suited for such applications. By the proper choice of detector materials and composition, one can meet the requirements for selectivity and sensitivity for widely different radiation fields. The use of many detectors, in combination with unfolding methods, makes it possible to obtain information also of the energy spectra of particles.

Activation detectors are first exposed in the radiation fields under study and then counted for the induced activities. In many cases, especially with a single activation reaction, simple counting techniques can be used. The use of high resolution gamma spectroscopy with Ge(Li) detectors in measuring induced activities makes it possible to develop more elaborate activation detectors and considerably increase the information obtained. Advanced computer techniques can be used to identify the induced radionuclides and determine their activities.

Such detectors and their gamma spectrum analysis are discussed in the following. Examples are chosen both in reactor and accelerator problems. It can be emphasized, however, that the applicability of the gamma spectrum analysis extends from activation detectors also to direct measurements of induced and natural radioactivity, fission product monitoring and other applications, such as activation analysis.

ACTIVATION DETECTORS

General Considerations

Activation detectors and their counting techniques must be chosen to suit the radiation field under study. Different detector sets and applications are described in some detail later. General requirements for the detector materials can be defined, however, to be the following /1/.

- All materials used should be available with high purity to minimize perturbing reactions with impurities.
- The materials should be commercially available and not too expensive.
- The materials should be easy to use in a simple form.
- The materials should have possibilities for preparing thin foils or for making dilute alloys or mixtures with a suitable carrier to minimize self-shielding corrections.
- The materials should have well defined $\sigma(E)$ curves.
- The product nuclides should have suitable half lives.
- The product nuclides should be the only long-lived radioisotopes present at the time of counting and have clearly defined well separated gamma peaks that can easily be identified and calculated.

- The product nuclides must have possibilities for simple activity
  determination, preferably gamma counting or coincidence techniques
  so that absorption corrections are not required.

The Activation Formalism

The detector response in the irradiation environment is dependent on the cross-section of the activation reaction. In particular, the saturation activity induced in the detector material is

$$A_{sat} = N \int_0^{E_{max}} \sigma(E) \phi(E) dE, \qquad (1)$$

where N is the total number of target nuclei in the detector,

$\sigma(E)$ is the energy dependent microscopic cross-section for the reaction under consideration,

$\phi(E)$ is the neutron flux density per unit energy interval.

In practice the counting rate remains smaller than the saturation activity. It is reduced by the factors accounting for the incomplete saturation during a finite exposure time, the decay of activity after exposure and during counting, the branching ratio of the decay mode and the efficiency of the counting system. These corrections can easily be accounted for when the required nuclear and calibration data are available.

In some cases the saturation activities are of direct interest, for instance in measuring induced activities and comparing them to star densities of hadronic cascade calculations in accelerator problems. In some cases other integral quantities, such as dose rates or estimates of radiation damage, can be derived through functional representations of the activities. In general, however, unfolding techniques are required to derive spectral information of the inducing particles. These techniques are the subject of the companion paper /2/ and are not discussed here. Rather, the examples used are chosen from applications where more direct interpretation is possible.

Resonance Detectors

In the thermal and intermediate neutron energy range up to a few keV, many neutron capture cross sections exhibit resonance behaviour. With such detectors the dominant contribution to the total activity comes from the resonance region through the so called resonance integral. The flux value at the resonance can be obtained directly from the activity measurement. Often, however, corrections must be applied to account for the activity contribution outside the resonance area. In either case, the counting of resonance detectors is done by measuring the activities induced. Typical resonance reactions and their principal data are summarized in table 1.

The use of high resolution gamma spectroscopy makes it possible to measure several resonance detectors simultaneously in specially designed multicomponent detectors which are under development at our laboratory /3/. As an example, Fig. 1 shows the gamma spectrum from such a detector and at the same time illustrates the resolving power of the Ge(Li) gamma spectrometer.

Table 1. Typical resonance reactions.

	Reaction	Resonance energy(eV)	Resonance integral(barn)	Half-life
1.	$^{115}In(n,\gamma)^{116m}In$	1.457	1930	54.3 m
2.	$^{197}Au(n,\gamma)^{198}Au$	4.906	1490	64.7 h
3.	$^{109}Ag(n,\gamma)^{110m}Ag$	5.2	40	252 d
4.	$^{186}W(n,\gamma)^{187}W$	18.8	486	23.9 h
5.	$^{75}As(n,\gamma)^{76}As$	47	35.0	26.4 h
6.	$^{59}Co(n,\gamma)^{60}Co$	132	53.1	5.27 A
7.	$^{65}Cu(n,\gamma)^{66}Cu$	227	1.38	5.3 m
8.	$^{55}Mn(n,\gamma)^{56}Mn$	337	9.4	2.58 h
9.	$^{63}Cu(n,\gamma)^{64}Cu$	580	3.17	12.7 h
10.	$^{23}Na(n,\gamma)^{24}Na$	2850	0.0075	15.0 h

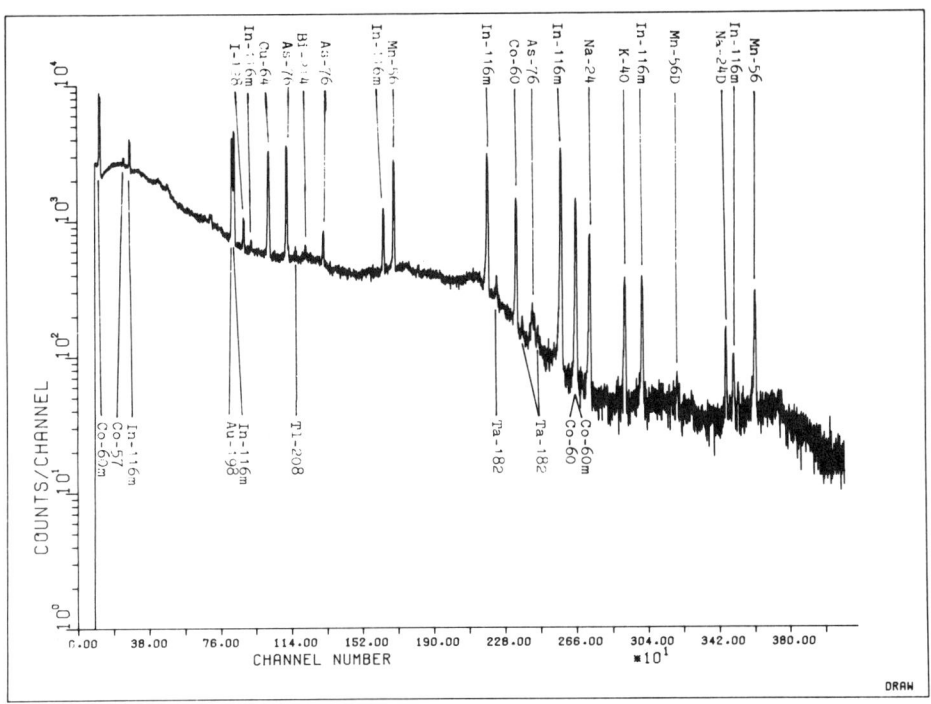

Fig. 1. The gamma spectrum of a multicomponent resonance detector.

Threshold reactions for reactor neutrons

In the MeV energy region, neutron induced (n,n'), (n,2n), (n,p), (n,α) and (n,fission) reactions can be used in addition to capture reactions. Many of these exhibit threshold behaviour and can be interpreted as flux integrators above their respective threshold energies. The saturation activity can then be written as

$$A = \int_{E_{thr}}^{E_{max}} \sigma(E) \phi(E) dE \qquad (2)$$

$$= \sigma_{eff} F_{thr}$$

where $\sigma_{eff} = \int_{E_{thr}}^{E_{max}} \sigma(E) \phi(E) dE / \int_{E_{thr}}^{E_{max}} \phi(E) de$

is an effective cross section and

$$F_{thr} = \int_{E_{thr}}^{E_{max}} \phi(E) de$$

is the energy integrated flux, both above the threshold energy. In some cases the cross section above the threshold can be assumed to be constant and σ_{eff} is thus independent of the flux spectrum. In others, knowledge of the spectrum is required to estimate the integrals.

Table 2 summarizes typical threshold reactions used in reactor measurements. As with resonance reactions, special composite detectors can be designed thus making it possible to measure up to 20 threshold reactions simultaneously using Ge(Li) gamma spectrometers /4/. The analysis methods applied to the gamma spectra are the same as those used in conjunction with the resonance detectors.

Activation detectors for accelerator studies

Studies of radiation fields near high energy accelerators require detectors of high threshold energies and high sensitivity. Techniques have been developed which use large detectors and special separation techniques enabling the detection of very small high energy fluxes. Different counting techniques are often employed to reach maximal sensitivities. Commonly used detectors include many of those also used in reactor studies, as summarized in Table 3 /5/.

Also, in addition to normal activation detectors, moderating or Bonner spheres are used in accelerator studies. Thermal neutron detectors are placed inside moderating polyethylene spheres of diameters up to 50 cm which extend the sensitivity of the detectors up to tens of MeV neutron energies. With these detectors, normal counting techniques are used but special unfolding methods are needed. Hence they are not discussed here in detail.

Table 2. Typical threshold reactions.

	Reaction	Half-life	Average σ(mbarn)	Threshold energy (MeV)	σ_{eff} (mbarn)
1.	$^{27}Al(n,\alpha)^{24}Na$	15,0 h	0,614	7,0	49,7
2.	$^{27}Al(n,p)^{27}Mg$	9,46 m	3,84	4,5	50,1
3.	$^{31}P(n,p)^{31}Si$	2,62 h	34,5	2,4	111
4.	$^{32}S(n,p)^{32}P$	14,3 d	60,8	3,1	306
5.	$^{46}Ti(n,p)^{46}Sc$	84,2 d	11,7	4,0	107
6.	$^{47}Ti(n,p)^{47}Sc$	3,4 d	17,2	2,2	48,8
7.	$^{48}Ti(n,p)^{48}Sc$	1,84 d	0,237	7,6	30,1
8.	$^{55}Mn(n,2n)^{54}Mn$	313 d	0,152	11,6	445
9.	$^{54}Fe(n,p)^{54}Mn$	313 d	70,4	3,3	404
10.	$^{56}Fe(n,p)^{56}Mn$	2,58 h	1,03	6,1	42,5
11.	$^{59}Co(n,\alpha)^{56}Mn$	2,58 h	0,156	6,8	10,1
12.	$^{58}Ni(n,p)^{58}Co$	71,3 d	114	2,8	470
13.	$^{63}Cu(n,\alpha)^{60}Co$	5,27 A	0,306	6,9	23,0
14.	$^{63}Cu(n,2n)^{62}Cu$	9,75 m	0,00864	12,4	482
15.	$^{65}Cu(n,2n)^{64}Cu$	12,7 h	0,292	11,2	621
16.	$^{64}Zn(n,p)^{64}Cu$	12,7 h	38,0	2,8	157
17.	$^{93}Nb(n,2n)^{92}Nb$	10,2 d	0,745	10,2	715
18.	$^{103}Rh(n,n')^{103m}Rh$	57 m	723	0,7	906
19.	$^{115}In(n,n')^{115}In$	4,48 h	174	1,3	294
20.	$^{127}I(n,2n)^{126}I$	12,8 d	0,686	10,5	835

Table 3. Characteristics of activation detectors.

Detector	Principal sensitivity	Reaction	Induced activity	Half-life	Counting
Indium foil	Thermal neutrons (0.02-20 MeV neutrons)	^{115}In(n,γ)^{116m}In	β^-,γ: 0.47 MeV (36%) 1.09 MeV (53%) 1.29 MeV (80%)	54 min	γ-spectrometer or β-counter
Gold foil (moderated)	Thermal neutrons (0.02-20 MeV neutrons)	^{197}Au(n,γ)^{198}Au	β^-,γ: 0.42 MeV (95%)	64.8 h	γ-spectrometer or β-counter
Sulphur pellet	3-25 MeV neutrons	^{32}S(n,p)^{32}P	β^- (no γ)	14.0 d	β-counter
Al $\rightarrow$ ^{24}Na	6-25 MeV neutrons	^{27}Al(n,α)^{24}Na	β^-,γ: 1.269 MeV (100%)	15.0 h	γ-spectrometer or β-counter
Carbon detector	Hadrons > 20 MeV	^{12}C(n,2n)^{11}C ^{12}C(spall)^{11}C	β^+ (100%)	20.3 min	γ-spectrometer or plastic scintillator counter
Al $\rightarrow$ ^{22}Na	Hadrons > 25 MeV	^{27}Al(spall)^{22}Na	β^+ (90%) γ: 1.275 MeV (100%)	2.6 y	γ-spectrometer or β-counter
Al $\rightarrow$ ^{18}F	Hadrons > 30 MeV	^{27}Al(spall)^{18}F	β^+ (97%)	109.8 min	γ-spectrometer or β-counter
Carbon $\rightarrow$ ^{7}Be	Hadrons > 50 MeV	^{12}C(spall)^{7}Be	γ: 0.477 MeV (12%)	53 d	γ-spectrometer
Terbium detector	Hadrons > 600 MeV	^{197}Au(spall)^{149}Tb Hg(spall)^{149}Tb	α: 3.95 MeV	4.1 h	α-counter
Cu-spallation (or Fe)	Hadrons > tens to hundreds of MeV	Cu(spall)^{52}Mn etc.	γ:		Ge(Li) spectrometer

Spallation detectors

In a spallation reaction, a nucleus far removed from the target nucleus is produced through fragmentation and evaporation processes caused by high energy particles. The corresponding threshold energies can be up to hundreds of MeV. High sensitivity detectors have been developed with such reactions, too, as illustrated by the Hg(spallation)^{149}Tb reaction in Table 3.

Spallation reactions also present a challenge for high resolution gamma spectroscopy. In principle, all nuclei lighter than the target are produced in the detector material and thus the number of activities can be large. Techniques using copper foils as multi-reaction activation detectors with up to 30 threshold reactions have been developed /6/.

Cross sections computed from empirical formulae /7/ for long lived spallation products are shown in Fig. 2. They are later used in interpretation of accelerator measurements.

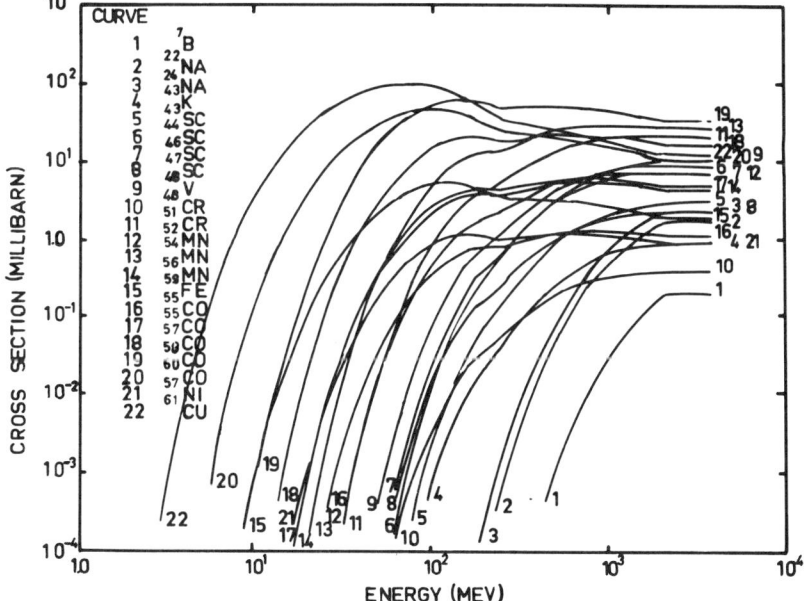

Fig. 2. Calculated spallation yields for different reaction products from Cu target /6/.

GAMMA SPECTRUM ANALYSIS

Counting and analysis procedures

Gamma spectra of irradiated activation detectors are measured with high resolution gamma spectrometers consisting of a Ge(Li) detector with its voltage supply and a preamplifier, a linear amplifier and a pulse height analyzer (PHA). Computer based analyzer systems using minicomputers have in recent years also become commercially available. In addition to data acquisition programs, they often incorporate spectrum analysis programs.

Several analysis schemes and programs have been developed and applied to activation detectors. These include both large computer programs applied to spectra measured with PHA systems and simpler codes which can be used directly on the data acquisition computers. In either case the analysis includes, after data acquisition and possible storage schemes, determination of peak centroids and areas, energy and efficiency calibrations, nuclide identification and calculation of saturation activities. Further interpretation of the results, such as unfolding, is done with separate codes.

In the following we briefly discuss mathematical techniques and their computer implementation for the analysis steps enumerated above. Since large numbers of programs have been reported, no effort is attempted for a comprehensive survey here. Rather, we illustrate the techniques by referring to different versions of a widely used SAMPO analysis method which we have previously reported /8/. This choice we justify, in addition to our familiarity with the program, with the fact that different versions extending from high accuracy large computer versions to reduced minicomputer programs of this analysis method have been extensively used with activation detectors.

Calibrations

Efficiency calibration of the Ge(Li) detector is required to relate the observed count rates to decay rates. The incident counting efficiency decreases with increasing photon energies and is further reduced by geometry factors.

Absolute efficiency calibrations are performed for each counting arrangement with standard sources. Interpolation and functional fitting of the calibration data are included in many analysis programs. They can also be easily included in minicomputer systems. An absolute accuracy of 3-10% is normally achieved.

The energy resolution of a Ge(Li) detector is typically about 1-3 keV full width at half maximum (FWHM) in the photon range up to 2 MeV. This resolution requires the use of a minimum of 2000 channels with 1-2 channels/keV for gamma energies up to 2 MeV. Peak centroids can typically be determined with an accuracy of 0.1-1 channel. This accuracy is often better than the integral linearity of the PHA system whose nonlinearity thus must be accounted for. Departures from a linear channel versus energy relationships can be up to several keV. Calibration procedures based on linear interpolation of polynomial fitting of points from standard peaks are used and can be incorporated also in minicomputer programs.

Typical results of efficiency and energy calibrations are shown in Fig. 3.

Spectrum Transformations

In many analysis methods smoothing or resolution improvement algorithms are employed to reduce the statistical fluctuations in the spectra. Mathematically the most elaborate of these methods is the one proposed by Inouye et al. /9/. Their procedure, although it significantly improves the appearance of the spectra, is difficult to use on small computers because of the significant computations. Simpler methods usually delete information useful in the error

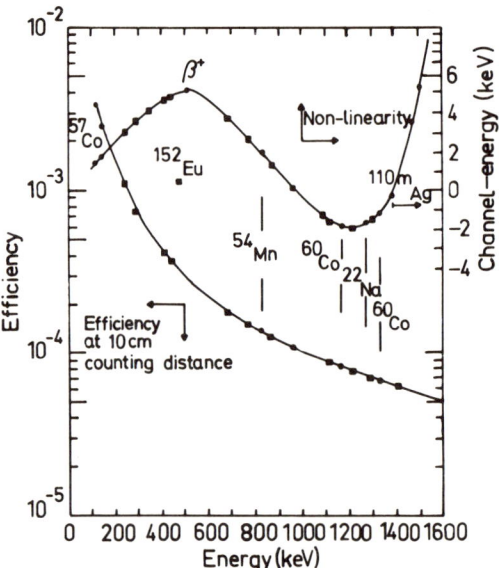

Fig. 3. The energy and efficiency calibrations for a Ge(Li)-spectrometer.

analysis but improve the reliability of the analysis if the method of peak detection and peak area calculation is simple.

Smoothing can also be incorporated into the peak search algorithms.

Peak Centroids

Peak search algorithms locate statistically significant peaks in the spectrum. The search methods can be based on visual inspection applicable in interactive analysis. Automated search uses computations of the zero crossings of the first derivative /10-14/, minima of the second derivative /8,15-17/ or maximal correlations with Gaussian shapes /18-21/. Most of these computations can be rather quickly performed even on small data acquisition computers.

In some algorithms (e.g., ref. 8), the peak centroids are determined only to the closest integral channel and later improved in nonlinear fitting of the peaks. In nonspectroscopy applications, such as measurement of activation detectors, it is more economical and sufficiently accurate to compute the peak centroids more accurately in the search algorithm and thus avoid nonlinear fitting later (e.g., ref. 26). Accuracies of the order of tenths of channel can be obtained for single peaks by the above mentioned search techniques or by computing centres of gravity for single peaks. With overlapping peaks, the accuracy obtained is worse, but still often adequate.

Peak Integrations

Several methods of varying complexity and accuracy can be used for computing peak areas. The simplest of them is the summation of the channel contents above the continuum which must always be subtracted when analyzing complex spectra. Variations occur in selection of integration limits and approximation of the continuum.

Resolving accurately overlapping peaks requires fitting many peaks simultaneously. In many programs, Gaussian peak shapes are used and fitted together with linear or higher order polynomial approximations for the continuum. A preferable technique is to use predetermined peak shape functions in the fitting and thus improve the stability of the results and reduce the number of fitting parameters. Peak shape functions are obtained by fitting single strong lines in the spectrum or a calibration spectrum measured under similar conditions. Typically, Gaussian shapes with corrections accounting for tailing are used as shape functions. Their parameters are determined by nonlinear least squares fitting and the resulting parameters, such as the Gaussian width and tailing parameters for different calibration peaks, are stored. Peak shapes in other parts of the spectrum are then determined by interpolating these shape parameters, and thus prescribed peak shapes can be used in actual fitting.

The actual fitting can use nonlinear least squares methods where peak centroids and heights are sought after. Both can be determined even for complex multiplets, as shown in Fig. 4. In routine applications, such as activation detectors, a linear least squares fitting without the free movement of the peak centroids typically produces results within 1-3% of those of the nonlinear method. A significant saving in computer time and core requirements can thus be achieved and it becomes possible to perform such a fitting even on minicomputers while still maintaining the resolving power of complex multiplets.

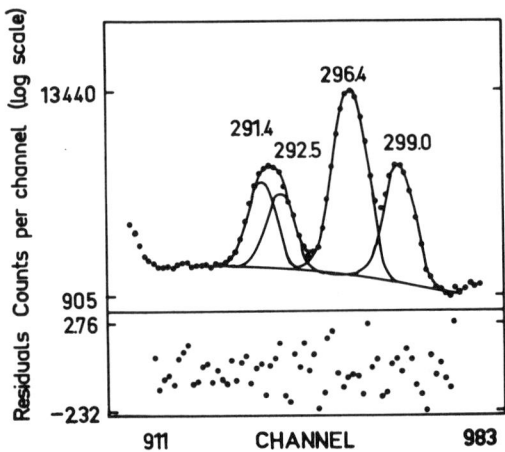

Fig. 4. A complete fit to a peak multiplet as done by the SAMPO program.

In the SAMPO programs Gaussian peak shapes with exponential tails are used as internally calibrated standards. In the high accuracy spectroscopy version /15/ nonlinear fitting is used for all peaks and interactive graphics can be optimally used. Linear fitting with the same peak shapes is used in an accelerated version /22/ with about 40 times faster fitting. A somewhat reduced version /23/ of the latter has also been installed on our data acquisition NOVA computer system with a 32 kiloword central memory and an additional disk memory, CAMAC interfaced pulse height conversion and a graphical display.

Nuclide Identification

Nuclide identification algorithms have been developed in recent years both for large and small computers. The simplest of them match peak centroids with potential reference energies and perform activity calculations. Since gamma energy libraries may contain thousands of lines, however, there are typically several possible nuclides for many of the fitted peaks and the user has to assure the right identifications.

In so doing he has to use information about several peaks belonging to the same nuclide. Peak energies, relative intensities and nuclide half lives are all useful. A graphical technique /24/ facilitating such an analysis method has been developed and used for spallation detectors.

The above considerations can be coded in identification algorithms, which typically develop identification criteria based on the closeness of the fitted and reference data of peak energies, relative intensities and feasibility of half lives. Combination on these criteria can then be incorporated into matrix formalism /25/, and nuclide identification with its confidence estimates can be performed.

More elaborate and formal pattern recognition techniques offer promising possibilities for nuclide identification or even combined fitting and identification /26/. Their application to gamma spectrum analysis is still under development, however.

The nuclide identification techniques with relatively compact reference libraries can be incorporated even into minicomputer analysis programs. Peak matching and matrix methods referred to have been included in our NOVA version of SAMPO /23/.

EXAMPLES OF ACTIVATION DETECTOR ANALYSIS

Multicomponent detectors for reactor neutron spectroscopy

Multicomponent resonance and threshold detectors have been succesfully used for measuring reactor neutron spectra in our Triga research reactor core. Analysis of gamma spectra, as shown in Fig. 1, have been done on a large computer with the linear version of SAMPO with nuclide identification. The interpretation of the results requires unfolding methods which are discussed in our second paper /2/. The gamma spectrum analysis can be performed also on the small data acquisition computer and eventually we intend to combine it with an unfolding algorithm also run on the data computer. This would allow direct neutron spectrum determination from the gamma spectrum of the multicomponent detector.

Hadron cascade measurements

Activation detector measurements have been used in the verification of hadron cascade calculations of accelerator radiation problems. An experiment requiring gamma spectrum analysis but no unfolding methods is used as an example.

Copper foils were placed around an external beam target at the CERN Proton Synchrotron bombarded by 24 GeV/c proton beam. Secondary particle fluxes from the target were determined by analyzing 12 long lived spallation products in the copper foils. The spallation reactions, whose cross sections were shown in Fig. 2, were interpreted as flux integrators above respective threshold energies ranging from 13 to 560 MeV. Experimental angular distributions of the particle fluxes above the same threshold energies could then be compared with hadron cascade computations of the same fluxes. Both results were found to be in good agreement /27/.

These spallation detector experiments as well as those requiring unfolding methods in their interpretation show the richness of information obtained with high resolution gamma spectroscopy and computerized analysis techniques as well as the good consistency of their interpretation with theoretical analysis of the radiation fields.

REFERENCES

1. Neutron Fluence Measurements, IAEA Technical Reports Series No. 107 (1970).
2. J.T. Routti and J.V. Sandberg, Unfolding Techniques for Activation Detector Analysis (see Lecture 24 in this volume).
3. K. Maunula, Multicomponent Activation Detectors for Reactor Neutron Spectroscopy (M.Sc thesis), Helsinki University of Technology, Department of Technical Physics (in Finnish) (1977).
4. J.T. Routti, Multicomponent Activation Detectors for Reactor Neutron Spectroscopy: Proposal and Optimization, Helsinki University of Technology, Department of Technical Physics Report (1975)(unpublished).
5. Radiation Problems Encountered in the Design of Multi-GeV Research Facilities, Report 71-21, CERN (1971).
6. J.T. Routti, Physics Scripta, $\underline{10}$, 107 (1974).
7. G. Rudstam, Z. Naturforsch., $\underline{21a}$, 1027 (1966).
8. J.T. Routti and S.G. Prussin, Nucl. Instr. Meth. $\underline{72}$, 125 (1969).
9. T. Inouye, T. Harper and N.C. Rasmussen, Nucl. Instr. Meth. $\underline{67}$, 125 (1969).
10. H. Weigel, J. Dauk, J. of Radioanal. Chem. $\underline{23}$, 171 (1974).
11. L.T. Felawka, J.G. Molnar, J.D. Chen and D.G. Boase, GAMAN, A Computer Program for the Qualitative and Quantitative Evaluation of Ge(Li) Gamma-Ray Spectra, Report AECL-4217, Atomic Energy of Canada Limited (1973).

12. V. Barnes, IEEE Trans. Nucl. Sci. NS-15, 377 (1968).
13. N. Sasamoto, K. Koyama, S. Tanaka, Nucl. Instr. & Meth. 125, 507 (1975).
14. H.P. Yule, Proceedings of the 1968 Int. Conf. on Modern Trends in Activation Analysis, Gaithersburg Maryland, USA, Vol. II, p. 1155 (1968).
15. J.T. Routti, SAMPO: Program for Computer Analysis of Gamma Spectra from Ge(Li) Detectors, Report UCRL-19452, Lawrence Berkeley Laboratory (1969).
16. M. Giannini, P.R. Oliva and M.C. Ramorino, Automatic Peak Identification in Analysis of Gamma-Ray Spectra Obtained with Ge(Li) Detectors, Report CNEN-RT/FI-72-14, Comitato Nazionale per l'Energia Nucleare (Rome)(1972).
17. B. Nyman, Nucl. Instr. & Meth. 108, 237 (1973).
18. R.G. Helmer and M.H. Putnam, GAUSS V: A Computer Program for the Analysis of Gamma-Ray Spectra from Ge(Li) Spectrometers, Report ANCR-1043, Aerojet Nuclear Company (1972).
19. A. Robertson, W.C. Prestwich and T.J. Kennett, Nucl. Instr. & Meth. 82, 141 (1970).
20. W.W. Black, Nucl. Instr. & Meth. 71, 317 (1969).
21. A.L. Connelly and W.W. Black, Nucl. Instr. & Meth. 82, 141 (1970).
22. G.C. Christensen, M.J. Koskelo and J.T. Routti, Gamma Spectrum Storage and Analysis Program SAMPO 76 with Nuclide Identification, Report HS-RP/015, CERN HS-Division (1977).
23. S. Toivonen, Analysis and Identification of Gamma Spectra Using a Minicomputer (M.Sc. thesis), Helsinki University of Technology, Department of Technical Physics (in Finnish)(1978).
24. J.T. Routti, Graphical Method for Nuclide Identification in Ge(Li) Gamma Spectra and Application to Spallation Studies, Report TKK-F-A227, Helsinki University of Technology, Department of Technical Physics (1974).
25. R. Gunnink and J.B. Niday, Computerized Quantitative Analysis by Gamma-Ray Spectrometry, Vols. I-IV, Report UCRL-51061, Lawrence Berkeley Laboratory (1972).

26. J.T. Routti, M.J. Koskelo, M.O. Enqvist, Gamma Spectrum Analysis Program SAMPO with Accelerated Peak Fitting and Nuclide Identification, presented at the ANS Topical Conference on Computers in Activation Analysis and Gamma-Ray Spectroscopy, May 1-4, 1978 Puerto Rico.
27. J.T. Routti, Nucl. Sci. Eng. <u>5</u>, 41 (1974).

INVITED PRESENTATIONS FROM STUDENTS AND SUMMARY LECTURE

MONTE CARLO CALCULATION OF EXPOSURE RATES IN DWELLING ROOMS[*]

Laszlo Koblinger

Central Research Institute for Physics (KFKI)
H-1525 Budapest
Hungary

Since all building materials contain natural radioactive isotopes, a fraction of the population's radiation burden originates from the walls of the buildings in which people live. A computer code is described here that can be used to predict the exposure rate of a room once the activity of the building material and the geometrical measurements of the room are known.

The Monte Carlo code REBEL-2 (Koblinger, 1976a) calculates the specific flux (normalized to unit source activity concentration) and exposure rate values at any given point in the room for the three most important sources: the uranium-radium series, the thorium series, and the potassium-40 isotope. The average energy and the spectrum of the photons are also computed by the code.

The code works by the Monte Carlo simulation of the adjoint equations; a new technique has been developed for the selection of the up-scattered pseudo-photon energies (Koblinger, 1976b).

The wall density dependence of the exposure rate is illustrated in Fig. 1 and a detailed description of the results is given in the paper by Koblinger (1978).

---

[*] Invited presentation.

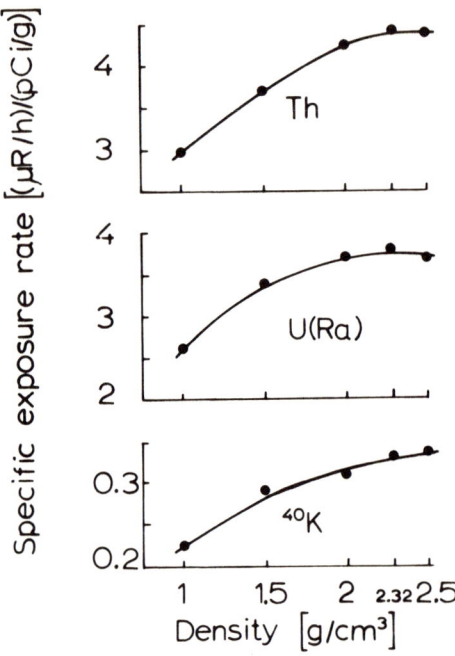

Fig. 1. Specific exposure rates versus density of 20 cm thick SiO_2 walls. Room measurements: $4 \times 5 \times 2.8$ m^3.

REFERENCES

Koblinger, L., 1976a, REBEL-2: An adjoint Monte Carlo code for the calculation of radiation in dwelling rooms, Central Research Institute for Physics (Budapest, Hungary) Report Number KFKI-76-65.

Koblinger, L., 1976b, A new energy sampling method, Central Research Institute for Physics (Budapest, Hungary) Report Number KFKI-76-57.

Koblinger, L., 1978, Calculation of exposure rates from gamma sources in walls of dwelling rooms, Health Physics, 34:459.

INTEGRAL EQUATION FOR RADIATION TRANSPORT - ASFIT [*]

V. Sundara Raman

Safety Research Laboratory[**]
Reactor Research Centre
Kalpakkam 603 102 India

ABSTRACT

ASFIT is a semi-analytical method for the transport of gamma-rays and neutrons. It is applicable to multi-region and multi-energy systems, and takes into account anisotropic scattering. The transport equation is written in the form of coupled integral equations separating the spatial and energy-angular transmissions. In the one-dimensional case, the basic equations are written as

$$\phi(x, E, \mu) = \int dx'\, S(x', E, \mu) T(E, x' \to x) , \quad (1)$$

$$S(x, E, \mu) = \int dE' \int d\mu'\, \phi(x, E', \mu') G(x, E' \to E, \mu' \to \mu)$$
$$+ S'(x, E, \mu) , \quad (2)$$

where $\phi(x, E, \mu)$ and $S(x, E, \mu)$ are the flux and source densities, at the space point x, for photons of energy E and with direction cosine μ; T and G are the kernels for space and energy-angle transmissions, respectively; S' is a non-scattering source term. Starting with a source S, ϕ is computed using Eq. (1) and then S is recomputed using Eq. (2), and this process (iteration) is continued

---

[*] Invited presentation
[**] This work was supported by the Technical Assistance Programme of the IAEA.

until convergence is obtained. In solving these equations, use is made of Legendre polynomial expansion in µ and discrete ordinate representation in x and E for S and ϕ. Constant and variable mesh widths are used in space coordinates. Linear and exponential interpolations are used for the source integral. The code also incorporates coarse mesh sizes to handle deep penetration problems.

AVAILABLE CODES

ASFIT-DS2 : One-dimensional transport code for gamma-rays using point cross-sections.
ASFIT-G : One-dimensional transport code for neutrons and gamma-rays using group cross-section data for computations.
ASFIT-3D : Three-dimensional transport code written for X, Y, Z geometry.

Using these codes, problems involving build-up factors for low and high atomic number materials, energy-angle distributions of ground-scattered radiation, and air contamination have been carried out.

A program, CHASFIT, is being developed for the transport of protons. Also the effect of bremsstrahlung on the scattered spectra is being investigated using this code. The photons generated by the secondary electrons are, in turn, tracked in an attempt to explain the low-energy tail of the measured spectra.

BIBLIOGRAPHY

Gopinath, D.V., and Santhanam, K., 1971, Radiation transport in one-dimensional finite systems - Part I, Nucl. Sci. Eng., 43:186.
Gopinath, D.V., and Santhanam, K., 1971, Radiation transport in one-dimensional finite systems - Part II, Nucl. Sci. Eng., 43:197.
Gopinath, D.V., Santhanam, K., and Burte, D.P., 1973, Some modifications in the anisotropic source-flux iteration technique, Nucl. Sci. Eng., 52:494.
Gopinath, D.V., and Sundararaman, V., 1976, Scattered radiation from extended sources, in "Proceedings of the National Symposium on Radiation Physics, Vol. XXVI, Mysore University, Mysore.
Gopinath, D.V., and Sundararaman, V., 1977, Solution on the ANS Benchmark problem for gamma-ray transport, Paper presented at the Fourth IARP Conference, Madras, India.
Gopinath, D.V., Natarajan, A., and Sundararaman, V., 1978, Higher-order interpolation schemes in ASFIT, Paper presented at the Fifth IARP Conference, Gwalior, India.
RSIC Computer code collection, CCC-153, 1975.

THERMAL EFFECTS INDUCED BY HIGH ENERGY PROTONS

IN TARGET AND ABSORBER MATERIALS*

P. Sievers

SPS Division
CERN
Geneva, Switzerland

ABSTRACT

With proton energies and intensities of about 400 GeV/c and several 10^{13} protons/pulse which are presently reached in high energy proton accelerators at FNAL and CERN, target and absorber materials which otherwise are radiation resistant, can be severely damaged by the elevated energy deposition densities and the resulting temperature rises induced by the incident protons. These are particularly high for proton beams of short duration of about 20 µs and small beam cross sections of about 2 mm diameter, incident on heavy materials, since therein a major amount of the beam energy is deposited via the π^0 initiated electron-gamma cascade.

Since no experimental data of the energy deposition density of high energy protons in various materials were available at the time when the CERN-SPS (Super Proton Sychrotron) was under construction, Monte Carlo cascade simulation programmes were used[1] in order to evaluate design parameters for targets and absorbers[2,4].

After the commissioning of the SPS, measurements were made of the energy deposition density in aluminium and copper and checked against the Monte Carlo programme FLUKA[1]. Good agreement was found

---
* Invited presentation

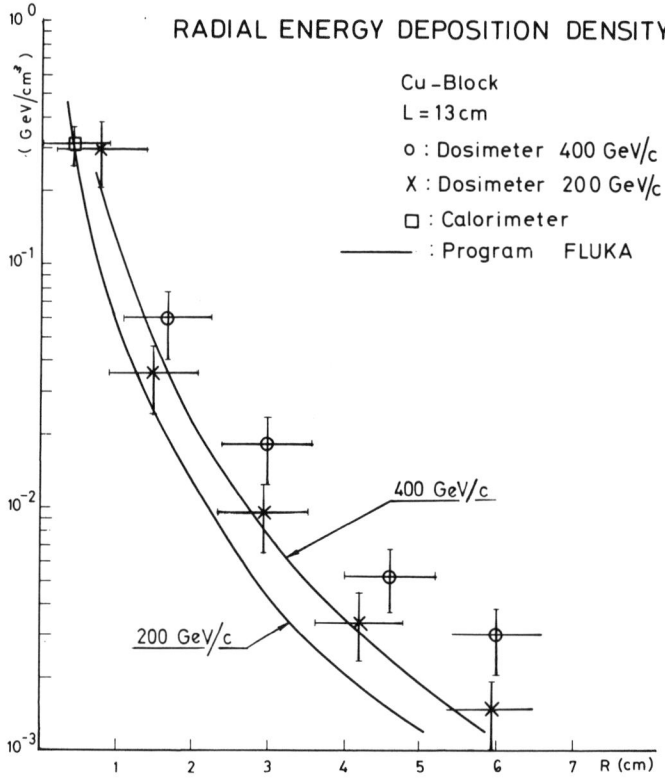

Fig. 1 Comparison of experimental (dosimeter, calorimeter) and computed (FLUKA) radial energy deposition densities in a copper cylinder at 13 cm distance from its front face.

along the central core of the cascade as well as laterally at depths above one to two hadron absorption lengths. Figure 1 shows, as an example, the radial energy deposition density in a copper cylinder at a depth of 13 cm.

As a result of the absorption of very short proton bursts, the temperature rises quasi-adiabatically deep inside massive blocks, which induces a thermal shock and subsequent radial and longitudinal stress waves inside the material[6]. Destruction of the core of a copper cylinder has already been experienced at moderate intensities of 5×10^{12} protons/pulse.

Further thermal stresses are induced by the rather inhomogeneous temperature fields, partially due to each proton pulse and partially due to the steady state temperature distribution inside the material. To evaluate these temperature fields, time and space dependent heat sources and boundary conditions in not trivial geometries have to be considered. These problems have been conveniently solved by a Monte Carlo heat diffusion code, as described in Ref. 7.

REFERENCES

1. J. Ranft and J.T. Routti, Monte Carlo Programs for Calculating Nuclear-Meson Cascades in Cylindrical Geometries, Report LAB-II/RA/71-4, CERN (1971);
   For further references see: H. Schoenbacher, Report LAB-II/RA/TM/74-5, CERN (1974).
2. W. Kalbreier, W.C. Middelkoop and P. Sievers, External Targets at the SPS, Report LAB-II/BT/74-1, CERN (1974).
3. K. Tentenberg, P. Sievers and W.C. Middelkoop, Absorber Blocks for Internal and External Beam Dumping at the SPS, Report LAB-II/BT/74-4, CERN (1974).
4. W. Kalbreier, A. Knesovic, G. Löhr, W.C. Middelkoop, P. Sievers and A. Warman, Target Stations and Beam Dumps for the CERN SPS, IEEE Trans Nucl. Sci. NS-24, No. 3, 1568 (1977).
5. P. Sievers, Measurements of the Energy Deposition of 200 and 400 GeV/c Protons in Aluminum and Copper, Report SPS/ABT/77-1, CERN (1977).
6. P. Sievers, Stress Waves in Matter due to the Rapid Heating by an Intense High Energy Particle Beam, Report LAB-II/BT/74-2, CERN (1974).
7. P. Sievers, Monte Carlo Treatment of the Non-Stationary Heat Transport, CERN Report LAB-II/BT/74-3.

SUMMARY LECTURE

Graham R. Stevenson

Health and Safety Division
CERN
Geneva, Switzerland

The success of a course such as this one must be based on two reactions: those of the persons actually attending the course and those of the wider population who might eventually read the proceedings of the course. The way in which the course was organized, by first discussing the physics of the topic, then the mechanics of the programs, and finally the use of the programs, certainly helped the participants to learn easily and should also help the interested reader to understand the contents of the course.

We, the participants, benefitted from the expertise of the lecturers, but this benefit was increased by the active participation of the students who demonstrated their own expertise in the subjects of the course. This enhanced the quality of the discussion immediately after the lectures and of the many informal exchanges that took place in the so-called free time. The participants in the course will therefore naturally feel that the course was more a success because they themselves helped to make it so. The reader can never experience this.

The practical consequences of the course will be judged by the increased ability of the participants to understand the problem they are faced with, to choose the program best suited to their problem and to understand the result of the calculation in drawing the correct conclusions. The participants were greatly aided in the aspect of practical application by attending a special workshop at CERN during the week following the course, where many of the programs discussed were actually run on real problems. If the reader is faced with similar difficulties he should not hesitate to contact the authors of the programs or the appropriate lecturers of this course.

On behalf of the students of the course and of the Course Directors I wish to thank the lecturers for all their efforts, which in my opinion made the course the success it was.

LIST OF PARTICIPANTS

Armstrong, Tony W., Science Applications Inc., La Jolla, CA, U.S.A.

Baldi, Antonio, CNEN, Bologna, Italy.

Cerchiari, Ugo, Istituto Nazionale Tumori, Milano, Italy.

Conte, Leopoldo, Servizio Fisica Sanitaria, Ospedale di Circolo, Varese, Italy.

Del Guerra, Alberto, Istituto di Fisica, Pisa, Italy.

Denk, Wolfgang, Gesellschaft für Reaktorsicherheit Forschungsgelande, Garching, Federal Republic of Germany.

Dinter, Herbert, Deutsches Elektronen-Synchrotron DESY, Hamburg, Federal Republic of Germany.

Engle, Ward W., Jr., Oak Ridge National Laboratory, Oak Ridge, TN, U.S.A.

Fasso, Alberto, HS Division, CERN, Geneva, Switzerland.

Gabriel, Tony A., Oak Ridge National Laboratory, Oak Ridge, TN, U.S.A.

Favale, Luciano, HS Division, CERN, Geneva, Switzerland.

Festag, Johannes, Gesellschaft für Schwerionenforschung mbH, Darmstadt, Federal Republic of Germany.

Gonzalez, J. A., Iberduero S.A., Bilbao, Spain.

Guaraldi, Rinaldo, CNEN, Bologna, Italy.

Hirayama, Ideo, KEK, National Laboratory for High-Energy Physics, Oho-Machi, Tsukuba-Gun, Ibaraki, Japan.

Jenkins, T. M., Radiation Physics Group, SLAC, Stanford, CA, U.S.A.

Koblinger, Laszlo, Central Research Institute for Physics, Budapest, Hungary.

Loewe, William E., Lawrence Livermore Laboratory, Livermore, CA, U.S.A.

Lundqvist, Hans, Gustaf Werner Institute, Uppsala, Sweden.

Markovich, Srpko, Radiation Protection Department, Nuclear Science Institute "Boris Kidric", Beograd, Yugoslavia.

Moritz, L. E., Safety Group, TRIUMF, University of British Columbia, Vancouver, B.C., Canada.

Nakamura, Takashi, Institute for Nuclear Study, University of Tokyo, Midori-cho,3-2-1, Tanashi, Tokyo, Japan.

Nelson, Walter R., HS Division, CERN, Geneva, Switzerland.

O'Brien, Keran, Environmental Measurements Laboratory, U.S. Department of Energy, New York, NY, U.S.A.

Orsini, Alberto, CNEN-DISP, Roma, Italy.

Pashoa, Anselmo S., Departmento de Fisica, Pontificia Universidade Catolica do Rio de Janeiro, Rio de Janeiro, Brazil.

Pasqui, Roberto, Istituto di Radiologia, Universita di Firenze, Firenze, Italy.

Pedroli, Guido, Servizio Fisica Sanitaria, Ospedale di Circolo, Varese, Italy.

Ponti, Carlo, CCR EURATOM, Ispra (VA), Italy.

Raman, Sundar V., Safety Research Laboratory, Reactor Research Center, Tamil Nadu, India.

Ranft, Johannes, Sektion Physik, Karl Marx Universität, Leipzig, German Democratic Republic.

Rindi, Alessandro, INFN, Frascati (Roma), Italy.

Roberti, Massimo, CNEN-DISP, Roma, Italy.

Routti, Jorma, Helsinki University of Technology, Department of Technical Physics, Otaniemi, Finland.

Sievers, Peter, SPS Division, CERN, Geneva, Switzerland.

Stevenson, G. R., HS Division, CERN, Geneva, Switzerland.

Tinti, Ranato, CNEN, Bologna, Italy.

Travaglini, Nedo, CNEN, Bologna, Italy.

Troyon, J. P., Institut de Radiophisique Appliquée, Lausanne, Switzerland.

Van Ginneken, A., Fermi National Accelerator Laboratory, Batavia, IL, U.S.A.

Von Gagern, Christoph, Max Planck Institut für Physik, Munchen, Federal Republic of Germany.

INDEX

Absorption length (see attenuation length)
Accelerator breeder, 99,110-115, 381
Activation, 99-100,269,337-339, 342,356,362-363,381,501-502
Activation detectors & methods, 389-497
Adjoint Boltzmann equation, 49-54, 75,78,82,95,100,104,501
AEGIS, 211-222,324,333-334
Age, 79-80,93-94
Albedo, 64,88,93 (also see boundary conditions)
AMPX system (see cross section data)
Angular convection, 61
Angular flux 18,127,134
Angular weight, 60
ANISN, 59,68,70,73,97-98,100,108, 121,123
Annihilation, 155,191,203,434,445
Antiproton (see radiation transport, hadron)
Approximation A, 144-145
Approximation B, 144-145
ASFIT, 503-505
Atomic displacement, 100,106,108-109
Attenuation length, 272-273,275, 343-344,361
Backscatter coefficient, 199-200
Barashenkov code (see "Dubna" code)
BCTIC (see Biomedical Computing Technology Information Center)
Bertini model (see Intranuclear-cascade-evaporation model)

BF_3 neutron detector, 438-439
Biomedical Computing Technology Information Center, 125-126
Boltzmann equation, 17-56,59-75, 77,95,104,128
Boltzmann operator, 103-104
BON, 398
Bonner sphere, 392-393,398,484
Boundary conditions, 36-37,50, 64-65
  albedo, 64,88,93
  method of Mark, 36-37
  method of Marshak, 36
  method of Yvon, 36-37
  periodic, 64
  vacuum, 64
Bragg peak, 273,360
Breeding, 99-100,110-115
Bremsstrahlung, 25-26,141-142, 146-148,150-151,157,160-161, 163,168-169,194,213-216,219-221,225,233,253-259
  spectrum, 395,409-441,443-447, 504
Build-up, 128-132,504

Calorimeter, 99,110,213,339,347, 381,508 (also see total absorption detector)
Capella-Krzywicki model, 306-307
CASCADE, 225
Cascades (see specific type)
CASIM, 277,284,292,323-338,374
Cerenkov radiation, 159-160
CHASFIT, 504
Chudakov effect, 155-156

CITATION, 123
Collided response, 90-91
Collision length biasing (see Monte Carlo methods)
Collisions (see interactions)
Combinatorial geometry, 375
Common Benchmark Program, 122
Compton scattering (see scattering)
Computer programs (indexed by name)

AEGIS	ANISN	ASFIT
BON	CASCADE	CASIM
CHASFIT	CITATION	COOLC
CRYSTAL-BALL		CYLKAZ
CYLKOZ	DIBRE	DOT
"Dubna"	EGS	ETRAN
EVAP-IV	FERDOR	FLUKA
FLUKOO	FORIST	HECC
HETC	HIC	KAPRYM
KAPRYZ	KASPRO	KENO
LOUHI	LUIN	LUNA-5
LYRA	MACK	MAGKA
MAGKO	MECC-7	MERCURE
MORSE	NMTC	NTC
PEGS	RDMM	REBEL-2
RECOIL	REFUM	RFSP
SABINE	SAMPO	SAND-II
SANDYL	SMUG	SPECTRA
SPUKST	STAY'SL	TAURUS
TRIPOLI	VENTURE	

Computer time (see Monte Carlo methods)
Contributions, 52-54
Conversion efficiency, 240-244
COOLC, 397
Cosmic rays, 27-29,269,273,381,389
Coulomb scattering (see scattering)
Coupling, 270-271
Coupling of codes, 251,272,355-356, 380
Critical energy, 143-144,168-169
Criticality, 66,68
Cross section data, 83,121-125, 403,431,435-439
  bibliography for photonuclear, 476-477
Cross section (also see scattering)
  effective, 437-441
  exclusive particle, 280-312
  inclusive particle, 280-310, 327-328, 344

Cross section (cont'd)
  empirical formula of Carey,328
  empirical formula of Ranft, 300-302, 339,344,347,350
CRYSTAL-Ball, 399-400,404
Current, 19
CYLKAZ, 343,346,348-350,356, 364-365
CYLKOZ, 343,346,359,364

Decay
  particle, 27,270,275,284,291-292,320,332,376-377
  radioactive, 20,22,31,427
Deep inelastic interaction, 299
Deep penetration, 66,68,82,88, 90,110,193,215
Delta ray, 225
Density effect, 154,225,377
Detector (also indexed by name of detector)
  design, 175,239-252,323,342, 356
  efficiency, 240-244,430-431, 489-490
  resolution, 247-250,358-359, 489-490
Diamond difference model (see flux models)
DIBRE, 409,421-428,436,441, 443-444,454,456
Diffusion equation, 37-38,127, 132-138
Discrete ordinates, 38-39,44-45, 59-75,77,97-98,128
  (also see ANISN, DOT)
Distributions (see cross section)
Dose, 100,130,206-207,211,223-237,253,260-266,339,342,356, 359,362-363,380-381,389,443, 481,501
DOT, 59,68-70,74,97-98,100,109, 121,123,380,382
DPA (see atomic displacement)
"Dubna" code, 277,321,350, 373-374

Effective energy range, 437-441
Effective cross section (see cross section)

INDEX 517

Efficiency (see detector or Monte Carlo methods)
EGS, 173-210, 239-266, 355-356, 380, 427-428, 449-452
Electromagnetic cascade, 141-266, 270-271, 280, 324, 332-333, 341-342, 344, 346, 355-356, 380, 382, 444-477, 507
Electron-photon shower (see electromagnetic cascade)
Electron (see radiation transport)
ENDF (see cross section data)
Energy balance (see hadronic cascade)
Energy loss, deposition and heating, 99-100, 103, 106-113, 130, 142-143, 153-154, 157, 211-213, 215, 221-222, 228, 248, 254, 261, 273-276, 323, 325, 332-333, 337, 339-371, 379, 427, 443, 507-509
(also see ionization loss)
Energy resolution (see detector)
EPR (see Experimental Power Reactor)
ESIS (see European Shielding Information Service)
ETRAN, 194, 253-260, 425, 427-428, 444, 449-452
EURACOS-II facility, 122
EURLIB (see cross section data)
European Shielding Information Service, 121-122
EVAP-IV, 321, 375, 377-378
Evaporation model, 318-319 (also see intranuclear-cascade-evapor- model)
Excitation (see ionization loss or nuclear excitation)
Exclusive distribution (see cross section)
Experimental Power Reactor (EPR), 100
Exponential transform (see Monte Carlo methods)
Extranuclear (see internuclear)

FERDOR, 397, 404
Feynman scaling, 285, 295, 298, 300, 339
Feynman x-distribution, 299

Fick's equation, 133
Finite difference method (see discrete ordinates)
Fission, 71-72, 82-84, 93, 110-115, 127, 135, 276, 314, 394, 427
"Fixup" model, 63-64 (see flux models)
Fluctuations
in Monte Carlo, 175, 212, 215, 248, 260, 292, 324-325
in unfolding, 391, 489
Fluence, 19, 106
FLUKA, 277, 284, 291, 300, 339-371, 507-508
FLUKOO, 343, 346, 348-350, 359, 361
Fluorescence, 193, 198, 209, 251, 254, 444
Flux density, 19, 72, 133, 501
Flux models, 59
linear, 62-68
step, 62-68
weighted difference, 66-70
Flux scale factor (see scale factor)
FORIST, 397
Fragmentation, 305
Fredholm integral equation, 390
Functional representation (see unfolding, solution methods)
Fusion, 99-109

Ge(Li) detector, 430-431, 479, 482-484, 488-490
Giant resonance, 271, 414, 439, 441, 456
Gluon, 299
Group structure, 34, 60, 78, 84, 104, 123, 134-135, 503-505

Hadron (see radiation transport)
Hadronic cascade, 251, 269-385, 481, 494-495
energy conservation in 284-287, 291-292, 306, 319, 327, 329
Hagedorn-Ranft (HR) model (see thermodynamic model)
Harder approximation, 25-26, 228-235
Heavy ions (see radiation transport)

HECC, 321
HETC, 110,112,116,277,321,373-385
HIC, 321

Impact parameter, 146-147,151-152, 303
Importance, 45-46,51,53
  (also see Monte Carlo methods)
Importance function (see Monte Carlo methods)
Inclusive distribution (see cross section)
Inelasticities, 277,285-286,293, 300,333-334,344-345,347
Interaction length (see attenuation length)
Interactions (see cross section or scattering or specific interaction by name (e.g., bremsstrahlung))
Internuclear cascade, 312,339,350
Intranuclear-cascade-evaporation model, 272,303,311-322,328, 350-351,373,376-377,379-380
Ionization loss, 142-143,153-154, 168,213,261,269,273-275,332,341, 346,376,445-446 (also see stopping power or energy loss)
Ionization potential, 154, 225

Kaon (see radiation transport)
KAPRYM, 346-349
KAPRYZ, 346-349,365,367
KASPRO, 251,277,284,292,300,339-371
KENO, 123
Kerma, 106,130,264-265
Kernel (see resolution function)

Landau-Pomeranchuk effect, 156-159
Least squares expansion (see unfolding, solution methods)
Lindenbaum-Sternheimer isobar model, 316
Linear estimation (see unfolding, solution methods)
Liquid-argon chanber, 249-250
LOUHI, 399-404
LUIN, 277

LUNA-5, 413
LYRA, 395,409-419,431,443,454, 456-457

MACK, 106
Macro (see MORTRAN)
MAGKA, 343,346,349,363-364
MAGKO, 343,346,349-350,359,364
Magnetic fields, (see radiation transport)
Marginal density function, 82
Mean excitation energy (see ionization potential)
Mean free path (see attenuation length
MECC-7, 321,375,377-378
Media polarization (see density effect)
Medical accelerators, 253-260, 265,269,443
MERCURE, 121,132
Method of moments, 130
Microdosimetry, 381
Monte Carlo methods, 39-49,51, 53,77-96,98-119,128,146, 173-266,277,312,318-319, 323-371,373-385,444-451, 501-502,507-509
  analog, 78,83,86,173-196,211-213,215-217,265,318-319, 324-325,328,379
  basic method (direct sampling), 42
  biasing & weighting, 103,106, 211-222,262,264,286-291, 323-337,379,447-448
  efficiency & computer time, 81-82,86,90,174,176,211-212, 218,240,254,260,324-325,330, 333,335-336,342,344,374,444-445
  estimation
    boundary crossing, 88-93
    collision density, 88-92
    statistical, 88-92
    track length, 88-92,106-108
  exponential transform, 88,106, 337

INDEX

Monte Carlo methods (cont'd)
  importance sampling, 46-49,80,
    82,175,193,212,240,290,292-294,
    339,347,374,379,447-448 (also
    see Monte Carlo methods, bias-
    ing & weighting)
  rejection technique, 42-43,80-82
  Russian roulette, 86,93,106
  sampling, 82-83,279-310,344-345,
    347,351
    correlated, 337
  splitting, 86,93,106,329,331-332
  statistical error, 92,94
MORSE, 77-119,121,123,175,240,380,
  382
MORTRAN
  language, 174
  macro, 178,182,190-191,265
Multigroup (see group structure)
Multiparticle events, 279-310
Multiple scattering (see scattering)
Multiplicity,271-273,279-281,283,
  293-296,303-304,306,311,324-325,
  333,351-353,373
Multiwire proportional counter,
  249-250
Muon (see radiation transport)

NaI detector, 247-249,430-431
Negatral (see radiation transport,
  electron)
Neutral beam injector, 100-101,
  103,106,108-109
Neutron (see radiation transport)
NMTC, 374-375
Nonelastic scattering (see scatter-
  ing, inelastic)
Nonnegativity, 391,399,401
Nonuniqueness, 391
NTC, 374-375
Nuclear density distribution,
  314-315
Nuclear excitation, 340,341,
  345-346
Nuclear material transmutation,381
Nuclear momentum distribution,
  314-315
Nuclear profile function, 306-307
Nuclear thickness, 303-304

Nuclear transparency, 313,319
Nuclide identification, 493-494

Orthonormal expansion (see unfolding, solution methods)

Pair production, 141,143-144,
  155,161-163,169,213-214,221,
  225,265,444-445
Particle production, 325 (also
  see cross section)
  models, 279-310,327-329
Path length stretching (see
  Monte Carlo methods,
  exponential transform)
Pauli exclusion principle, 313,
  317,319
Peak centroid, 490-493
Peak integration, 491-494
PEGS, 176,178,191-192,241,250
Phase space density, 20
Photoelectric effect, 166-167,
  193,198,204,209,225,444
Photon (see radiation transport)
Photonuclear Data Center (NBS),
  435
Photonuclear interaction, 271,
  410-441
Pion (see radiation transport)
Polarization (see density
  effect)
Potential energy distribution,
  315-316
Propagating particle, 330-331,
  333,336
Proton recoil emulsion, 392,
  397
Proton (see radiation trans-
  port)

Quark, 299
  recombination model, 299

Radiation damage, 100,103,106,
  211,213,323,325,443,481,507-
  509
Radiation density, 19
Radiation length, 142-144,148,
  153
Radiation Shielding Information
  Center,124-126,194,374-375

Radiation transport, 17-56,503-505,
  electron, 24-26 (also see electromagnetic cascade)
  muon, 26-28,110,270-271,275,347, 376-377,380
  hadron (proton, neutron, pion, kaon..), 27-30,34,44-45,59 75, 77-119,503-505
  (also see hadronic cascade)
  heavy ion, 375,381
  photon, 59-75,77-119,381,503-505
  (also see electromagnetic cascade)
  in magnetic fields, 191,244-247, 251,323,333,337,346,366
Radiative interaction (see bremsstrahlung)
Radiative losses (see bremsstrahlung)
Radioactivity (see activation)
Radiobiology, 381
Radionuclide production (see activation)
Range
  electron, 25,445-446,471-472
  (also see Harder approximation)
  straggling, 375-377
Rapidity, 301, 304-308
RDMM, 395
Reactions (see interactions)
REBEL-2, 501-502
RECOIL, 106
Recombination model (see quark)
Recording particle, 330-333,335
Reflection coefficient, 198-199
REFUM, 409,431
Reggeon theory, 306
Rejection technique (see Monte Carlo methods)
Resolution function, 390-393,404
Resolution of detectors (see detector, design)
Resonance detector, 392,398,482-483,494
RFSP, 399-400
RSIC (see Radiation Shielding Information Center)
Russian roulette (see Monte Carlo methods)

SABINE, 121,132
SAMPO, 488,492-494
SAND-II, 397-398,400,404,443, 454,456-457
SANDYL, 194,254
Scale factor, 72,73,75
Scaler flux (see flux density)
Scattering, 21-23,35,53,60,66, 71,91,127,135
  Compton, 163-166,198,204, 214-216,220-221,225,257, 444
  Coulomb, 148-153,213,225, 269,276,325,375-377,445
  down scattering, 21,23,134
  elastic, 270-273,344,377-378
  electron, 314,422
  inelastic, 270-272,311-322, 377-378
  low energy photon, 198-199, 239-240
  multiple, 146,188-189,193, 225,254,256,325,341,375-377,424,444-445
  up scattering, 21,23,71,501
Scintillation detector, 240-244, 356,410
  liquid, 392,397,430-431
Screening parameter, 147
Secondaries, 27
Semiempirical method, 272-273
Shower counter, 193,239-252
Silicon detector, 261-264,381
Slowing down, 20-21,24-26,34-35, 38,41-42,137,225,325,421-422, 444-445
Smoothness, 391,399,489-490
SMUG, 106
Sn theory, 33
Source term, 20,22,50-51,60, 71-72,79-80,82,91,93,95,104, 110,127-129,320, 379-380
  (also see albedo)
Spacecraft shielding, 381
Space-dependent rebalance (see scale factor)
Space-dependent scaling (see scale factor)
Space radiation, 381

INDEX   521

Spallation detector, 393,403,487,
  493-495
SPECTRA, 399-400,404
Spherical harmonics, 30-37,44-45
Splitting (see Monte Carlo methods)
SPUKST, 279
Star density, 323,325,329-332,
  339-371,379,481
Statistical bootstrap model, 298
STAY'SL, 399,403
Stopping power, 21,25,156,207,209,
  213,232,234-235,377
  collision, 142-143,153-154,168,
    225,445-446
  radiative, 142,168,446
  restricted, 207-208
Straggling, 444 (also see range)
Straight-ahead approximation,
  24-30,46,136
Streamer chamber, 244-247
Stress waves, 509
Superconducting magnets, 364-365
Synchrotron radiation, 193,197-199,
  239,251

TAURUS, 443-444,449-454,462-473
TFTR (see Tokamak)
Thermodynamic model, 298-299,306-
  307,328,335
Thermoluminescent dosimeter,
  223-237,360
Threshold detector (see activa-
  tion detectors & methods)
Tissue equivalence, 381
TLD (see thermoluminescent dosi-
  meter)
Tokamak, 99,100,101,109
Total absorption detector, 346,
  358
Track length, 187-188
Transition radiation, 159-160
TRIPOLI, 121

Uncollided response, 90-91,128
Unfolding, 389-407,413-427,484,
  488,494
  solution methods, 394-404
    functional representation, 394
    iterative, 397

Unfolding, (cont'd)
  linear estimation, 391,
    396-397
  regularization (functional
    & parameter), 398-399
  series expansion, 394-395,
    413,415,417-419

Variance reduction techniques
  (see Monte Carlo methods,
  importance sampling)
VENTURE, 123

Weighting factor, 398
Weighting function, 395,400,
  413,416-417,419
Woods-Saxon density distribu-
  tion, 303
Wyard's formula, 414
"Zero" model, 66,68 (also see
  flux models)